P9-CKY-677

MULTI-CARRIER SPREAD SPECTRUM & RELATED TOPICS

Multi-Carrier Spread Spectrum & Related Topics

Edited by

Khaled Fazel
Bosch-Telecom GmbH

and

Stefan Kaiser
German Aerospace Centre (DLR)

KLUWER ACADEMIC PUBLISHERS
BOSTON / DORDRECHT / LONDON

A C.I.P. Catalogue record for this book is available from the Library of Congress.

ISBN 0-7923-7740-0

Published by Kluwer Academic Publishers,
P.O. Box 17, 3300 AA Dordrecht, The Netherlands.

Sold and distributed in North, Central and South America
by Kluwer Academic Publishers,
101 Philip Drive, Norwell, MA 02061, U.S.A.

In all other countries, sold and distributed
by Kluwer Academic Publishers,
P.O. Box 322, 3300 AH Dordrecht, The Netherlands.

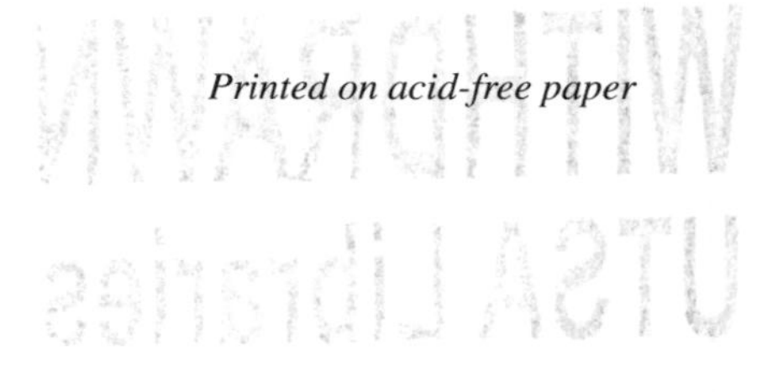

Printed on acid-free paper

Printed in the Netherlands.

TABLE OF CONTENTS

*Invited paper

EDITORIAL INTRODUCTION

Khaled Fazel
Digital Microwave Systems
Bosch Telecom GmbH
D-71522 Backnang, Germany

Stefan Kaiser
German Aerospace Center (DLR)
Institute for Communications Technology
D-82234 Wessling, Germany

In this last decade of this millennium the technique of multi-carrier transmission for wireless broadband multimedia applications has been receiving wide interests. Its first great success was in 1990 as it was selected in the European Digital Audio Broadcasting (DAB) standard. Its further prominent successes were in 1995 and 1998 as it was selected as modulation scheme in the European Digital Video Broadcasting (DVB-T) and in three broadband wireless indoor standards, namely ETSI-Hiperlan-II, American IEEE-802.11 and Japanese MMAC, respectively.

The benefits and success of multi-carrier (MC) modulation in one side and the flexibility offered by spread spectrum (SS) technique in other hand motivated many researchers to investigate the combination of both techniques, known as ***multi-carrier spread-spectrum*** (MC-SS). This combination benefits from the main advantages of both systems and offers high flexibility, high spectral efficiency, simple detection strategies, narrow-band interference rejection capability, etc.. The basic principle of this combination is straightforward: The spreading is performed as direct sequence SS (DS-SS) but instead of transmitting the chips over a single carrier, several sub-carriers could be employed. As depicted in Figure 1, after spreading with assigned user specific code of processing gain G the frequency mapping and multi-carrier modulation is applied. In the receiver side after multi-carrier demodulation and frequency de-mapping, the corresponding detection algorithm will be performed. The MC modulation & demodulation could be easily done in the digital domain by performing IFFT and FFT operations.

In 1993 different multiple access concepts based on this combination for mobile and wireless indoor communications have been introduced, called OFDM/CDMA or MC-CDMA, MC-DS-CDMA and MT-CDMA. The main differences between these schemes are in the spreading, frequency mapping and the detection strategies.
The MC-CDMA or OFDM/CDMA is based on a serial concatenation of DS spreading with MCM. The high rate DS spread data stream is MC modulated in that way that the chips of a spread data symbol are transmitted in parallel on each sub-carrier. As for DS-CDMA a user may occupy the total bandwidth for the transmission of a single data symbol. The separation of the users' signals is performed in the code domain. This means that the MC-CDMA system performs the spreading in the frequency domain, which allows for simple signal detection strategies. This concept was proposed with OFDM for optimum use of the available bandwidth for the downlink of a cellular system using orthogonal Walsh-Hadamard codes.

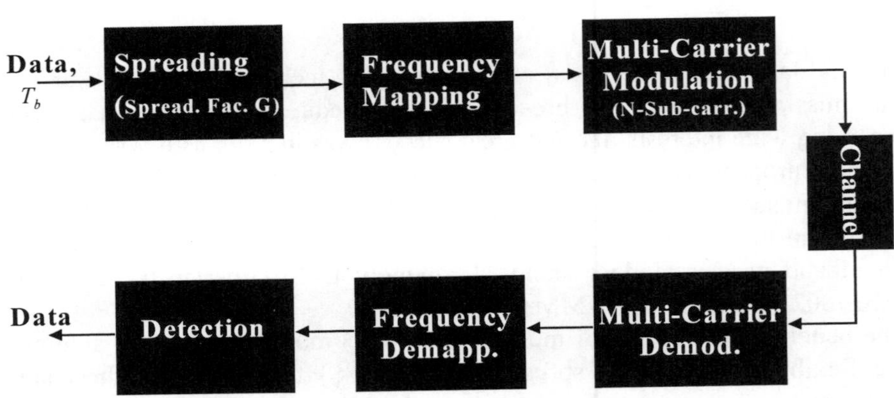

Figure 1. Principle of Multi-Carrier Spread Spectrum

The MC-DS-CDMA modulates the sub-streams on sub-carriers with a carrier spacing proportional to the inverse of the chip rate. First the data stream is converted onto parallel low rate sub-streams before applying the DS spreading on each sub-stream in the time domain and modulating onto each sub-carrier. This will guarantee the orthogonality between the spectra of the sub-streams. If the spreading code length is smaller or equal to the number of sub-carriers a single data symbol is not spread in the frequency, instead in time domain. Spread spectrum is obtained by modulating the spread data symbols on parallel sub-carriers. This concept by using high number of sub-carriers benefits from the time diversity. However, due to the frequency non-selective fading per sub-channel, frequency diversity could only be exploited if channel coding with interleaving or sub-carrier hopping

is employed. Furthermore, here the sub-carrier spacing might be chosen larger than the chip rate that may provide a higher frequency diversity.
Finally, the multi-tone-CDMA (MT-CDMA) applies the same data mapping and spreading as MC-DS-CDMA. However, its sub-carrier spacing is much smaller than the inverse of the chip rate that results in inter-carrier-interference (ICI). On the other hand, the tight sub-carrier spacing enables the use of spreading codes which are longer than the spreading code of a DS-CDMA. Therefore, at the expense of higher ICI, under certain conditions the system can supply more users than the DS-CDMA system. In Table 1 the main characteristics of these three schemes are illustrated.

MA-Schemes	**MC-CDMA**	**MC-DS-CDMA**	**MT-CDMA**
Spreading & Code-Division	Frequency Domain	Time Domain	Time Domain
Sub-carrier Spacing	1 / (NxTb /G)	≥ 1 / (NxTb /G)	1/ (N x Tb)
Detection	MRC, EGC, MMSE, MLD, Iterative	Coherent, No Rake (Correlation)	Rake or MIMO Equalizer
Applications	- Downlink - Uplink synch.	- Downlink - Uplink asynch.	- Downlink - Uplink asynch.
Specific characteristics	Very efficient for the downlink by using ⊥ codes	Designed especially for the uplink	Designed especially for the uplink

*Table 1.*Main Characteristics of different Multi-Carrier Spread-Spectrum Concepts

Since 1993, especially the first and the second schemes have been deeply studied and new alternative solutions have been proposed. Meanwhile, deep system analysis and comparison with DS-CDMA have been performed that show the superiority of MC-CDMA. In addition, new application fields have been proposed such as high rate cellular mobile, high rate wireless indoor and LMDS. In addition, a multitude of research activities has been addressed to develop appropriate strategies on detection and interference cancellation. The application of channel coding in MC-SS schemes, especially studying the use of convolutional and Turbo codes, has been another research subject during the last years. The synchronization and channel estimation aspects for the up- and downlink and the sensitivity of MC-SS schemes to synchronization imperfections have been also treated by many researchers. However, regarding the implementation and demonstration issue few works have been done.

SCOPE OF THIS ISSUE

The aim of this issue, consisting of seven parts, is to edit the ensemble of 41 contributions presented during three days of the Second International Workshop on *Multi-Carrier Spread-Spectrum (MC-SS) & Related Topics*, held from Sept. 15-17, 1999 in Oberpfaffenhofen, Germany.

The first part is devoted to the ***general issues*** of MC-SS & related topics. First, Sari, Vanhaverbeke and Moeneclaey give an overview of different multiple-access techniques. At the expense of slightly higher interference they propose to overlap the signals issued from different multiple-access schemes in order to increase the system capacity. Jankiraman and Prasad propose a hybrid CDMA/OFDMA with SFH for wide-band multimedia applications, where they emphasize that a hybrid scheme provides many advantages comparing to a single multiple access scheme. The idea of a generalized multi-carrier CDMA has been proposed by Giannakis, Anghel, Wang and Scaglione. They analyze the system performance in the uplink of a wireless cellular system suffering from multipath propagation. Finally, Kaiser, Kryzmien and Fazel propose the general idea of an asynchronous spread spectrum multi-carrier multiple access scheme. Different detection strategies in the case of shorter guard time have been analyzed.

In the second part of this book the different ***applications*** of multi-carrier and multi-carrier spread-spectrum for cellular mobile/fixed, underwater acoustic communications and HFC networks have been proposed. First Steendam and Moeneclaey propose and analyze the performance of a flexible form of MC-CDMA in a cellular mobile system. Then, Ormondroyd, Lam and Davies describe the application of multi-carrier spread spectrum for broadband underwater acoustic transmission. The results showed that by applying spreading together with multi-carrier transmission the system capacity will be doubled comparing to the conventional scheme. Ping presents a new combined MC-CDMA scheme for cellular mobile communications. Fazel and Engels present the concept of a transmission scheme based on multi-carrier FDMA for micro-waves point-to-multi-point system, where by employing adaptive modulation, dynamic bandwidth allocation and directive antenna the system capacity will be highly increased. Finally, Cherubini analyzes the performance of a hybrid TDMA/CDMA system based on multi-tone modulation for upstream transmission of hybrid fiber & coaxial (HFC) networks.

The third part of this issue is devoted to ***coding and modulation***. A general overview on coding and spreading for MC-CDMA is given by Lindner. He shows that by combing channel coding and spreading, high performance can be achieved in a frequency selective fading channel. Ziemer and Welch analyze the concept of a multi-carrier modulated DS-SS with differential PSK modulation using equal gain combing detection in Rician fading channel in the presence of Doppler and delay spread. The influence of code selection on the performance of MC-SS is analyzed by Kühne, Nahler and Fettweis. The use of complex spreading code in a MC-CDMA scheme is proposed by Dekorsey and Kammeyer that provides a higher gain compared to the conventional spreading codes. Haas and Kaiser study the performance of a two dimensional differential demodulation of OFDM frames. The application of block Turbo coding for Hiperlan-II based on OFDM is proposed by Maynou, Hinrichs and Mann Pelz. Finally, the performance of several spreading codes for MCSS in a frequency selective fading channel is analyzed by Xing, Rinne and Renfors.

The fourth part is devoted to the ***detection and multiplexing*** techniques. Here first Jarot and Nakagawa analyze the performance of a detection technique for MC-DS-CDMA in the presence of antenna diversity and power control. A wavelet-based MC-SS system with constant power is proposed by Attallah and Lim. For blind channel identification in multi-carrier CDMA systems a new method is proposed by Iglesia, Escudero and Castedo. A multiuser detection with iterative channel estimation in the case of multipath propagation is suggested by Lampe and Windpassinger. Bader, Zazo and Paez-Borrallo present a multi-user detection using a hexagonal pilot distribution for channel acquisition and tracking in MC-CDMA systems. Finally, Zong and Bar-Ness propose an adaptive multi-shot multi-user detection for an asynchronous MC-CDMA transmission, where as detection algorithm the bootstrap algorithm is employed.

In the fifth part of this book different ***interference cancellation*** strategies for combating multi-user interference and narrow-band interference are proposed. First Visser and Bar-Ness introduce a blind adaptive interference cancellation method for multi-carrier modulated systems. Then Ochiai and Imai present the performance of a simple interference cancellation strategy for the downlink of a cellular system based on MC-CDMA. McCormick, Grant and Povey present a wide- and narrow-band interference cancellation method for asynchronous MC-CDMA. Using a sub-space projection technique, Iglesia, Escudero and Castedo develop a multiple access interference cancellation for MC-CDMA. An interference cancellation technique for MC-SS system is presented by Nahler, Kühne and Fettweis. A Narrow-band interference mitigation technique for wireless OFDM-based

LANs is presented by Ness, Thoen, Van der Perre, Gyselinckx and Engels. Finally Baudais, Helard and Citerne analyze the performance of different interference cancellations strategies for MC-CDMA.

The ***synchronization and channel estimation*** aspects for MC and MC-SS transmission are discussed in the sixth part of this issue. An overview of MC-CDMA sensitivity to synchronization imperfection is given by Steendam and Moeneclaey. Hara presents a blind frequency offset/ symbol timing and symbol period estimation and sub-carrier recovery for OFDM signals in fading channels. The proposed algorithm is quite fast and needs only few OFDM symbols for its convergence. Robertson and Kaiser analyze the effect of Doppler spread in OFDM systems and the aspects of simulations for multi-carrier mobile radio and broadcast systems. A reduced stated joint detection, channel estimation and equalization for multi-carrier transmission is presented by De Broeck, Kullmann and Sorger. Hoeher describes the channel estimation strategies with superimposed pilot sequences for multi-carrier systems. A comparison of different channel estimation techniques for MC-CDMA and MC/JD-CDMA is done by Steiner. Said, Prasetyo and Aghvami present the performance of an OFDM burst synchronization method using Turbo codes. Finally, Matic, Petrochilos, Coenen, Schoute and Prasad propose an algorithm for simultaneous acquisition of frame, carrier frequency and symbol timing offsets and for channel estimation using a single OFDM symbol.

The last part of this book is devoted to the ***realization and implementation*** aspects. First, Matheus, Kammeyer and Tuisel present the implementation of multi-carrier systems using poly-phase filterbanks. Zazo, Bader and Paez-Borrallo analyze the improvement of a HF multi-carrier modem by using spread spectrum techniques. Then, Bury and Lindner propose the realization of an improved MMSE based block decision feedback equalizer for MC-CDMA. Finally, Fazel and Kaiser make a comparison of the performance of MC-CDMA and DS-CDMA in the presence of non-linear distortions using BPSK modulation. When using pre-distortions, both techniques achieve quite the same performance.

In conclusions, we wish to thank all of the authors who have contributed to this issue, and all those in general who responded enthusiastically to the call. We also hope that this book may serve to promote further research in this new area.

ACKNOWLEDGMENTS

The editors wish to express their sincere thanks for the support of the chairmen of the different sessions of the workshop namely, Prof. S. Hara from University of Osaka, Prof. P. Hoeher from University of Kiel, Prof. H. Imai from University of Tokyo, Prof. W.A. Krzymien from University of Alberta/TRLabs, Prof. J. Lindner from University of Ulm, Prof. R. Prasad from University of Aalborg, Prof. H. Sari from Alactel and Prof. R. Ziemer from University of Colorado. Many thanks to all invited authors whose contributions made the workshop successful. Furthermore, many thanks to the panelists, namely Dr. E. Auer from Bosch Telecom, Prof. M. Nakagawa from Keio University, Dr. R. van Nee from Luncent Technologies and Prof. H. Sari from Alcatel that have kindly accepted our invitations. We would like also to thank J. Uelner from DLR for her active support for the local organisation of the workshop.

This second international workshop on Multi-Carrier Spread-Spectrum & Related Topics could not be realised without the

- assistance of the ***TPC members***:

P. W. Baier (Germany)
Y. Bar-Ness (USA)
K. Fazel (Germany)
G. P. Fettweis (Germany)
J. Hagenauer (Germany)
S. Hara (Japan)
H. Imai (Japan)
S. Kaiser (Germany)
W. A. Krzymien (Canada)
J. Lindner (Germany)
L. B. Milstein (USA)
M. Moeneclaey (Belgium)
M. Nakagawa (Japan)
S. Pasupathy (Canada)
R. Prasad (Denmark)
M. Renfors (Finland)
H. Sari (France)
G. L. Stuber (USA)
L. Vandendorpe (Bel.)

- technical and financial support of:

German Aerospace Center (DLR)
Bosch Telecom GmbH

- and technical support of:

IEEE Communications Society, German Section
Information Technology Society (ITG) within VDE

Section I

GENERAL ISSUES ON MULTI-CARRIER SPREAD-SPECTRUM & RELATED TOPICS

Some Novel Concepts in Multiplexing and Multiple Access

Hikmet Sari[1], Frederik Vanhaverbeke[2], and Marc Moeneclaey[2]
[1]*Alcatel Radio Communications Division, 5 rue Noel Pons, 92734 Nanterre Cedex, France*
[2]*TELIN Dept., University of Ghent, St.-Pietersnieuwstraat 41, B-9000 Gent, Belgium*

Key words: Multiplexing, Multiple Access, TDMA, CDMA

Abstract: This paper introduces some novel multiplexing and multiple access concepts which increase the number of users that can be accommodated on a given channel. Specifically, N users are accommodated without any interference on a channel whose bandwidth is N times the bandwidth of the individual user signals, and additional users are accommodated at the expense of some signal-to-noise ratio (SNR) penalty. This breaks the hard capacity limit of N users that is specific to orthogonal-waveform multiple access (OWMA) schemes, and the multiuser interference (MUI) that is specific to CDMA with pseudo-noise spreading sequences (PN-CDMA) does not appear in the proposed multiple access techniques until the number of users exceeds N.

1. INTRODUCTION

From the standpoint of the number of users that can be accommodated on a given channel, multiple access techniques can be separated into two basic categories with distinct operating principles: The first category, which can be referred to as orthogonal-waveform multiple access (OWMA), includes the classical frequency-division multiple access (FDMA), time-division multiple access (TDMA), code-division multiple access with orthogonal spreading sequences (called Orthogonal CDMA, or OCDMA) [1], [2], orthogonal frequency-division multiple access (OFDMA) [3], and any other multiple access scheme which assigns orthogonal signal waveforms. These techniques can accommodate N users without any interference if the channel bandwidth is N times the bandwidth of the individual user signals, but N appears as a hard limit that can not be exceeded without reducing the individual bit rates. The other category includes CDMA with pseudonoise (PN) spreading sequences (PN-CDMA) and some other related schemes. Since the spreading sequences are not orthogonal in this technique, different user signals interfere with each other, and the interference power grows linearly with the number of simultaneous users. Consequently, PN-CDMA has a soft capacity which depends on the required performance.

In this paper, we describe several new concepts which allow to increase the number of users K beyond N which represents the hard capacity limit of OWMA, while ensuring that there is no multiuser interference (MUI) for $K \leq N$. The first one consists of augmenting OCDMA with PN sequences when the N orthogonal sequences are all assigned. The excess users which use PN sequences are subjected to mutual interference in addition to interference from orthogonal sequence users, while orthogonal sequence users only get interference from PN sequence users. The second introduced concept consists of using two sets of orthogonal signal waveforms. Several examples are given using TDMA, OCDMA, OFDMA, and multi-carrier OCDMA (MC-OCDMA) [4] as the component signal sets.

The paper is organized as follows: First, in the next section, we formulate the capacity problem of existing multiple access techniques. Then, in Section 3, we present a CDMA scheme which uses a combination of orthogonal and PN spreading sequences along with an iterative multistage detection technique. Section 4 describes a multiple access concept that uses two sets of orthogonal signal waveforms and gives several examples. Finally, we give our conclusions in Section 5.

2. PROBLEM FORMULATION

The number of users that can be accommodated on a multiple access channel is quite straightforward in OWMA. Suppose that the channel bandwidth is NW Hz, and that each user wants to transmit a data rate which requires a bandwidth of W Hz in the single-user case. The maximum number of users is then N for all OWMA techniques. A simple way to view this is to consider a TDMA system in which each frame is composed of N time slots and each of the K users gets one time slot per frame. The channel resources are thus entirely used for $K = N$. A similar reasoning holds for OCDMA. Consider an OCDMA scheme in which the spectrum of each user signal is spread by a factor of N through a multiplication by a Walsh-Hadamard (WH) [2] sequence of length N (N chips per symbol period). The number of WH sequences of this length being precisely equal to N, only N users can be accommodated in this system. An equivalent OFDMA scheme uses N carriers and assigns one carrier to each user. Here too, the maximum number of users N appears as the ratio of the total channel bandwidth to the bandwidth of the individual user signals. The same applies to any other OWMA scheme.

In PN-CDMA, spreading sequences are uncorrelated PN sequences and all users interfere with each other. The despreading operation at the receiver improves the signal-to-interference ratio (SIR) by a factor of 10log(N), where N denotes the spreading factor. Therefore, if the useful signal power is normalized by 1 and the spreading factor is N, the interference power from each other user is 1/N at the threshold detector input. The total interference power is proportional to the number of interfering users, and the number of users which can be supported is a function of the performance degradation that can be tolerated. In other words, PN-CDMA has a soft capacity that is determined by the required performance, which may be perceived as a desirable feature, but this technique also has the undesirable feature that MUI arises as soon as there are two active users. Fig. 1 gives an illustration of the interference power in OWMA and PN-CDMA for $N = 8$.

From the above capacity review of the existing multiple access techniques, it is clearly desirable to devise multiple access schemes which allow to accommodate a higher number of users than the spreading factor N while also ensuring interference-free transmission as long as the number of users K does not exceed N. This would combine the advantages of OWMA and PN-CDMA while avoiding their shortcomings. In search for such schemes, we have devised several new multiple access techniques which we describe below.

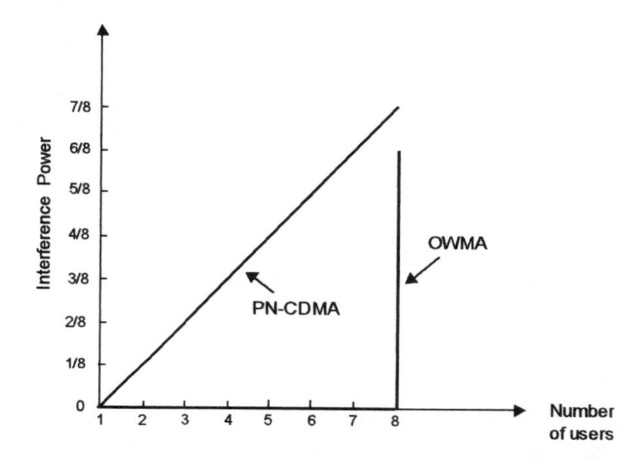

Figure 1. Interference power in OWMA and PN-CDMA for N=8

3. COMBINING OCDMA WITH PN-CDMA

The first idea toward extending the cell capacity beyond the spreading factor N while maintaining interference-free transmission for $K \leq N$, is to use OCDMA and augment it with PN-CDMA when $K > N$ [5]. More specifically, the base station in this scheme assigns WH spreading sequences to the first N users and PN sequences to the additional users. With $K = N + M$ users, a set of M users will be employing PN sequences. Signal detection is carried out iteratively, each iteration consisting of two separate stages, one for the set of users with WH sequences, and one for the set of users with PN sequences.

First Iteration

Stage 1: Since they use orthogonal spreading sequences, the first set of N users do not have any mutual interference. They only get an interference power of M/N from the second set of users. As long as M remains small, the correlator output for these users can be sent to a threshold detector to make symbol decisions with good reliability. For example, $M = N/4$ leads to an SIR of 6 dB, and since the interference is the sum of a large number of random variables, it tends to be gaussian. This is equivalent to operating the receiver at a signal-to-noise ratio (SNR) of 6 dB on an additive white gaussian noise (AWGN) channel. For a binary phase-shift keying (BPSK) modulation, this leads to a bit error rate (BER) on the order of 10^{-3}. This BER is not sufficiently low to consider the first-stage decisions as final, but they can be used as preliminary decisions in the next stage for interference cancellation.

Stage 2: Considering next the set of M users with PN sequences, they get an interference power of $N.1/N = 1$ from the first set of users in addition to their mutual interference power which amounts to $(M-1)/N$. The total interference power for these users is therefore 1+(M-1)/N which is clearly prohibitive for a threshold decision device, but the preliminary decisions of the first stage can be used to synthesize and subtract the interference of WH sequence users.

Let W_1, W_2, ..., W_N designate the N orthogonal sequences assigned to the first set of users. For i = 1, 2, ..., N, we write $W_i = (w_{i,1}, w_{i,2}, ..., w_{i,N})$. That is, $w_{i,j}$ designates the *j*th chip of the sequence W_i. Next, suppose that P_1, P_2, ..., P_M designate the portion corresponding to the current symbol of the PN sequences allocated to the second set of users. Note that the W_i sequences are independent of the symbol index, and although the PN sequences do not repeat from one symbol to the next, we also drop their symbol index, because the signal processing we consider in the receiver is memoryless. That is, detection of the current symbol does not involve signal samples from previous and future symbols. Consequently, we write $P_i = (p_{i,1}, p_{i,2}, ..., p_{i,N})$ for i = 1, 2, ..., M.

After despreading at the receiver, the total interference from the WH sequence users on the *k*th PN sequence user (the user with index N+k) signal is

$$I_{N+k} = \sum_{i=1}^{N} a_i \sum_{j=1}^{N} p_{k,j} w_{i,j} = \sum_{i=1}^{N} a_i P_k W_i^T \tag{1}$$

where a_i is the data symbol transmitted by the *i*th user during the current symbol interval, and the superscript T denotes transpose. Each term in this sum represents the interference from one user. Since P_k and W_i, i = 1, 2, ..., N are known to the receiver, I_{N+k} can be estimated once the symbol decisions are made by the receivers of users 1 to N. This estimate of I_{N+k} is then subtracted from the corresponding correlator output before sending this signal to the threshold detector. If all decisions are correct, interference cancellation is perfect, and the only remaining interference is the mutual interference of PN sequence users. The power level of this interference is (M-1)/N, and as for the first set of users, signal detection is possible at least to obtain preliminary decisions

Second iteration: The symbol decisions made for PN sequence users in the first iteration are used to synthesize their interference on the WH sequence users. The interference corrupting the *k*th user signal (k = 1, 2, ..., N) is given by

$$I_k = \sum_{i=1}^{M} a_{N+i} \sum_{j=1}^{N} w_{k,j} p_{i,j} = \sum_{i=1}^{M} a_{N+i} W_k P_i^T \,. \tag{2}$$

This interference is synthesized by substituting the second-stage decisions of the first iteration for the symbols actually transmitted. Since the decisions are correct with a probability close to 1, the synthesized replica is virtually identical to the actual interference. The synthesized interference is subtracted from the *k*th WH sequence user signal at the correlator output and the resulting signal is passed to the threshold detector. This process is repeated for all WH sequence user signals.

The second-iteration decisions for PN sequence users are made after subtracting their mutual interference based on the first-iteration decisions in addition to subtracting the interference of WH sequence users based on the second iteration decisions. The total interference corrupting the *k*th PN sequence user is given by

$$I'_{N+k} = I_{N+k} + \sum_{k' \neq k} a_{N+k'} \sum_{j=1}^{N} p_{k,j} p_{k',j} = I_{N+k} + \sum_{k' \neq k} a_{N+k'} P_k P_{k'}^T \,. \tag{3}$$

The first term in this expression is the interference from the set of WH sequence users given by (1), and the second represents the interference from the other PN sequence users. After subtracting the best available estimate of this interference, the correlator output for the kth PN sequence user is sent to the threshold detector which makes the second-iteration decision for that user. The results indicate that if the number of excess users M is not too large, the second iteration gives sufficiently good performance and the detection process stops at this iteration. But for larger values of M, further improvements are still possible from additional iterations.

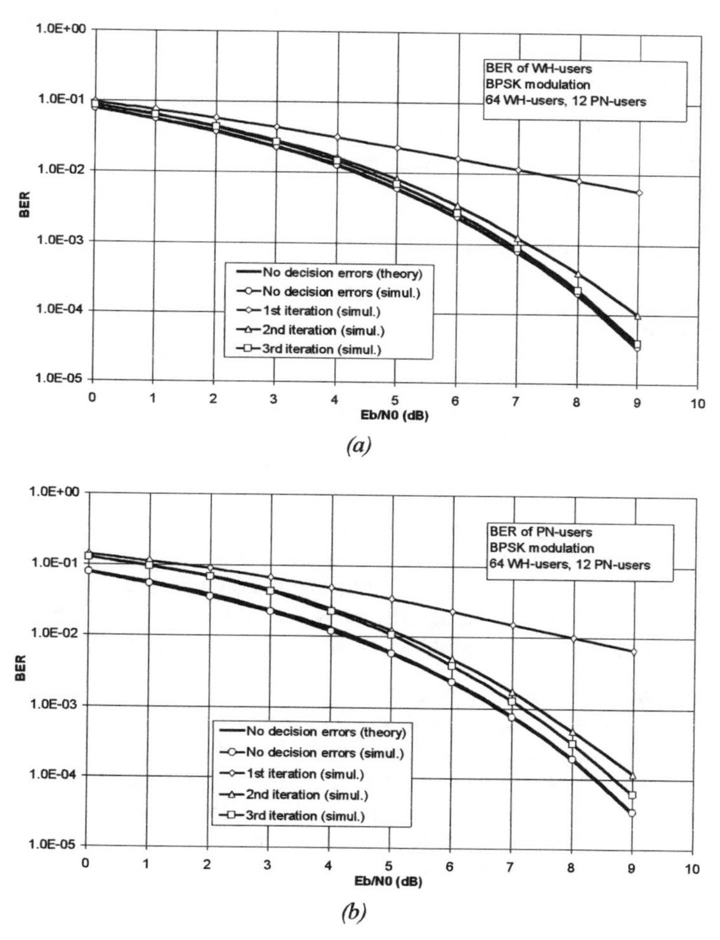

Figure 2. BER of OCDMA augmented with PN-CDMA.
a) BER of WH sequence users, b) BER of PN sequence users

Performance of this multiple access technique was evaluated analytically (assuming no decision errors in the interference cancellation steps) and by means of computer simulations. Fig. 2 shows the results obtained using a BPSK modulation, a spreading factor N = 64, and a number of PN sequence users M = 12. The first (resp. the second) plot gives the BER curves corresponding to the WH sequence users (resp. the PN sequence users) after the first, the second, and the third iterations. We observe that (as expected) the first iteration does not give sufficiently reliable decisions, but convergence quickly occurs after the second or the third iteration. We can also see that for a given number of iterations, performance is slightly better for WH sequence users which are free of mutual interference.

4. USING TWO SETS OF ORTHOGONAL SIGNAL WAVEFORMS

The multiple access technique described in the previous section makes use of two sets of signal waveforms, but only one of them is orthogonal, the other being a set of uncorrelated PN sequences. Analyzing the properties and performance results of this technique, it appeared more desirable to use two sets of orthogonal signal waveforms. In such a multiple access scheme, a given user will not interfere with other users which make use of resources from the same set, but only with users whose resources are from the other set. We will start the description of this concept using OCDMA and TDMA as the two sets of orthogonal signal waveforms.

4.1 Combined TDMA/OCDMA

For the sake of simplicity, we assume that the TDMA frame consists of N time slots of one symbol each. Suppose that resource allocation starts with length-N WH sequences. Up to N users, the considered scheme coincides with OCDMA. Once all WH sequences are assigned, the base station assigns TDMA time slots to additional users. All signal waveforms are assumed to have equal energy, denoted $E_S = PT_S$, where T_S designates the OCDMA symbol period. Fig. 3 shows the instantaneous power of the waveforms used for N = 8. The energy of OCDMA waveforms is uniformly distributed over the symbol period $T_S = 8T_C$, where T_C is the chip period (which is also the TDMA symbol period). The TDMA waveform in this figure interferes with the first chip of all OCDMA users, but all other chip periods are free of interference. Since all users have the same symbol energy, the correlator output in the OCDMA receivers will have an SIR of N, or 10.log(N) in the dB scale, when only one TDMA user is active. With M TDMA users, the SIR is N/M.

First iteration: Provided that M remains moderately small with respect to N, the correlator output in the N OCDMA receivers can be sent to a threshold detector to make preliminary decisions. This is the first stage of the detection process. The second stage starts with the synthesis of the interference of OCDMA users on TDMA users using the preliminary decisions obtained from the first stage. Omitting the time index as previously, the interference on the *k*th TDMA user (the user with index N+k) can be written as follows:

$$I_{N+k} = \sum_{i=1}^{N} a_i w_{i,k} \tag{4}$$

where a_i is the present symbol of the *i*th OCDMA user, and $w_{i,k}$ is the *k*th chip of the WH sequence assigned to that user. This interference is estimated by simply substituting the first-stage decisions $\hat{a}_i$ (i = 1, 2, …., N) for the symbols actually transmitted a_i. The estimated interference is subtracted from the received TDMA signals, and after this operation, the TDMA signals are passed to a threshold detector. As in the previous technique, M = N/4 leads to an SIR of 6 dB and a BER of 10^{-3} for the first-stage preliminary decisions assuming BPSK signaling. This means that interference cancellation in stage 2 is close to ideal even with such a high value of M. In other words, the second-stage decisions on the transmitted TDMA symbols will be very reliable, and the corresponding BER curve will be close to the ideal curve corresponding to interference-free transmission.

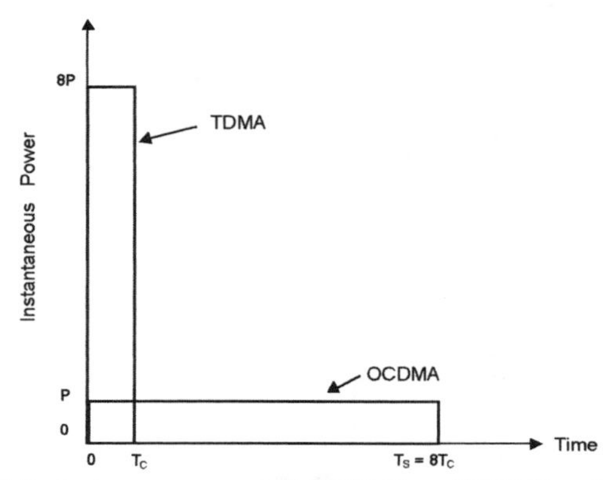

Figure 3. Instantaneous power of OCDMA and TDMA pulses for N=8

Second iteration: In the second iteration, the interference of TDMA users on OCDMA users is synthesized using the first-iteration symbol decisions for these users. The interference from M TDMA users at the correlator output of OCDMA receivers is expressed as

$$I_k = \sum_{i=1}^{M} w_{k,i} a_{N+i} \tag{5}$$

for the *k*th user. The estimated interference is subtracted from the correlator output, and an improved decision is made for the symbol transmitted by each one of the N OCDMA users. Since most TDMA symbol decisions made in the first iteration are correct, interference cancellation from OCDMA user signals in the second iteration will be close to perfect, and this step will hopefully give the final receiver decisions for these symbols. Also, there is little need in the second iteration to use these decisions for canceling in a second stage the interference caused by the OCDMA users on the TDMA users as most of the TDMA user signals have been detected correctly during the first iteration. The simulation results indicate that this is indeed the case for small values of M, but that further iterations are required in the detection process for larger values of M.

For this multiple access technique too, the user BER values were evaluated by means of computer simulations and the results were checked against the theoretical results obtained in the absence of decision errors in the interference cancellation steps. Fig. 4 shows the results obtained using BPSK, N = 64, and M = 12. The first (resp. the second) plot gives the results corresponding to the OCDMA users (resp. TDMA users) after one and two iterations. We observe that the first-iteration decisions are not reliable for OCDMA users as they are made in the presence of interference from 12 TDMA users. In contrast, the first-iteration decisions of TDMA users are much more reliable since they are made after subtracting the interference from the 64 OCDMA users. But the most remarkable result is that for both sets of users, the BER curve after the second iteration virtually coincides with the ideal curve which corresponds to interference-free transmission. This result implies that no further iterations are needed at least with this number of excess users.

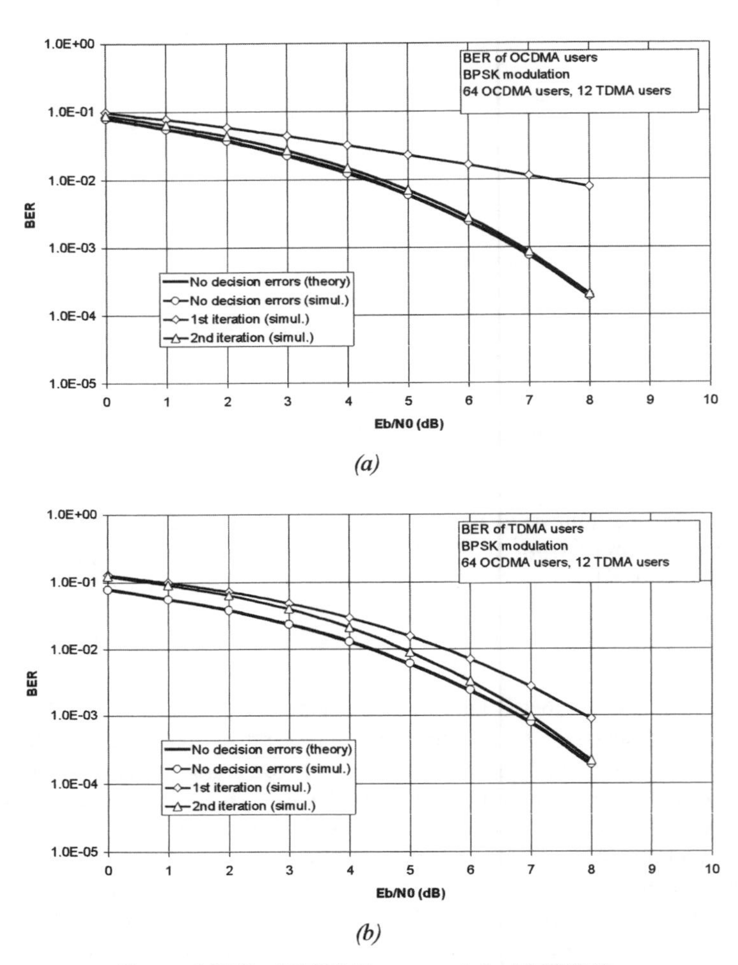

Figure 4. BER of OCDMA augmented with TDMA.
a) BER of OCDMA sequence users, b) BER of TDMA sequence users

Instead of starting with WH spreading sequences for the first N users and assigning TDMA time slots to the additional users, exactly the opposite can be made. In this case, the first N users get N TDMA time slots and additional users get length-N WH sequences. The interference problem and its cancellation remain the same as previously, although performance may differ. Specifically, each TDMA user is corrupted by one chip of each OCDMA user, and the total interference power from the set of M OCDMA users is M/N. Again, as long as M does not take excessive values, preliminary decisions can be made on the symbols transmitted by the N TDMA users. Next, those decisions are used to synthesize and cancel the interference of TDMA users on OCDMA users and make decisions on the symbols transmitted by the latter users. Further iterations continue as described above.

4.2 Other Signal Set Combinations

The TDMA and OCDMA waveform sets are only one example, and any other sets of orthogonal signal waveforms can be used. Note that both TDMA and OCDMA are time-domain techniques. Their frequency-domain counterparts are OFDMA and multicarrier OCDMA (MC-OCDMA) [4], respectively. In an OFDMA system with N carriers, each assigned to one user, the component OFDMA signals occupy (1/N)th of the channel bandwidth, and the symbol period is N times the symbol period of TDMA. That is, the symbol period in OFDMA is the same as in OCDMA. As for MC-OCDMA, it consists of spreading the transmitted signal spectrum in the frequency domain instead of spreading it in the time domain. This is performed by entering to an inverse discrete Fourier transform the baseband signal after spectral spreading. The combination of OFDMA with MC-OCDMA has the same power density representation as combined TDMA/OCDMA except that the time axis in Fig. 3 must be replaced by the frequency axis. It is quite straightforward to see that at the threshold detector input, each OFDMA user gives an interference of 1/N to MC-OCDMA users, and conversely, each MC-OCDMA user gives the same amount of interference to OFDMA users. The new multiple access concept is therefore readily applicable to combined OFDMA/MC-OCDMA, and resource allocation to the first N users can start with either of the two sets of signal waveforms.

We can also combine time-domain and frequency-domain techniques. To view this, it is useful to look at Fig. 5 which illustrates the 3-D power density of the basis functions in TDMA, OFDMA, OCDMA, and MC-OCDMA. We can see that TDMA and OFDMA are easily distinguished through their distinct power distributions in the (time, frequency) plane. In a combined TDMA/OFDMA system, each OFDMA (resp. TDMA) user gives an interference power of 1/N to each TDMA (resp. OFDMA) symbol, and as in the combined TDMA/OCDMA described in the previous section, either of the two sets can be assigned to the first N users, while the other set is assigned to the next M users. Users from the first set will be corrupted by an interference power that is equal to M/N, whereas users from the second set will be corrupted by an interference power of 1. Iterative multistage detection follows the same basic principle as previously, the first stage making decisions for the first set of users and the second stage making decisions for the second set of users.

Fig. 5 also suggests that due to their distinct power density representations, our technique is equally applicable to the TDMA/MC-OCDMA and OFDMA/OCDMA signal set combinations, because in each case the mutual interference between two users from different signal sets is 1/N when the number of basis functions per set is N. But the power density representation alone is misleading, because it would suggest that the new multiple access concept is not applicable to the OCDMA/MC-OCDMA combination as these signal waveform sets have identical power densities. A closer analysis of these multiple access techniques indicates, however, that despite the identical power density of the signal waveforms involved, the mutual interference at the threshold detector input is small, and the new concept is also applicable to this combination.

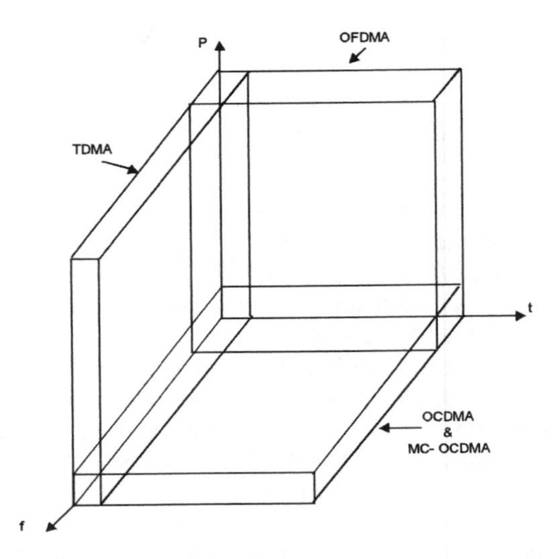

Figure 5. Power density of OWMA techniques in the (time, frequency) plane.

5. SUMMARY AND CONCLUSIONS

We have introduced two novel multiplexing and multiple access concepts which extend the number of users beyond the spreading factor N (the total channel bandwidth divided by the bandwidth required by each user in the single-user case). The first one is a full CDMA concept which assigns orthogonal sequences to the first N users and PN sequences to the additional users. The second concept uses two sets of orthogonal signal waveforms, and avoids any interference between users with resources from the same set. The second technique was described using TDMA and OCDMA, but its generalization to other signal sets was also briefly outlined. In both techniques, iterative multistage detection was used to compensate for MUI. Computer simulation results indicated that the presented techniques increase the number of users on a multiple access channel by 30 to 40 % at the expense of a negligible SNR penalty. This opens up some interesting perspectives in the fields of multiplexing and multiple access.

REFERENCES

[1] A. J. Viterbi, CDMA: Principles of Spread Spectrum Communication, Addison-Wesley Wireless Communications Series, 1995, Reading, Massachusetts.

[2] T. S. Rappaport, "Wireless Communications: Principle & Practice," IEEE Press & Prentice Hall, 1996.

[3] H. Sari and G. Karam, "Orthogonal Frequency-Division Multiple Access and its Application to CATV Networks," European Transactions on Telecommunications (ETT), vol. 9, no. 6, pp. 507-516, November-December 1998.

[4] N. Yee, J.-P. Linnartz, and G. Fettweis, "Multicarrier CDMA for Indoor Wireless Radio Networks," Proc. PIMRC '93, pp. 109-113, September 1993, Yokohama, Japan.

[5] H. Sari, F. Vanhaverbeke, and M. Moeneclaey, "Increasing the Capacity of CDMA Using Hybrid Spreading Sequences and Iterative Multistage Detection," Proc. the 1999 IEEE 50th Vehicular Technology Conf. (VTC '99 – Fall), September 1999, Amsterdam.

Wideband Multimedia Solution Using Hybrid CDMA/OFDM/SFH Techniques

M. Jankiraman[+] and Ramjee Prasad[*]

+Centre for Wireless Personal Communication, IRCTR,
Telecommunications and Traffic Systems Group, Faculty of Information Technology and Systems,
Delft University of Technology, P.O. Box 5031, 2600 GA Delft, The Netherlands
Wireless Information and Multimedia Chair, Co-Director Center for Personkommunikation, Aalborg University, Institute of Electronic Systems, Frederik Bajers Vej 7A5, DK-9220, Aalborg, Denmark
Email : m.jankiraman@its.tudelft.nl , prasad@cpk.auc.dk

Key Words : CDMA/OFDM/SFH, Wideband Multimedia

Abstract : This paper investigates the combination of CDMA with OFDM and SFH as a possible solution for a multiple-access system. The CDMA component provides the inherent advantage of DS-CDMA systems incorporating a spreading signal based on PN code sequence, by providing user discrimination based on coding at the same carrier frequency and simultaneously. The OFDM component provides resistance to multipath effects making it unnecessary to use RAKE receivers for CDMA and thus avoid hardware complexity. The SFH component provides resistance to the "near-far" problem inherent in CDMA. Thus we have a system that retains the greatest advantage of CDMA viz. multiple access without the disadvantages of "near-far" problems and complexity of RAKE receivers.

1. INTRODUCTION

Multicarrier systems have gained an increased interest during the last years. This has been fuelled by a large demand on frequency allocation resulting in a crowded spectrum as well as a large number of users requiring simultaneous access. The quest received a fillip with the onset of CDMA systems. CDMA protocols do not achieve their multiple access property by a division of the transmissions of different users in either time or frequency and it is already getting too crowded in these domains, but instead make a division by assigning each user a different code. This code is used to transform a user's signal to a wideband signal (spread spectrum signal). If a receiver receives multiple wideband signals, it will use the code assigned to a particular user to transform the wideband signal received from that user back to the original signal. All other codewords will appear as noise due to decorrelation.

Each user, therefore, operates independently with no knowledge of the other users. There is, however, a problem. The power of multiple users at a receiver determines the noise floor after decorrelation. If this power of a near user is not controlled as compared to a the power of a far user, then these signals will not appear equal at the base station. Then the stronger received signal levels raise the noise floor at the base station demodulators for the weaker signals, thereby decreasing the probability that weaker signals will be received. This is called the "near-far" problem. Therefore, an elaborate power control scheme is provided by each base station by rapidly sampling the radio signal strength indicator (RSSI) levels of each mobile and then sending a power change command over the forward link. The second problem in CDMA systems is that the signals travel by different paths to the receiver. It is, therefore, preferred to use a RAKE receiver to utilise maximum ratio combining techniques to take advantage of all the multipath delays, in order to get a strong signal. These RAKE receivers are extremely complex.

In order to combat the "near-far" problem, DS-CDMA was combined with Slow Frequency Hopped (SFH) systems. This hybrid technique consists of a direct sequence modulated signal whose centre frequency is made to hop periodically in a pseudorandom fashion. This ensures that, even within the same cell, no two mobiles are operating at the same frequency.

There was, however, an industrial demand for very high bit rates, irrespective of the type of access scheme used. This was fuelled by multimedia considerations requiring bit rates as high as 155 Mbps. This gave rise to OFDM systems. In such systems very high data rates are converted to very low parallel data rates using a series-to-parallel converter. This ensures flat fading for all the sub-carriers , i.e. , a wideband signal becomes a packet of narrowband signals. This will automatically combat multipath effects removing the need for equalisers and RAKE receivers. A variant of this approach was earlier introduced as Multicarrier - CDMA or MC-CDMA. This proposal envisages interfacing a DS-CDMA system with a system of orthogonal coding (not OFDM) using only Walsh coding [1][11][4]. We shall briefly examine the advantages and disadvantages of the prevailing systems before proceeding to examine the new proposal. It is pointed out that the advantages/disadvantages listed are not comprehensive, but only those relevant to this paper.

It can be seen in table 1, that each multiple access approach has its advantages and disadvantages. It is especially to be noted that MC-CDMA achieves the same result as OFDM, but in a more complex manner. This paper proposes a comprehensive approach maximising the merits and minimising the demerits of the individual schemes.

TYPES OF MULTICARRIER ACCESS SCHEMES

TYPE	ADVANTAGES	DISADVANTAGES	REMARKS
DS-CDMA	i) Can address multiple users simultaneously and at same frequency. ii) Interference rejection	i) Problems due to "near-far" effect. ii) Complex Time Domain RAKE receivers. iii)Synchronisation within fraction of chip time becomes difficult.	Refer [1] [2] [3]
SFH-CDMA	i)Reduces "near-far" effect. ii) Synchronisation within fraction of hop time is easier. iii)No need for contiguous bandwidths	i)Complex Frequency Domain RAKE receivers. ii)Coherent demodulation difficult because of phase relationship during hops.	Refer [1] [2] [3]
MC-CDMA	i) Double the processing gain of DS-CDMA. ii)Higher number of users as full bandwidth is utilised unlike in DS-CDMA	i) Peak-to-Average ratio problem. ii) Synchronisation problems. iii)Overcrowding of the spectrum as each bit is spread across the available bandwidth based on Walsh coding.	Refer [4] [1]
OFDM-SFH	i) Anti-multipath capability. ii)Multiple access due to FH requires very wide bandwidths depending upon the number of users.	i) Peak-to-Average ratio problem. ii) Synchronisation problems.	Refer[5] [6]

Table 1 : Types of Multicarrier Access Schemes

2. OVERALL CONCEPT

The overall concept is shown in the schematic in fig 1 :

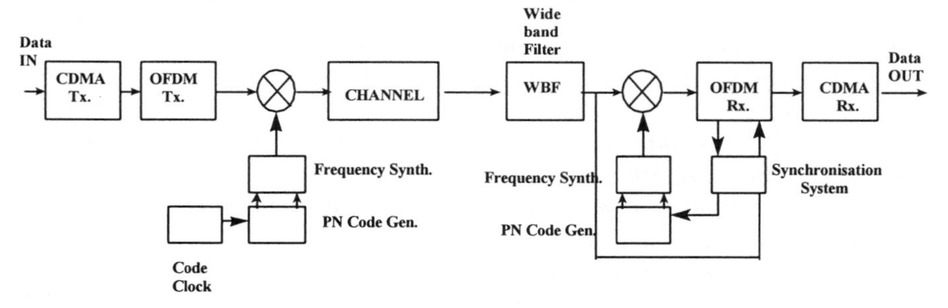

OVERALL SYSTEM SCHEMATIC

Figure 1 : Overall System Schematic

The schematic is self-explanatory. However, there are a few salient points to be noted :

i) **Bandwidth and other considerations :** In this paper, we are assuming Rayleigh fading conditions and AWGN. We are also assuming perfect OFDM synchronisation with no carrier offset. However, in reality, with a multimedia requirement of high bit rates, typically 155 Mbps, we need wide bandwidths of around 100 MHz or higher. The CDMA systems till now pertain to voice channels with a bandspread factor of 128 for a data rate which is at the most 9600 bps. This is necessary due to the adverse transmission conditions and also due to a lot of users at that frequency (around 850 MHz). In our case, however, we intend to operate at around 60 GHz, where larger bandwidths are available. This means Rician fading conditions and line-of-sight transmissions, conditions which are not so severe. Therefore, a bandspread factor of 128 is unnecessary. More likely, a bandspread factor of 10 or less will prove sufficient. This, however, has to be verified by extensive simulations. If we assume a bandspread factor of 10, then we would require a bandwidth of at most 1GHz for a 100 MHz data rate. In such a case the proposal shown in fig 2 is a better approach. In this case, we hop bandslots of subcarriers within the bandwidth of the OFDM system. If the bandspread factor is much less than 10, we can adopt the approach fig 3 which is easier to implement.

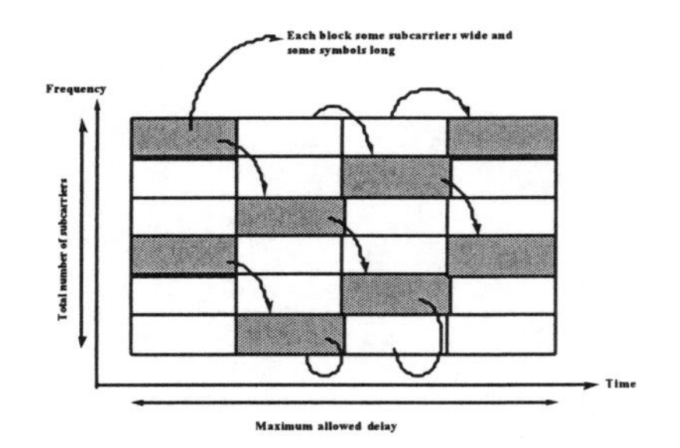

Figure 2: Hopping within Bandwidth

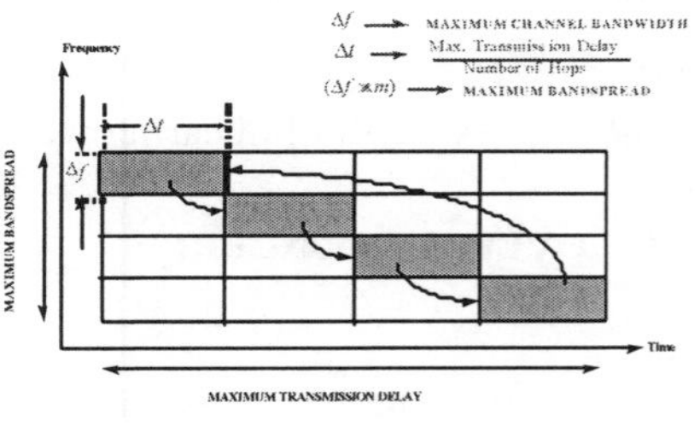

Figure 3 : Hopping between Bandwidths

The basic motivation for the scheme in fig 1 above, is not so much having a robust design, for multipath rejection as having a design that can incorporate a lot of users. There is no need for such a robust design at such frequencies.

But we can expect a steep rise in the number of users when high data rates become realisable, especially with regard to video-telephones. It is the CDMA aspect (code diversity) which really gives rise to a lot of users. The frequency hopping has been introduced to obtain frequency diversity, to reduce the "near-far" problem. This limits the number of users in order to avoid "collisions". However, we still wish to incorporate a lot of users. The CDMA aspect makes up for this limitation by introducing a larger number of users due to code diversity. Interleaving and error correction coding may be dispensed with if the need so arises , i.e. , if the channel is not severe. In case, in the foreseeable future, the channel does pose problems, we can increase the spread factor of the CDMA transmission and/or introduce FEC and interleaving. The OFDM aspect is required, because it eliminate the need for RAKE receivers and allows us to use coherent modulation inspite of frequency hopping which makes coherent signal processing difficult. It also helps reduce the burden of synchronisation related to CDMA systems (see (v) below).

ii) **Error correction coding :** In the CDMA transmitter, there are two levels of coding viz. Convolution encoding (which is an error correcting coding) and Walsh encoding (which is a spreading code) . The latter is an orthogonal coverage, since PN sequences by themselves are insufficient to ensure channel isolation. This orthogonal coverage should not be confused with the orthogonal sub-carrier in OFDM transmitter. The Walsh coding ensures orthogonality *between* users. The OFDM orthogonality ensures that the sub-carriers for each data frame are orthogonal. The convolutional encoding ensures robustness of data as discussed above.

iii) **Modulation :** In our case, the CDMA sequence after Walsh coding, does not get converted to RF via QPSK, DPSK or any other type of modulation, but instead is fed as an input to the OFDM transmitter. In the OFDM transmitter it gets modulated as M-QAM or any other type of modulation.

iv) **CDMA Receiver :** Similarly, in the CDMA receiver, the OFDM receiver gives it a sequence after M-QAM demodulation. Thereafter, the CDMA signal processing is carried out, in that there is a *digital* correlator (correlated to the respective PN sequence) which ensures channel isolation. The output of the correlator is then given to a Viterbi decoder. The output from this decoder is the required data sequence. RAKE receivers are not required in this case at the matched filter level.

v) **Synchronisation :** Stringency of synchronisation is, however, still required as the PN sequences need to be synchronised. However, in such a hybrid system, the burden of synchronisation is transferred to the OFDM system. The OFDM system has a more sophisticated synchronisation system [5] than in CDMA synchronisation systems, as the OFDM system utilises the cyclic prefixes to synchronisation. Hence, the PN sequences emerging from the OFDM system and going to the CDMA system, are already better synchronised than in a pure CDMA system. The reader will recall that synchronisation is one of the limiting factors in CDMA systems for high data rates. It is expected that in our system, such problems will be considerably reduced.

vi) **Bit Error Probabilities :** The proposed system is essentially an OFDM-FH system. This is, because, the transmission and reception is carried out by the OFDM-FH system. The CDMA aspect only generates the data stream, but in a more complicated way. Hence, the same analysis applies to this system as for OFDM-FH systems, for determining Bit Error Probabilities.

vii) **Trade-off between OFDM and CDMA :** The OFDM-FH system by itself does not solve the multimedia requirement. This is, because, multimedia requires very high bit rates, typically, 155 Mbps. This requires large bandwidths of typically

100 MHz. By using FH among users, our number of users comes down drastically, being limited by the bandwidth available. By adding CDMA to this, we have enhanced the number of users, since CDMA supports a large number of users working at the same frequency and bandwidth hopping is used to obtain frequency diversity with a view to reducing the "near-far" effect. Hence, there is a trade-off.

viii) **CDMA Signal Processing :** It will be argued that the present data rate of 1.2288 Mbps in IS-95 systems, is woefully inadequate for multimedia applications. This is acknowledged. However, it must be borne in mind that if we want a large number of users, we need to use CDMA techniques. Hence, efforts must be made to increase data rates using better PLLs for synchronisation and high speed digital electronics. It is pointed out that in this proposal, the entire CDMA signal processing is carried out in the digital domain both for the transmitter as well as for the receiver. At no time is matched filtering carried out at RF level for CDMA signals nor is there any operation at RF level including transmission and reception for CDMA signals. This is a big plus point, as signal processing is easier in the digital domain.

3. PERFORMANCE EVALUATION

A Comparison of DS-CDMA and DS-CDMA-SFH (DS-SFH) Systems

It is assumed that there are K active users, each having a transmitter-receiver pair. Each user employs a channel encoder. We denote by $b_K(t)$ the modulating sequence of the Kth user, which randomly takes values from the set {+1, -1} with equal probability. The coded bit duration is indicated by T and the transmitted power by P.

$$h_K(t) = \sum_{l=1}^{L} h_{Kl}\delta(t - \tau_{Kl})\exp(j\phi_{Kl}) \text{ --------------------(1)}$$

For each path, the gain h_{Kl} is Rayleigh distributed, the delay τ_{Kl} is uniformly distributed over [0,2π]. We note that the average power for each path and each user is $E[h_{Kl}^2] = \sigma_0/2$. The receiver introduces AWGN *n(t)* with two sided power spectral density *N0/2*.

1. DS-CDMA

The transmitted signal $x_K(t)$ is :

$$x_K(t) = \text{Re}\{\sqrt{2P}b_K(t)c_K(t)\exp[j(2\pi f_c t + \theta_K)]\} \text{ ------------------(2)}$$

where f_c is the carrier frequency, θ_K is the phase introduced by the DPSK modulator, $c_K(t)$ is a PN sequence with a rate $1/T_c$ and rectangular pulse.

The received signal *r(t)* is given by :

$$r(t)=\sum_{K=1}^{K}\sum_{l=1}^{L}h_{Kl}x_K(t-\tau_{Kl})\cos\phi_{Kl}+n(t) \text{ --------------------(3)}$$

At the ith receiver, the signal corrupted by noise and interference, is filtered, downconverted and recovered at the output of a filter matched to the spreading code $c_i(t)$. It is sampled at the instants $\lambda T+\tau_{il}$ and is then given to the DPSK demodulator. The demodulated output is denoted by $\Psi_l[\lambda]=\mathrm{Re}\{d_l[\lambda]d_l^*[\lambda-1]\}$ where $d_l[\lambda]$ is the complex notation of the matched filter output. This demodulator output, is then given to a Maximum Ratio Combiner(MRC) to yield the final output $\Psi[\lambda]=\frac{1}{M}\sum_{l=1}^{M}\Psi_l[\lambda]$, where M is the diversity,
$M \le L$.

The matched filter output $d_l[\lambda]$ comprises,

$$d_l[\lambda]=W_l[\lambda]+I_l^{MP}[\lambda]+I_l^{MA}[\lambda]+N_l[\lambda] \text{ ------------------(4)}$$

where $W_l[\lambda]$ is the required signal,

$$W_l[\lambda]=\sqrt{P/2}h_{il}b_i[\lambda]\exp(j\phi_{Kl}) \text{ ----------------------------(5)}$$

The interference contributions are zero - mean complex Gaussian random variables and comprise I_l^{MP} which is the multipath interference term caused due to L-1 other path signals and I_l^{MA} is the multiaccess interference term caused due to K-1 other users.
The average SNR is given by the approximation [7] :

$$SNR=\gamma=\left(\left(\frac{2E_b}{N_0}\right)^{-1}+\frac{LK-1}{N}\times\frac{1}{3}\right)^{-\frac{1}{2}} \text{ ----------------------------(6)}$$

which is the SNR per bit. The constant 1/3 is due to the rectangular chip pulse. This equation is valid for binary DS/SSMA systems. In case of MRC with diversity of order M, given that the received signal is Rayleigh distributed, the bit error probability P_b at the decoder input is given by [3],

$$P_b=\frac{1}{2^{(2M-1)}(M-1)!(1+\gamma)^M}\sum_{m=0}^{M-1}C_m(M-1+m)!\left(\frac{\gamma}{\gamma+1}\right)^m \text{ ---------------(7)}$$

where

$$C_m=\frac{1}{m!}\sum_{n=0}^{M-m-1}\binom{2M-1}{n} \text{ ----------------------------(8)}$$

At the output of the (n, k) block decoder, the probability of bit error is given by [3],

$$P_e = \sum_{j=t+1}^{n} \binom{n}{j} P_b^j (1-P_b)^{n-j} \quad \text{------(9)}$$

where t denotes the error correction capability of the code.

2. DS-CDMA-SFH

The general derivation for this system is similar, except that the system uses Binary Frequency Shift Keying (BFSK) modulation followed by non - coherent demodulation, because, of the nature of slow frequency hopping as discussed earlier. The considered system is asynchronous as this is the more realistic case. Even when synchronism can be achieved between individual user clocks, radio signals will not arrive synchronously to each user due to propagation delays.
The probability of bit error for BFSK is given by [3],

$$Pb = \frac{1}{2^{(2M-1)}} \exp\left(-\frac{\gamma}{2}\right) \sum_{K=0}^{M-1} C_K \left(\frac{\gamma}{2}\right)^K \quad \text{------(10)}$$

where,

$$C_K = \frac{1}{K!} \sum_{n=0}^{M-1-K} \binom{2M-1}{n} \quad \text{------(11)}$$

However, if two users transmit simultaneously in the same frequency band, a collision or "hit" occurs. In this case, we assume the probability of error as 0.5. Hence, the overall probability of bit error for frequency hopped BFSK signal is ,

$$P_b = Pb(1-P_{hit}) + \frac{1}{2} P_{hit} \quad \text{------(12)}$$

where P_{hit} is the probability of hit, derived as discussed below.

If there are q possible hopping channels, there is a $\frac{1}{q}$ probability that a given interfering user will be present in the desired user's slot. Hence, for K-1 interfering users, the probability that at least one is present in the desired frequency slot is,

$$P_{hit} = 1 - \left\{1 - \frac{1}{q}\left(1 + \frac{1}{N_b}\right)\right\}^{K-1} \quad \text{------(13)}$$

where N_b is the number of bits/hop. We take this value as 1 , i.e. , 1 hop/bit for slow hopping.

At the output of the (n, k) block decoder, the probability of bit error is given by [3],

$$P_e = \sum_{j=t+1}^{n} \binom{n}{j} P_b^j (1-P_b)^{n-j} \quad \text{------------------------(14)}$$

where t denotes the error correction capability of the code.

Numerical Results

We now examine the performance of the DS-CDMA and DS-SFH systems.

In fig 4, we compare the DS-CDMA with DS-SFH for various diversity values ranging from 1 to 4. We have assumed q=1000 i.e. a very large number of frequency hopped channels for the DS-SFH system. In such a case, the limiting error will be due to other users. It can be seen that at high SNRs, the DS-SFH is superior . At a diversity of 4, the DS-CDMA performs well at SNRs below 10 dB. The asymptotic behaviour of the DS-SFH curve is due to multiple access interference as expected.

B. OFDM - FH

The probability of bit error for M- QAM is given by [2]:

$$P_b = 4\left(1-\frac{1}{\sqrt{M}}\right)\frac{1}{2}erfc\left(\sqrt{\frac{E_b}{N_0}}\right) \quad \text{-------------------------------(15)}$$

In our case, M=16. We correct this, for probability of hit using (8) and then correct for error correction coding using (10).

Numerical Results

We now compare DS-SFH which has already shown itself to be a superior system compared to DS-CDMA with OFDM - FH.

We note from fig 5, that the OFDM system performs better due to the coherent nature of 16 QAM as compared to DS-SFH, operating with a diversity of 4. The OFDM can perform even better if we take into account the interference and frequency diversity caused due to interleaving and frequency hopping. We have used a strong error correcting code like the Reed-Solomon (32,12) code in both cases as compared to BCH(15,7) in fig 4. We have also assumed a q of 1000 with 10 users for both cases and also a cyclic prefix of 20% ($\eta_g = 0.8$) for the OFDM system.

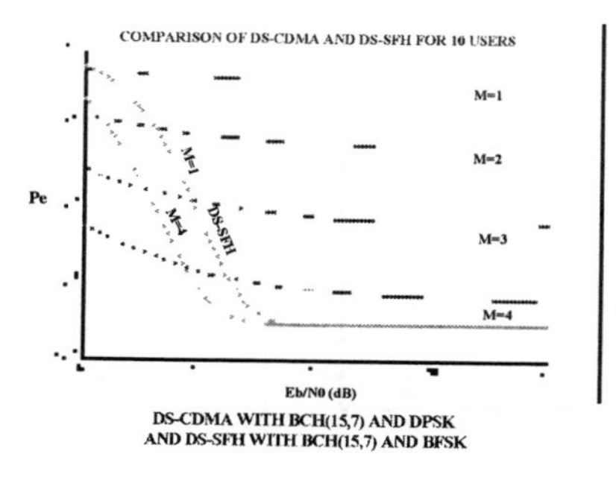

DS-CDMA WITH BCH(15,7) AND DPSK AND DS-SFH WITH BCH(15,7) AND BFSK

Figure 4

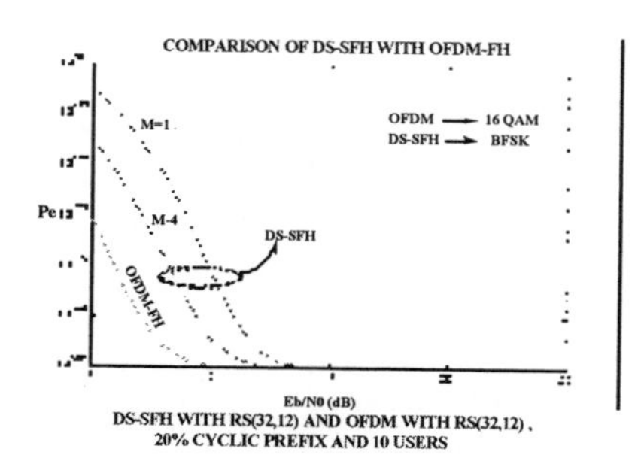

DS-SFH WITH RS(32,12) AND OFDM WITH RS(32,12), 20% CYCLIC PREFIX AND 10 USERS

Figure 5

If we reduce the cyclic prefix further, the SNR improves, but at the cost of poor synchronisation [5][6]. Hence, we can definitely conclude that OFDM - FH is a better choice. The bit error probability for the OFDM-FH system appears to be extremely low. The explanation for this is, that unlike for DS-SFH, we have not only taken 16 QAM as the coherent modulation technique, but we have also used a strong coding like Reed-Solomon coding. This is apparent in fig 6. In the figure, we note that the curve for bit error without RS coding and with 16 QAM, is inferior to the one with RS coding. The former is similar to the one derived by Proakis [3] for 16 QAM modulation. This result is due to the effect of equation (10) with RS coding. In reality, due to residual "near-far" effect, RS coding will help to improve the system performance, but the OFDM-FH result will not be so exceptional.
Furthermore, in fig 5, both the curves tend to zero for SNRs in excess of 10 dB. This is, because, if we take a large q, equation (9) reduces to $P_{hit} = (K-1)/q$. If we now take a large SNR, the P_b (equation(6) and (8)) tends to $(1/2) \times (K-1)/q$. For large q this value tends to zero i.e. multiple access interference limited. It now remains for us to identify as to what should be the typical value of q for such an OFDM -FH system. This is shown in fig 7 below.
We can see that Reed-Solomon coding performs better as compared to BCH coding. A q value of 20 hops appears to be an optimum value. This is, because, the chances of collisions and consequent errors appear to be less with a q of 20. However, a larger spread of q is preferred.
Finally, we examine the effect of multiple users on the performance of the system.
We note from fig 8, that a maximum of 10 users is ideal. Users in excess of this number, will cause collision and consequent poor bit error capability. However, this is where the advantage of our approach pays off. We can pass the burden of the additional users to the CDMA system!!! This means that we are looking at a scenario of 10 users operating simultaneously using the OFDM-FH concept, but in actuality there will be many more, being limited only by the capability of the respective CDMA systems.

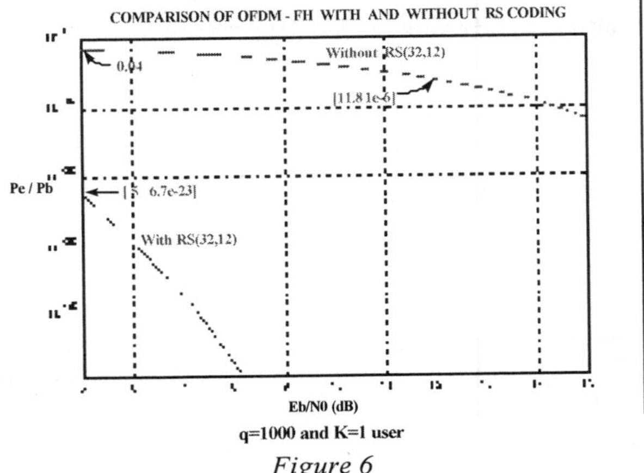

q=1000 and K=1 user

Figure 6

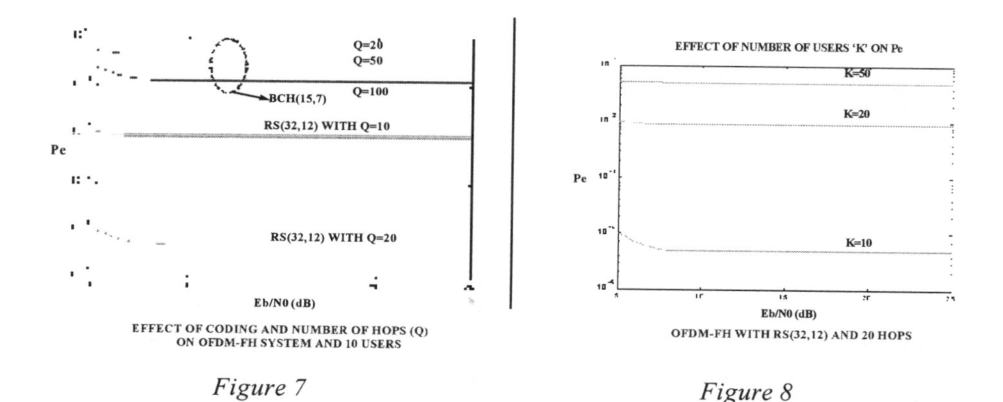

Figure 7

Figure 8

4. CONCLUSION AND RECOMMENDATIONS

This paper discusses a novel concept of integrating CDMA with OFDM-FH with a view to reducing the main drawbacks of CDMA viz. "near-far" effect and complex RAKE receivers. This approach as given in this paper, essentially applies to the downlink. It is, however, also applicable in the uplink provided synchronisation problems are addressed in the OFDM systems. This is driven by a need to maintain orthogonality between mobile receivers. This is essentially an OFDM problem area and does not in any way restrict our suggested approach. This approach yields the greatest advantage of CDMA viz large number of users without the drawbacks of CDMA like power control, multipath effects etc. The suggested system makes this approach a promising solution for multimedia applications. It is recommended that efforts be made to increase the bit rates of CDMA systems for multimedia, but this proposed system is independent of data rate and can handle data rates as high as 155 Mbps due to OFDM. It is proposed to conduct further simulations using this approach, leading to a formal proposal for a fourth generation system for wideband multimedia.

5. REFERENCES

[1] CDMA for Wireless Personal Communications, Ramjee Prasad, Artech House, Boston, London,1996

[2]Wireless Communications - Principles & Practice, Theodore S. Rappaport, Prentice-Hall,1996

[3]Digital Communications, John G. Proakis, McGraw-Hill Book Co.,1995, Third Edition

[4] *Overview of Multicarrier CDMA*, S. Hara and R. Prasad, IEEE Comm. mag., no. 12, vol 35, pp 126-133, Dec 1997.

[5]*A Novel Algorithmic Synchronisation Technique for OFDM based Wireless Multimedia Communications*, M.Jankiraman and Ramjee Prasad, ICC '99, Vancouver,Canada.

[6]*ML Estimation of Timing and Frequency Offset in Multicarrier Systems*, Magnus Sandell et al, Research report, Div. of Signal Processing, Lulea University of Technology, Sweden

[7] *Coherent Hybrid DS-SFH Spread Spectrum Multiple Access Communications*, Evaggelos A. Geraniotis, IEEE JSAC, vol SAC-3, No. 5, September 1985, pp 695-705.

[8] *A Conceptual Study of OFDM - based Multiple Access Schemes* : Part 1 - Air Interface Requirements, Mattias Wahlqvist et al, Technical Report Tdoc 117/96, ETSI STC SMG2 meeting no 18, Helsinki, Finland, May 1996

[9] *A Conceptual Study of OFDM - based Multiple Access Schemes* : Part 2 - Channel estimation in the Uplink, Jan-Jaap van de Beek et al, Technical Report Tdoc 116/96, ETSI STC SMG2 meeting no 18, Helsinki, Finland, May 1996

[10] *A Conceptual Study of OFDM - based Multiple Access Schemes* : Part 3 - Performance Evaluation of a Coded System, Jan-Jaap van de Beek et al, Technical Report Tdoc 166/96, ETSI STC SMG2 meeting no 19, Dusseldorf, Germany, September 1996

[11] Universal Wireless Personal Communication, Ramjee Prasad, Artech House, Boston, London, 1998

GENERALIZED MULTI-CARRIER CDMA FOR MUI/ISI-RESILIENT UPLINK TRANSMISSIONS IRRESPECTIVE OF FREQUENCY-SELECTIVE MULTIPATH

Georgios Giannakis, Paul A. Anghel, Zhengdao Wang, Anna Scaglione
Dept. of ECE, Univ. of Minnesota, 200 Union Street., Minneapolis, MN 55455, U.S.A.

Abstract Relying on symbol blocking and judicious design of user codes, this paper develops a Generalized Multicarrier (GMC) quasi-synchronous CDMA system capable of multiuser interference (MUI) elimination and intersymbol interference (ISI) suppression, irrespective of the wireless frequency selective channels encountered in the uplink. As the term reveals, GMC-CDMA provides a unifying framework for multicarrier (MC) CDMA systems and as this paper shows, it offers flexibility in full load (maximum number of users allowed by the available bandwidth) and in reduced load settings. A blind channel estimation algorithm is also derived. Analytic evaluation and simulations illustrate that GMC-CDMA outperforms competing MC-CDMA alternatives especially in the presence of uplink multipath channels.

1. INTRODUCTION

Mitigation of frequency selective multipath and elimination of multiuser interference have received considerable attention as they constitute the main limiting performance factors in wireless CDMA systems. Multicarrier (MC) CDMA systems have been introduced to mitigate both MUI and ISI caused by frequency selective channel effects, but they do not guarantee (blind or not) recovery of the transmitted symbols in the uplink without imposing constraints on the unknown multipath channel nulls [1, 6, 9]. In [7, 8] a spread-spectrum multicarrier multiple access scheme was developed and shown to achieve MUI

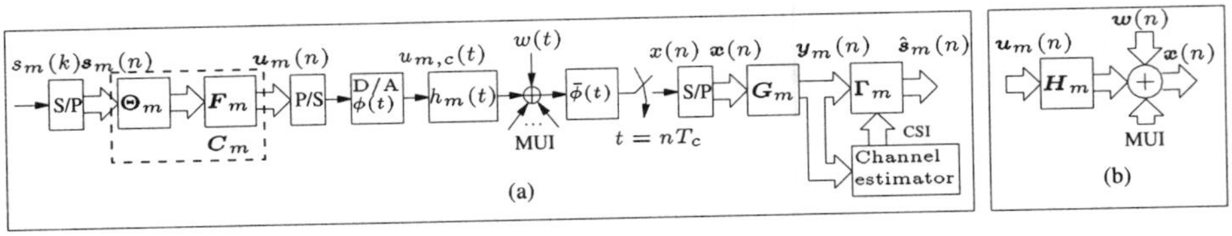

Figure 1 (a) Baseband transceiver model (b) Discrete-time equivalent channel model.

elimination irrespective of the frequency selective uplink channels. Relative to [7, 8], the generalized multicarrier (GMC) CDMA system designed in this paper, offers the following distinct features: (i) it quantifies the minimum redundancy needed for uplink bandwidth efficient transmissions; (ii) without channel coding and symbol interleaving, it establishes conditions that guarantee FIR-channel-irrespective symbol recovery with FIR linear equalizers; (iii) it offers capabilities for blind channel estimation by exploiting the redundant GMC-precoded transmission; (iv) it has low (linear) complexity and does not trade-off bandwidth efficiency in order to lower the exponential complexity of MLSE receivers. Features i)-iv) are present also in the so called AMOUR system of [4, 5, 10], which however, was designed for fully loaded systems. GMC-CDMA retains AMOUR's low complexity and bandwidth efficiency while at the same combines spread-spectrum with multicarrier features to improve performance when the system is not fully loaded.

2. SYSTEM MODEL

The baseband equivalent transmitter and receiver model for the mth user is depicted in Fig. 1(a), where $m \in [0, M_a - 1]$ and M_a is the number of active users out of a maximum M users that can be accommodated by the available bandwidth. The information stream $s_m(k)$ with symbol rate $1/T_s$ is first serial-to-parallel (S/P) converted to blocks[1] $\boldsymbol{s}_m(n)$ of length $K \times 1$ with the kth entry of the nth block denoted as: $s_{m,k}(n) := s_m(k + nK)$, $k \in [0, K-1]$. The $\boldsymbol{s}_m(n)$ blocks are multiplied by a $J \times K$ $(J > K)$ tall matrix $\boldsymbol{\Theta}_m$, which introduces redundancy and spreads the K symbols in $\boldsymbol{s}_m(n)$ by J-long codes. Precoder $\boldsymbol{\Theta}_m$ will facilitate ISI suppression, while the subsequent redundant precoder described by the tall $P \times J$ matrix $\boldsymbol{F}_m$ will accomplish MUI elimination. The precoded $P \times 1$ output vector $\boldsymbol{u}_m(n)$ is first parallel to serial (P/S) and then digital to analog converted using a chip waveform $\phi(t)$ of duration T_c with Nyquist characteristics, before being transmitted through the frequency selective channel $h_m(t)$. Although we focus on the uplink, the downlink scenario is subsumed by our model (it corresponds to having $h_m(t) = h(t)\ \forall m$). The resulting aggregate signal $x(t)$ from all active users is filtered with a receive-filter $\bar{\phi}(t)$ matched to $\phi(t)$ and then sampled at the chip rate $1/T_c$. Next, the sampled signal $x(n)$ is serial to parallel converted and processed by the digital multichannel receiver.

[1]Throughout this paper, k is the symbol index and n will be used to index blocks-of-symbols.

From the multichannel input-output viewpoint depicted in Fig. 1(b), the $K \times 1$ vector $\boldsymbol{u}_m(n)$ propagates through an equivalent channel described by the $P \times P$ lower triangular Toeplitz (convolution) matrix $\boldsymbol{H}_m$ with (i,j)th entry $h_m(i-j)$, where $h_m(l)$, $l \in [0, L]$, $m \in [0, M_a - 1]$ are the taps of the discrete chip-rate sampled FIR channels assumed to have maximum order L. In addition to transmit-receive filters, each channel $h_m(l)$, includes user quasi-synchronism in the form of delay factors (in this case $L = L_d + L_m$, where L_d captures asynchronism ($\ll P$ chips) relative to a reference user, and L_m expresses (in chips) the maximum multipath delay spread). To avoid channel-induced inter-block interference (IBI), we pad our transmitted blocks $\boldsymbol{u}_m(n)$ with L zeros (guard bits). Specifically, we design our $P \times J$ precoders $\boldsymbol{F}_m$ such that:
d1) $P = M_a J + L$ and the $L \times J$ lower submatrix of $\boldsymbol{F}_m$ is set to zero.
Under d1), the $P \times 1$ data vector $\boldsymbol{x}(n)$ received in AGN $\boldsymbol{w}(n)$ is given by:

$$\boldsymbol{x}(n) = \sum_{m=0}^{M_a-1} \boldsymbol{H}_m \boldsymbol{C}_m \boldsymbol{s}_m(n) + \boldsymbol{w}(n), \qquad \boldsymbol{C}_m := \boldsymbol{F}_m \boldsymbol{\Theta}_m. \tag{1}$$

Processing the multichannel data $\boldsymbol{x}(n)$ by the mth user's receiver amounts to multiplying it with the $J \times P$ receiver matrix $\boldsymbol{G}_m$ that yields $\boldsymbol{y}_m(n) = \boldsymbol{G}_m \boldsymbol{x}(n)$. Similar to [4], the precoder/decoder matrices $\{\boldsymbol{F}_m, \boldsymbol{G}_m\}$ will be judiciously designed such that MUI is eliminated from $\boldsymbol{x}(n)$ independent of the channels $\boldsymbol{H}_m$. Channel status information (CSI) acquired from the channel estimator (see Fig. 1) will be used to specify the linear equalizer $\boldsymbol{\Gamma}_m$ which removes ISI from the MUI-free signals $\boldsymbol{y}_m(n)$ to obtain the estimated symbols $\hat{\boldsymbol{s}}_m(n) = \boldsymbol{\Gamma}_m \boldsymbol{G}_m \boldsymbol{x}(n)$ that are passed on to the decision device.

3. MUI/ISI ELIMINATING CODES

We pursue MUI elimination from $\boldsymbol{x}(n)$ in the Z-domain [4]. Let us define $\boldsymbol{v}_P(z) := [1\ z^{-1}\ ...\ z^{-P+1}]^{\mathrm{T}}$ (^{T} denotes transpose), and Z-transform the entries of $\boldsymbol{x}(n)$ in (1) to obtain: $X(n;z) := \boldsymbol{v}_P^{\mathrm{T}}(z)\boldsymbol{x}(n)$. Substituting $\boldsymbol{x}(n)$ from (1), we find:

$$X(n;z) = \sum_{m=0}^{M_a-1} H_m(z)[F_{m,0}(z)\ ...\ F_{m,J-1}(z)]\boldsymbol{\Theta}_m \boldsymbol{s}_m(n) + \boldsymbol{v}_P^{\mathrm{T}}(z)\boldsymbol{w}(n), \tag{2}$$

where $H_m(z) := \sum_{l=0}^{L} h_m(l) z^{-l}$ and $F_{m,j}(z)$ is the Z-transform of matrix $\boldsymbol{F}_m$'s jth column. Note that evaluating $X(n;z)$ at $z = \rho_{\mu,i}$ amounts to using a receiver $\boldsymbol{v}_P(\rho_{\mu,i})$ that performs a simple inner product operation $\boldsymbol{v}_P^{\mathrm{T}}(\rho_{\mu,i})\boldsymbol{x}(n)$. Hence, forming $\boldsymbol{y}_\mu := [X(n;\rho_{\mu,0})\ X(n;\rho_{\mu,1})\ \ldots\ X(n;\rho_{\mu,J-1})]^{\mathrm{T}}$ requires a receiver $\boldsymbol{G}_\mu := [\boldsymbol{v}_P(\rho_{\mu,0})\ ...\ \boldsymbol{v}_P(\rho_{\mu,J-1})]^{\mathrm{T}}$ to obtain:

$$\boldsymbol{y}_\mu(n) = \boldsymbol{G}_\mu \boldsymbol{x}(n). \tag{3}$$

The principle behind designing MUI-free precoders $\boldsymbol{F}_m$ is to seek J points $\{\rho_{\mu,i}\}_{i=0}^{J-1}$ for every active user $\mu \in [0, M_a - 1]$ on which $X(n; z = \rho_{\mu,i})$ contains the μth user's signal of interest, while MUI from the remaining $M-1$ users is eliminated. If in addition to MUI we also want to cancel the inter-chip interference from precoder $\boldsymbol{F}_m$, we must select:

$$F_{m,j}(\rho_{\mu,i}) = \delta(j-i)\delta(m-\mu), \qquad \forall m,\mu \in [0, M_a-1], \;\; \forall j,i \in [0, J-1]. \quad (4)$$

The minimum degree polynomial $F_{m,j}(z)$ that satisfies (4) can be uniquely computed by Lagrange interpolation through the M_aJ points $\rho_{\mu,i}$ as follows [4]:

$$F_{m,j}(z) = \prod_{\substack{\mu=0 \\ (\mu,i)\neq(m,j)}}^{M_a-1} \prod_{i=0}^{J-1} \frac{1-\rho_{\mu,i}z^{-1}}{1-\rho_{\mu,i}\rho_{m,j}^{-1}}. \quad (5)$$

Because manipulation of circulant matrices can be performed with FFT, low-complexity transceivers result if $\boldsymbol{F}_m$ is formed by FFT exponentials, which corresponds to choosing $\{\rho_{\mu,i}\}_{i=0}^{J-1}$ in (5) equispaced on the unit circle as: $\rho_{\mu,i} = \exp(j2\pi(\mu + iM_a)/M_aJ)$ $\forall \mu, i$. Accounting for the L trailing zeros as per d1), such a choice leads to:

$$\boldsymbol{F}_\mu = \frac{e^{j2\pi\mu/M_aJ}}{M_aJ}\begin{bmatrix} \boldsymbol{v}^*_{M_aJ}(1) & \boldsymbol{v}^*_{M_aJ}(e^{j2\pi/J}) & \ldots & \boldsymbol{v}^*_{M_aJ}(e^{j2\pi(J-1)/J}) \\ & \boldsymbol{0}_{L\times J} & & \end{bmatrix}, \quad (6)$$

where * denotes conjugation. The degree of the mth user's jth code polynomial in (5) is M_aJ-1. Adding the L guard chips to $F_{m,j}(z)$'s inverse Z-transform, sets the number of rows for precoder $\boldsymbol{F}_m$ to $P = M_aJ + L$, which explains our choice in d1).

Next, we substitute (2) into (3) and take account of (4) to obtain the MUI-free:

$$\boldsymbol{y}_\mu(n) = \boldsymbol{D}_{H_\mu}\boldsymbol{\Theta}_\mu \boldsymbol{s}_\mu(n) + \boldsymbol{\eta}_\mu(n) \;\;\Rightarrow\;\; \hat{\boldsymbol{s}}_\mu(n) = \boldsymbol{\Gamma}_\mu^{zf}\boldsymbol{y}_\mu(n) := \boldsymbol{\Theta}_\mu^\dagger \boldsymbol{D}_{H_\mu}^\dagger \boldsymbol{y}_\mu(n), \quad (7)$$

where $\boldsymbol{D}_{H_\mu} := \text{diag}[H_\mu(\rho_{\mu,0}) \ldots H_\mu(\rho_{\mu,J-1})]$ is a $J \times J$ diagonal matrix with entries $H_\mu(\rho_{\mu,j})$, $\boldsymbol{\eta}_\mu(n) := \boldsymbol{G}_\mu \boldsymbol{w}(n)$, and † denotes pseudoinverse. With $\mathcal{C}^K$ denoting the vector space of complex K-tuples, suppose we design $\boldsymbol{\Theta}_\mu$ in (7) to satisfy:

d2) $J \geq K + L$ and *any* $J - L$ rows of $\boldsymbol{\Theta}_\mu$ span the $\mathcal{C}^K$ row vector space;

Notice that d2) can always be checked and enforced at the transmitter. Under d2), $\boldsymbol{D}_{H_\mu}\boldsymbol{\Theta}_\mu$ in (7) will always be full rank, because the added redundancy ($\geq L$) can afford even L diagonal entries of $\boldsymbol{D}_{H_\mu}$ to be zero (recall that $H_\mu(z)$ has maximum order L and thus at most L nulls). Therefore, identifiability of $\boldsymbol{s}_\mu(n)$ can be guaranteed irrespective of the multipath channel $H_\mu(z)$. Possible choices for $\boldsymbol{\Theta}_\mu$ that are flexible enough for our design include: (a)the $J \times K$ Vandermonde matrix $\boldsymbol{\Theta}_\mu := [\boldsymbol{v}(\rho_{\mu,0}, K) \ldots \boldsymbol{v}(\rho_{\mu,J-1}, K)]^{\mathrm{T}}$ used in the AMOUR system [4], which for $\rho_{\mu,i} = \exp(j2\pi(\mu + iM_a)/(M_aJ))$ becomes $\exp(j2\pi\mu/(M_aJ))$ times a truncated $J \times K$ FFT matrix; (b)a truncated $J \times K$ Walsh-Hadamard (WH) matrix; (c)a $J \times K$ matrix with equiprobable ± 1 random entries.

Channel estimation, blind or pilot-based, is needed to build a ZF-equalizer $\boldsymbol{\Gamma}_\mu^{zf}$ in (7), which will guarantee ISI-free detection (MMSE equalizers are also possible). For Θ_μ's selected as in (a), a blind channel estimation method was developed in [4]. In the sequel, we will establish the identifiability conditions and derive a more general blind channel estimation algorithm allowing spread-spectrum precoders Θ_μ to be chosen as in (b) or (c).

4. BLIND CHANNEL ESTIMATION

We will suppose here that instead of d2), we design Θ_μ such that:
d2′) $J \geq K + L$ and *any* K rows of Θ_μ span the $\mathcal{C}^K$ row vector space.
Note that when $J = K + L$, d2′) is equivalent to d2). To estimate $H_\mu(z)$ under d2′) in the noiseless case, user μ collects N blocks of $\boldsymbol{y}_\mu(n)$ in a $J \times N$ matrix $\boldsymbol{Y}_\mu := [\boldsymbol{y}_\mu(0) \cdots \boldsymbol{y}_\mu(N-1)]$ and forms $\boldsymbol{Y}_\mu \boldsymbol{Y}_\mu^{\mathcal{H}} = \boldsymbol{D}_{H_\mu} \Theta_\mu \boldsymbol{S}_\mu \boldsymbol{S}_\mu^{\mathcal{H}} \Theta_\mu^{\mathcal{H}} \boldsymbol{D}_{H_\mu}^{\mathcal{H}}$, where $\boldsymbol{S}_\mu := [\boldsymbol{s}_\mu(0) \cdots \boldsymbol{s}_\mu(N-1)]_{K \times N}$. User μ also chooses:
d3) N large enough so that $\boldsymbol{S}_\mu \boldsymbol{S}_\mu^{\mathcal{H}}$ is of full rank K.
Under d2′) and d3), we have $\text{rank}(\boldsymbol{Y}_\mu \boldsymbol{Y}_\mu^{\mathcal{H}}) = K$ and range space $\mathcal{R}(\boldsymbol{Y}_\mu \boldsymbol{Y}_\mu^{\mathcal{H}}) = \mathcal{R}(\boldsymbol{D}_{H_\mu} \Theta_\mu)$. Thus, the nullity of $\boldsymbol{Y}_\mu \boldsymbol{Y}_\mu^{\mathcal{H}}$ is $\nu(\boldsymbol{Y}_\mu \boldsymbol{Y}_\mu^{\mathcal{H}}) = J - K$. Further, the eigen-decomposition

$$\boldsymbol{Y}_\mu \boldsymbol{Y}_\mu^{\mathcal{H}} = [\boldsymbol{U}\ \tilde{\boldsymbol{U}}] \begin{bmatrix} \boldsymbol{\Sigma}_{K \times K} & \boldsymbol{0}_{K \times (J-K)} \\ \boldsymbol{0}_{(J-K) \times K} & \boldsymbol{0}_{(J-K) \times (J-K)} \end{bmatrix} \begin{bmatrix} \boldsymbol{U}^{\mathcal{H}} \\ \tilde{\boldsymbol{U}}^{\mathcal{H}} \end{bmatrix} \tag{8}$$

yields the $J \times (J-K)$ matrix $\tilde{\boldsymbol{U}}$ whose columns span the null space $\mathcal{N}(\boldsymbol{Y}_\mu \boldsymbol{Y}_\mu^{\mathcal{H}})$. Because the latter is orthogonal to $\mathcal{R}(\boldsymbol{Y}_\mu \boldsymbol{Y}_\mu^{\mathcal{H}}) = \mathcal{R}(\boldsymbol{D}_{H_\mu} \Theta_\mu)$, it follows that $\tilde{\boldsymbol{u}}_l^{\mathcal{H}} \boldsymbol{D}_{H_\mu} \Theta_\mu = \boldsymbol{0}_{1 \times K}^{\mathcal{H}}$, $l \in [1, J-K]$, where $\tilde{\boldsymbol{u}}_l$ denotes the lth column of $\tilde{\boldsymbol{U}}$. With $\boldsymbol{D}_{u_l}$ denoting the diagonal matrix $\boldsymbol{D}_{u_l} := \text{diag}[\tilde{\boldsymbol{u}}_l^{\mathcal{H}}]$ and $\boldsymbol{d}_{H_\mu}^T := [H(\rho_{\mu,0}), \ldots, H(\rho_{\mu,J-1})]$, we can write $\tilde{\boldsymbol{u}}_l^{\mathcal{H}} \boldsymbol{D}_{H_\mu} = \boldsymbol{d}_{H_\mu}^T \boldsymbol{D}_{u_l}$. It can be easily verified that with $\boldsymbol{h}_\mu^T := [h_\mu(0), \ldots, h_\mu(L)]$ and with $\boldsymbol{V}_\mu$ being a $(L+1) \times J$ matrix whose $(l+1, j+1)$st entry is $\rho_{\mu,j}^{-l}$, one can write $\boldsymbol{d}_{H_\mu}^T = \boldsymbol{h}_\mu^T \boldsymbol{V}_\mu$. This yields

$$\boldsymbol{h}_\mu^T \boldsymbol{V}_\mu [\boldsymbol{D}_{u_0} \Theta_\mu \vdots \cdots \vdots \boldsymbol{D}_{u_{J-K}} \Theta_\mu] = \boldsymbol{0}_{1 \times K(J-K)}^{\mathrm{T}}, \tag{9}$$

from which one can solve for $\boldsymbol{h}_\mu$. We have established uniqueness (within a scale) in solving for $\boldsymbol{h}_\mu$, but we omit the proof due to lack of space. In the noisy case, if the covariance matrix of $\boldsymbol{\eta}_\mu$ is known, we can prewhiten $\boldsymbol{Y}_\mu$ before SVD, and a similar blind channel algorithm can be devised. We summarize our results in the following:

Theorem: *i) Design a GMC-CDMA system according to d1) and d2), and suppose that CSI is available at the receiver using pilot symbols. User symbols $\boldsymbol{s}_\mu(n)$ can then be always recovered with linear processing as in (7), irrespective of frequency selective multipath channels up to order L. ii) A GMC-CDMA system designed according to d1), d2′), and d3) guarantees blind identifiability (within a scale) of channels $h_\mu(l)$ with maximum order L, and the channel estimate is found by solving (9) for the null eigenvector.*

Note that unlike [7, 8], even blind channel-irrespective symbol recovery is assured by the Theorem, without bandwidth consuming channel coding/interleaving, and with linear receiver processing (as opposed to the exponentially complex MLSE used in [7, 8]).

5. UNIFYING FRAMEWORK

A number of multiuser multicarrier schemes fall under the model of Fig. 1(a). We outline some of them in this section by describing their baseband discrete-time equivalent models in order to illustrate the generality of GMC-CDMA.

Multicarrier CDMA (MC-CDMA), [3, 11]: For this multicarrier scheme, no blocking of data symbols occurs at the receiver and the first precoding matrix $\boldsymbol{\Theta}_m$ is a $Q \times 1$ vector $\boldsymbol{\theta}_m$ (the spreading filter). The $\boldsymbol{F}_m$ matrix is no longer user dependent, as it is selected to be a $Q \times Q$ IFFT matrix augmented either by an $L \times Q$ all-zero matrix as in d1), or, by an $L \times Q$ cyclic prefix matrix to yield a $(Q + L) \times Q$ matrix corresponding to an OFDM precoder (modulator). At the OFDM receiver, the cyclic prefix is discarded. For MC-CDMA to have the same bandwidth as GMC-CDMA, Q must be chosen as: $Q = \lfloor P/K \rfloor - L$. The frequency selective channel matrix $\boldsymbol{H}_m$ is $(Q + L) \times (Q + L)$ and for the zero-padded transmissions we have instead of (1): $\boldsymbol{x}(n) = \sum_{m=0}^{M_a-1} \boldsymbol{H}_m \boldsymbol{F} \boldsymbol{\theta}_m s_m(n) + \boldsymbol{w}(n)$. At the receiver end, the matrix $\boldsymbol{G}$ is selected to be a $Q \times (Q+L)$ extended FFT matrix with entries given by powers of $\exp(j2\pi/Q)$, or, a $Q \times Q$ FFT matrix increased by the $L \times Q$ leading zeros to discard the cyclic prefix used at the transmitter. Matrix $\boldsymbol{\Gamma}_m$ becomes now an $1 \times Q$ vector to be chosen according to the selected multiuser detection technique. Although simpler than GMC-CDMA, because matrix $[\boldsymbol{H}_0 \boldsymbol{F}_0 \boldsymbol{\theta}_0 \ \ldots \ \boldsymbol{H}_{M_a-1} \boldsymbol{F}_{M_a-1} \boldsymbol{\theta}_{M_a-1}]$ is not guaranteed to be invertible, symbol recovery is not assured in MC-CDMA even when CSI is available.

Multicarrier Direct-Sequence CDMA (MC-DS-CDMA), [1, 2]: Both GMC-CDMA and MS-DS-CDMA are multicarrier techniques using a block precoder $\boldsymbol{\Theta}_m$, except that for MC-DS-CDMA the precoder is particularized to be a $QK \times K$ block diagonal matrix with blocks of size $Q \times 1$. The dimensionality of the OFDM transmitter-receiver pair $\{\boldsymbol{F}_m, \boldsymbol{G}_m\}$ is $(QK + L) \times QK$ and $QK \times (QK + L)$, respectively. Keeping in mind the bandwidth constraint we choose $Q = \lfloor (P - L)/K \rfloor$. Each user transmits K symbols in parallel, spreads them with codes of length Q, and modulates each spread symbol with a specific set of Q subcarriers. Same as the $\boldsymbol{\Theta}_m$ precoder, $\boldsymbol{\Gamma}_m$ has a block diagonal structure with K blocks of size $1 \times Q$. If $K = 1$, then $Q = P - L$ and MC-DS-CDMA reduces to MC-CDMA.

Multitone CDMA (MT-CDMA), [9]: This multicarrier technique applies the same data mapping as MC-DS-CDMA. However, in contrast with MC-DS-CDMA, MT-CDMA uses a block diagonal precoder $\boldsymbol{\Theta}_m$ with K blocks of size $QQ' \times 1$. The $KQ \times KQQ'$ precoder is given by: $\boldsymbol{F}_m = [\boldsymbol{v}^*_{KQ}(1) \ \boldsymbol{v}^*_{KQ}(j2\pi/(KQQ') \ \ldots \boldsymbol{v}^*_{KQ}(j2\pi(KQQ'-1)/(KQQ'))]$, while the receiver is: $\boldsymbol{G}_m = [\boldsymbol{0}_{KQ \times L} \ \boldsymbol{v}_{KQ}(1) \ \boldsymbol{v}_{KQ}(j2\pi/(KQQ')) \ldots \boldsymbol{v}_{KQ}(j2\pi(KQQ'-1)/(KQQ'))]^{\mathrm{T}}$. The system has the advantage of allowing a bigger size $\boldsymbol{\Theta}_m$ matrix (longer spreading codes), which reduces self-interference and MUI at the expense of introducing subcarrier interference.

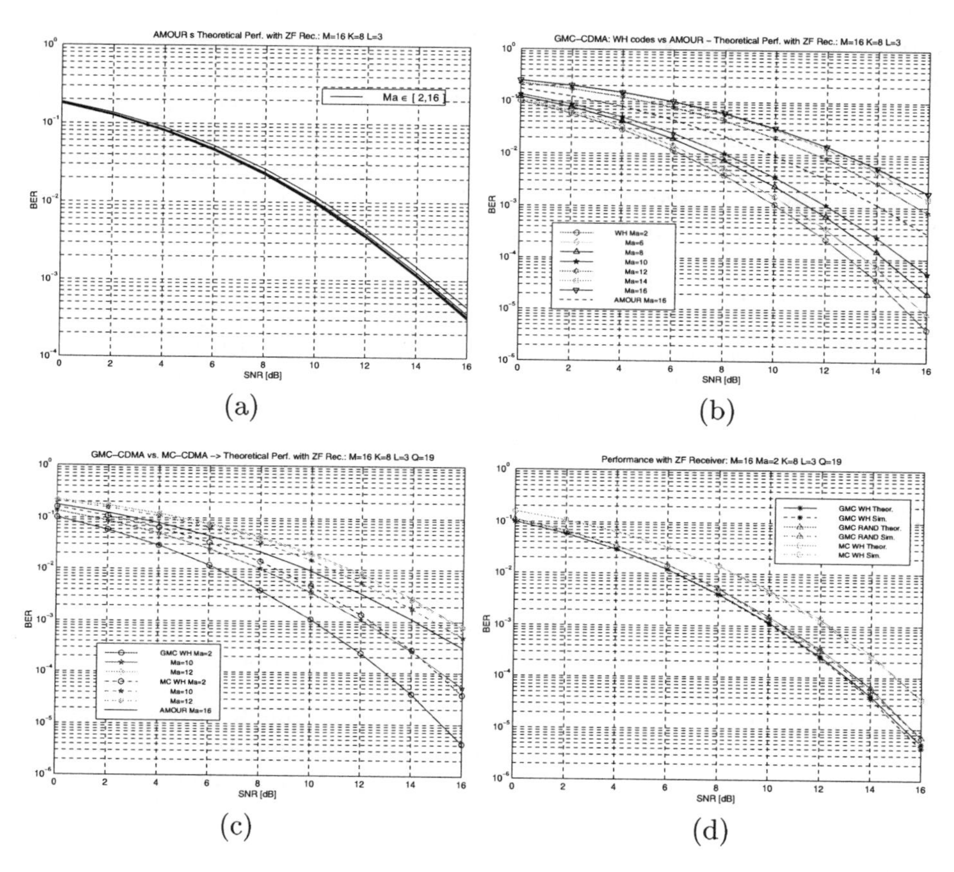

Figure 2 (a) AMOUR: different number of active users. (b) GMC-CDMA: Theoretical WH vs. AMOUR. (c) Theoretical: GMC-CDMA vs. MC-CDMA; (d) Simulation $M_a = 2$: GMC-CDMA vs. MC-CDMA.

6. PERFORMANCE ANALYSIS

The theoretical BER evaluation for an AMOUR system [4] that uses a ZF equalizer can be extended to a GMC-CDMA system with $\mathbf{\Gamma}_{\mu}^{zf}$ as in (7). Perfect knowledge of the channel is assumed. Similar to [4], we choose for simplicity a BPSK constellation to obtain in terms of the Q-function an average bit error rate (BER) $\bar{P}e$:

$$\bar{P}e = \frac{1}{MK} \sum_{m=0}^{M_a-1} \sum_{k=0}^{K-1} \mathrm{Q}\left(\sqrt{\frac{1}{\bar{\boldsymbol{g}}_{m,k}^{\mathrm{H}} \bar{\boldsymbol{g}}_{m,k} E_{m,k}}} \sqrt{\frac{2E_b}{N_0}}\right), \tag{10}$$

where $\bar{\boldsymbol{g}}_{m,k}^{\mathcal{H}}$ is the kth row of matrix $\boldsymbol{\Gamma}_m \boldsymbol{G}_m$, $E_{m,k} := \sum_{i=0}^{P-1} |c_{m,k}(i)|^2$ is the energy of the mth user's kth code, and E_b/N_0 is the bit SNR.

First, we plot (10) for an AMOUR system designed for $M = 16$ users with data symbols drawn from a BPSK constellation, each one experiencing a Rayleigh fading channel of order $L = 3$. The length of the $\boldsymbol{s}_m(n)$ blocks is $K = 8$. To avoid channel dependent performance, we averaged (10) over 100 Monte Carlo channel realizations. We decrease gradually the number of active users in the system from 16 down to 2 and each time we redesign AMOUR to incorporate the available bandwidth. The theoretical BER curves from Fig. 2(a) show that under different load conditions there is practically no difference in performance. Next, keeping the same set up we compare GMC-CDMA using WH codes versus AMOUR under different load conditions. It is clear from Fig. 2(b) that under 65% load the WH codes outperform the AMOUR codes. In Fig. 2(c) the theoretical performance of GMC-CDMA with WH codes and same parameters as before is compared with an equivalent MC-CDMA system using OFDM transceivers with trailing zeros. GMC-CDMA has a lower BER than MC-CDMA independent of the number of active users if the WH codes are replaced by AMOUR codes when the system load increases over 65%. Code redesign/switching has the drawback of requiring knowledge of the system load at the mobile transmitter. To verify our theoretical claims, we simulated GMC (both with WH and random codes) and MC with maximum number of users $M = 16$, of which $M_a = 2$ were active and used block size $K = 8$. CSI was assumed to be perfect ($L = 3$). The results plotted in Fig. 2(d) show that at 12% load (value in our region of interest) both WH and random codes exhibit improved performance over MC-CDMA.
Acknowledgments: The work in this paper was supported by NSF CCR grant no. 98-05350 and the NSF Wireless Initiative grant no. 99-79443. The authors also wish to thank prof. S. Barbarossa for discussions on this and related subjects.

References

[1] Q. Chen, E. S. Sousa, and S. Pasupathy, "Performance of a coded multi-carrier DS-CDMA system in multipath fading channels," *Wireless Personal Comm.*, vol. 2, pp. 167–183, 1995.

[2] V. M. DaSilva and E. S. Sousa, "Multicarrier orthogonal CDMA signals for quasi-synchronous communication systems," *Journal on Select. Areas on Comm.*, pp. 842–852, June 1994.

[3] K. Fazel, "Performance of CDMA/OFDM for mobile communication system," in *Proceedings IEEE Intern. Conf. on Univ. Personal Comm.*, Ottawa, Canada, Oct. 1993, pp. 975–979.

[4] G. B. Giannakis, Z. Wang, A. Scaglione, and S. Barbarossa, "Mutually orthogonal transceivers for blind uplink CDMA irrespective of multipath channel nulls," in *Proc. of Intl. Conf. on ASSP*, Phoenix, AZ, Mar. 1999, vol. 5, pp. 2741–2744.

[5] G. B. Giannakis, Z. Wang, A. Scaglione, and S. Barbarossa, "AMOUR - generalized multicarrier transceivers for blind CDMA irrespective of multipath," *IEEE Trans. on Comm.*, (submitted December, 1998);, see also, Proc. of GLOBECOM, Rio de Janeiro, Brazil, Dec. 5-9, 1999 (to appear).

[6] S. Hara and R. Prasad, "Overview of multicarrier CDMA," *IEEE Comm. Mag.*, pp. 126–133, Dec. 1997.

[7] S. Kaiser and K. Fazel, "A spread spectrum multi-carrier multiple access system for mobile communications," in *First International Workshop in Multicarrier Spread-Spectrum*, Oberpfaffenhofen, Germany, Apr. 1997, pp. 49–56.

[8] S. Kaiser and K. Fazel, "A flexible spread-spectrum multi-carrier multiple access system for multimedia applications," in *IEEE International Symposium on Personal, Indoor and Mobile Radio Comm.*, Helsinki, Findland, Sept. 1997, pp. 100–104.

[9] L.. Vandendorpe, "Multitone direct sequence CDMA system in an indoor wireless environment," in *Proceedings of IEEE First Symposium of Communications and Vehicular Technology*, Delft, The Netherlands, Oct. 1993, pp. 4.1.1–4.1.8.

[10] Z. Wang, A. Scaglione, G. B. Giannakis, and S. Barbarossa, "Vandermonde-lagrange mutually orthogonal flexible transceivers for blind CDMA in unknown multipath," in *Proc. of IEEE-SP Workshop on Signal Proc. Advances in Wireless Comm.*, May 1999, pp. 42–45.

[11] N. Yee, J-P. Linnartz, and G. Fettweis, "Multicarrier CDMA in indoor wireless radio networks," in *Proc. of IEEE PIMRC '93*, Sept. 1993, pp. 109–13.

PERFORMANCE TRADEOFFS IN ASYNCHRONOUS SPREAD SPECTRUM MULTI-CARRIER MULTIPLE ACCESS[1]

Stefan Kaiser
German Aerospace Center (DLR), Institute for Communications Technology
82234 Oberpfaffenhofen, Germany
Stefan.Kaiser@dlr.de

Witold A. Krzymien
University of Alberta / TRLabs, 800 Park Plaza, 10611 - 98th Avenue
Edmonton, Alberta, Canada T5K 2P7
wak@edm.trlabs.ca

Khaled Fazel
Bosch Telecom GmbH, Digital Microwave Systems, 71522 Backnang, Germany
Khaled.Fazel@de.bosch.com

Abstract The paper presents simple detection algorithms to mitigate interference due to asynchronism in the uplink of a mobile radio system using spread spectrum multi-carrier multiple access (SS-MC-MA). The effects of a pilot-symbol-aided uplink channel estimation are taken into account in the analysis. The presented system allows some intersymbol interference (ISI) and intersubchannel interference (ICI) in the uplink but in return reduces the loss in bandwidth efficiency due to a shortened guard interval and avoids the complexity of a dedicated uplink synchronization scheme.

1. INTRODUCTION

Future mobile radio systems have to be highly bandwidth efficient due to limited radio spectrum available for allocation, and on the other hand have to use transmission techniques which enable low complexity

[1]This work was funded by the German Aerospace Center (DLR), the Telecommunications Research Laboratories (TR*Labs*), and the Natural Sciences & Engineering Research Council (NSERC) of Canada.

implementation suitable for low-cost mobile consumer products. Taking into account these requirements, a multiple access scheme called spread spectrum multi-carrier multiple access (SS-MC-MA) has been proposed [1], in which user separation is achieved through frequency division multiple access (FDMA) at the subcarrier level. Each user performs code division multiplexing (CDM) of its own data on its exclusive subset of subcarriers. The use of the spread spectrum technique for multiplexing and interleaving of the subcarriers of different users introduces frequency diversity, the advantages of which outweigh the drawback of self-interference due to the CDM. The multi-carrier modulation can efficiently be realized through orthogonal frequency division multiplexing (OFDM) [2]. The proposed multiple access approach is suitable for the uplink of a mobile radio system to avoid multiple access interference (MAI), and hence simplify the uplink channel estimation.

In this paper we demonstrate the ability of SS-MC-MA to cope with asynchronism in the uplink. The SS-MC-MA system under investigation uses only the frame structure received on the synchronous downlink to synchronize uplink transmissions from different users, requiring no additional synchronization measures in the uplink [3]. Frequency division duplex (FDD) is used to separate uplink and downlink transmission. A guard interval that is smaller than the sum of the maximum time offset between the users and the maximum excess delay of the frequency selective multipath radio channel is used. It minimizes the loss in bandwidth efficiency by allowing residual interference. The residual interference is minimized by proper positioning of the detection interval in the receiver. We consider the performance of the asynchronous SS-MC-MA uplink with perfect knowledge of the radio channel parameters, and with pilot-symbol-aided channel estimation.

2. SS-MC-MA UPLINK STRUCTURE

The SS-MC-MA system investigated in this paper accommodates K simultaneously active users in the uplink. It is an FDMA scheme on subcarrier level, and each user k, $k = 1, \ldots, K$, transmits exclusively on a subset of L subcarriers of N_c available subcarriers. The total number of subcarriers is $N_c = KL$. The SS-MC-MA transmitter is shown in Fig. 1. After channel coding, code bit interleaving (with sufficient interleaver size to introduce both time and frequency diversity), and QPSK symbol mapping, L complex-valued data symbols of user k are spread by multiplication with orthogonal Hadamard codes of size L, and superimposed on the same subset of L subcarriers [1], corresponding to the Hadamard transform (CDM). The resulting sequence modulates in parallel the subcarriers assigned to user k. In order to exploit optimally the

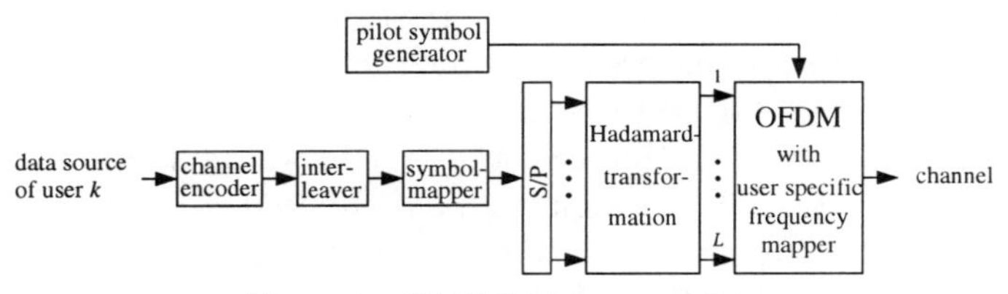

Figure 1 SS-MC-MA transmitter

frequency diversity offered by the wideband mobile radio channel, the subcarriers assigned to different users are interleaved. To achieve MC modulation/demodulation the OFDM operation is applied [2]. Since L symbols of one user after spreading are superimposed on a subset of L subcarriers, self-interference occurs in non-ideal channels.

The block diagram of the SS-MC-MA receiver is shown in Fig. 2. In

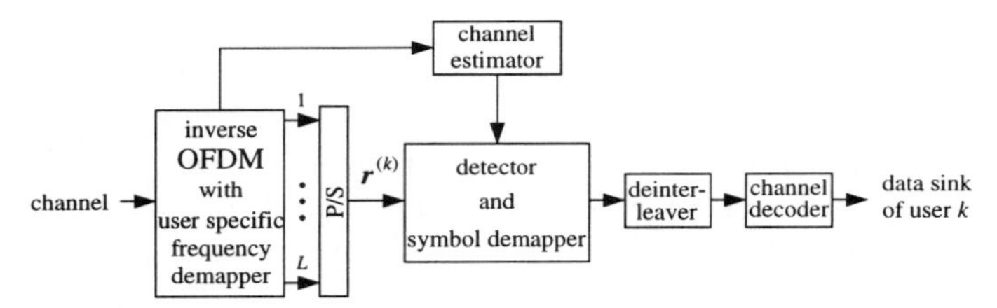

Figure 2 SS-MC-MA receiver

the receiver, MC demodulation is performed with the inverse OFDM operation, and de-interleaving is applied. We consider a maximum likelihood detector for the detection of the data of user k. After symbol demapping, code bit deinterleaving, and channel decoding, the detected source bits of user k are obtained.

Pilot-symbol-aided channel estimation is used for coherent detection. For the uplink, one-dimensional FIR filters in time dimension are applied for subchannel estimation.

3. INTERFERENCE MITIGATION

In the uplink, significant time offsets between the signals arriving at the base station occur due to different propagation distances between the mobile stations and the base station. With the assumption of uniform surface distribution of users within the cell, the resulting probability density function $p_\delta(\delta^{(k)})$ of the signal delays at the base station is linear for $\delta^{(k)}$ from 0 to $\delta_{\max}$, where $\delta_{\max}$ is the maximum time offset between signals from different users within a cell arriving at the base station and $\delta^{(k)} \in [0, \delta_{\max}]$, $k = 1, \ldots, K$, are the delays of the signals of the K users. Without compensation of different propagation delays of the signals in the uplink and an insufficient guard interval, the user synchronism is lost and interference results, which can significantly deteriorate the performance of an OFDM system.

In the following the resulting ISI and ICI in an asynchronous multi-carrier link is described. We assign each subchannel m its own delay δ_m, $m = 0, \ldots, N_c - 1$. The delays on subcarriers assigned to the same user are determined by the location of the user, and are the same. One arbitrary symbol $S_{m,i}$ of the sequence obtained after the Hadamard transformation is transmitted per subcarrier in one OFDM symbol slot. The index m is the subchannel index and i is the OFDM symbol index (discrete time).

We considering multipath propagation with N_p paths, where $\rho_{m,p}$ is the complex-valued attenuation and $\tau_{m,p}$ the excess delay of the pth path, $p = 1, \ldots, N_p$, of subchannel m. The total delay $\epsilon_{m,p}$ per arriving path p on subchannel m at the base station is the sum of the propagation delay δ_m and path delay $\tau_{m,p}$, i.e., $\epsilon_{m,p} = \delta_m + \tau_{m,p}$, where $\epsilon_{m,p} \in [0, \epsilon_{\max}]$, $\epsilon_{\max} = \delta_{\max} + \tau_{\max}$, and $\tau_{\max}$ is the maximum excess delay of any subchannel. The probability density function $p_\epsilon(\epsilon)$ is obtained by convolution of the probability density function $p_\delta(\delta^{(k)})$ and the probability density function $p_\tau(\tau)$ of the multipath propagation delay.

The received symbol after inverse OFDM operation in the receiver is

$$R_{m,i} = Y_{m,i} + N_{m,i}. \tag{1}$$

The component $N_{m,i}$ is the additive noise component. The output $Y_{m,i}$ of the demodulator is the noise-free decision variable that includes ISI and ICI. $Y_{m,i}$ can be written as [3]

$$\begin{aligned} Y_{m,i} &= \sum_{p=0}^{N_p-1} \Big[\rho_{m,p} \left(S_{m,i}\, \mu_{m,m}(\epsilon_{m,p}) + S_{m,i-1}\, \lambda_{m,m}(\epsilon_{m,p}) \right) \\ &\quad + \sum_{\substack{n=0 \\ n \neq m}}^{N_c-1} \rho_{n,p} \left(S_{n,i}\, \mu_{m,n}(\epsilon_{n,p}) + S_{n,i-1}\, \lambda_{m,n}(\epsilon_{n,p}) \right) \Big], \end{aligned} \tag{2}$$

where $\lambda_{m,n}(\epsilon_{n,p})$ and $\mu_{m,n}(\epsilon_{n,p})$ are defined in [3]. The desired component in (2) on subcarrier m is $\rho_{m,p}\, S_{m,i}\, \mu_{m,m}(\epsilon_{m,p})$, and the ISI on the same subcarrier is $\rho_{m,p}\, S_{m,i-1}\, \lambda_{m,m}(\epsilon_{m,p})$. ICI from the subcarriers $n \neq m$ is given by $\rho_{n,p}\, (S_{n,i}\, \mu_{m,n}(\epsilon_{n,p}) + S_{n,i-1}\, \lambda_{m,n}(\epsilon_{n,p}))$. The average signal-to-interference power ratio (SIR) on subcarrier m is

$$(SIR)_m = \frac{P_s}{P_{ICI} + P_{ISI}}, \tag{3}$$

where P_s is the power of the desired component and $P_{ICI} + P_{ISI}$ is the interference power.

Two detection algorithms differing in the choice of the position of the detection interval in the receiver are investigated and introduced in the following.

Algorithm I: In order to keep the loss in bandwidth efficiency due to the insertion of guard intervals at a tolerable level, we consider an uplink scheme which allows ISI and ICI by choosing $T_g < \epsilon_{\max}$, i.e., we use a short guard interval. At the same time the detection interval in the receiver is chosen such that the residual interference is minimized. Principle features of the Algorithm I are illustrated in Fig. 3. $\Delta + T_g$ is the beginning of the integration interval for demodulation (of duration T_s), which is the same for all users. The time shift Δ can take on values in the interval $[0, \epsilon_{\max} - T_g)$. It is explained in [3] how to find the optimum Δ.

We can define the worst case user as the user which is next to the base station and has a delay of $\delta^{(k)} \approx 0\,\mu$s. Due to the fixed detection interval starting at $\Delta + T_g$, the worst case user in addition to the ICI from other users suffers from the maximum possible ISI.

Algorithm II: This is a straightforward solution where the integration interval for demodulation of the data of user k starts at the time instant $\delta^{(k)} + T_g$. The integration interval can thus be different for each user in order to minimize the interference originating from its own subset of subcarriers. Hence, this approach does not take into account the probability density function $p_\delta(\delta^{(k)})$ of the delays and, thus, of the interference from other asynchronous users. Principle features of the Algorithm II are illustrated in Fig. 3.

For Algorithm II we can define the worst case user as the user with maximum distance R to the base station and a maximum delay of $\delta^{(k)} = \delta_{\max}$. Thus, each user with $\delta^{(k)} < \delta_{\max}$ interferes with the desired user, and the guard interval gives no advantage.

4. RESULTS AND DISCUSSION

The investigated SS-MC-MA system is designed for medium size cells with a radius between 1 km and 3 km, typical for future outdoor cellular mobile radio systems. This results in maximum time offsets $\delta_{\max}$ between the users of between 6.7 μs and 20 μs. The transmission bandwidth is 2 MHz, the carrier frequency $f_c = 2$ GHz and the number of subcarriers $N_c = 256$. The Hadamard codes used for spreading are of length $L = 8$. QPSK symbol mapping is used and convolutional codes with rate 1/2 and memory 6 are applied. The total number of active users is 128 [3] and the net bit rate per user results in 10.8 kbit/s.

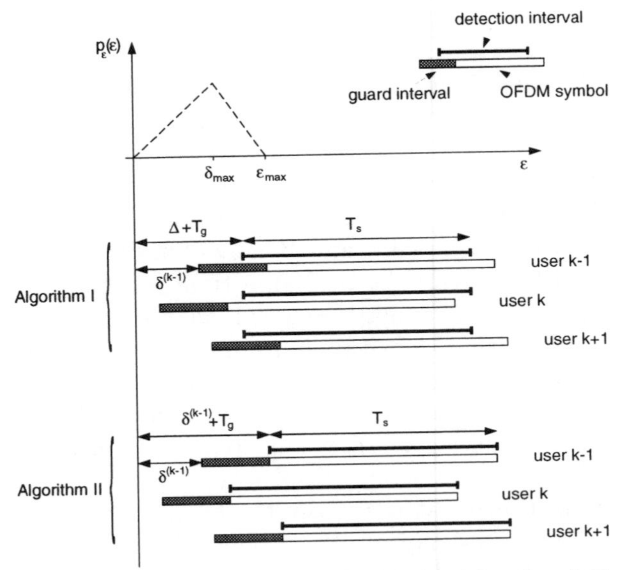

Figure 3 Principle of interference mitigation with Algorithm I and Algorithm II

The FIR filters for channel estimation are designed to become optimal Wiener filters in the worst case condition when the channel's Doppler power spectrum is rectangular with the maximum Doppler shift $f_{D_{max}} = 333.3$ Hz. The pilot symbol spacing is chosen such that the channel transfer function is two-times oversampled. The one-dimensional FIR filters are realized with 5 taps, which is a good compromise between performance and complexity [4]. The separation of pilot symbols is 5 symbols.

The mobile radio channel model is taken from [5]. The 'Outdoor Residential -High Antenna' (Channel B) channel model with maximum excess delay $\tau_{\max} = 15$ μs is chosen. Velocity of the mobile user is 30 km/h, resulting in the maximum Doppler frequency of 55.6 Hz, and the classical Doppler spectrum is assumed [5, 6]. All Monte Carlo simulation results shown in the following are for the most critical case of a fully loaded system. Moreover, the signal-to-noise ratio (SNR) degradation due to the guard interval and pilot symbols is taken into account in the results.

In Fig. 4 the SIR versus the guard time T_g is depicted. The SIR is shown with interference mitigation according to Algorithm I and Algorithm II for different cell sizes. The worst case user of Algorithm I and Algorithm II is considered. It is obvious that only Algorithm I is able to deal with the interference since the straightforward solution of

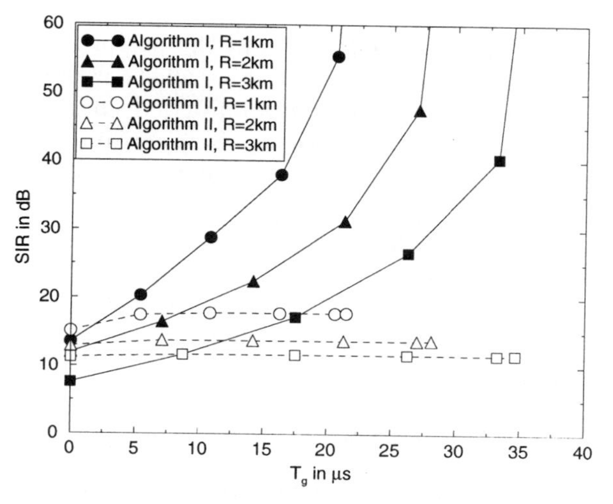

Figure 4 SIR for different cell sizes; worst case user; fully loaded system

Algorithm II does not take into account interference from asynchronous users. Thus, in the following, the focus is only on Algorithm I.

In Fig. 5 the bit error rate (BER) versus E_b/N_0 is shown for different cell sizes. The guard interval duration T_g is 15 μs. Results with pilot-

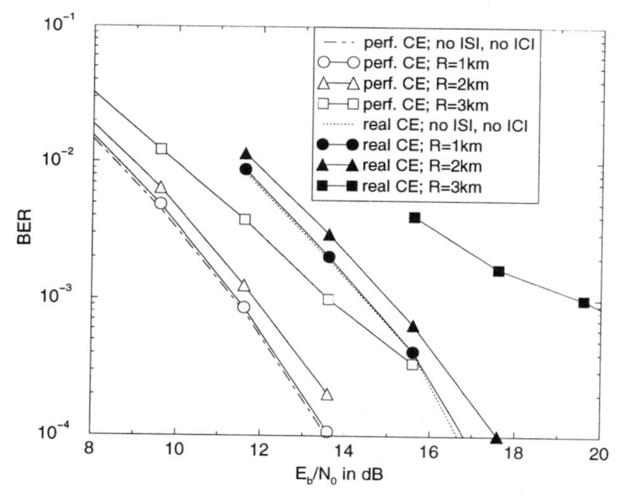

Figure 5 Performance of Algorithm I with channel coding and pilot-symbol-aided channel estimation (real CE); worst case user; fully loaded system

symbol-aided channel estimation (real CE) and with perfect channel knowledge (perf. CE) are shown. Moreover, as a lower bound, the BER performance of SS-MC-MA without ISI and ICI is plotted. Pilot-symbol-aided channel estimation produces tolerable BER performance in cells with up to 2 km radius. When $R = 2$ km, the performance degradation due to the presence of ISI and ICI is the same whether the perfect channel knowledge is available, or the pilot-symbol-aided channel estimation is used. However, if the interference exceeds a certain amount (e.g. for $R = 3$ km) the SNR degradation with pilot-symbol-aided CE becomes more severe.

Finally, it can be seen that with a radius of 2 km the total delay $\epsilon_{\max}$ is 28.3 μs and a guard time duration of only 15 μs is used. I.e., the guard interval can be reduced by about 50% with the proposed system, resulting in increased bandwidth efficiency and only moderate SNR degradation of less than 1 dB.

5. CONCLUSIONS

The performance tradeoffs in asynchronous SS-MC-MA in the uplink of a mobile radio system have been investigated. The presented system allows ISI and ICI in the uplink but in return reduces the loss in bandwidth efficiency due to shortened guard interval and avoids the complexity of a dedicated uplink synchronization scheme. The residual interference is minimized by proper positioning of the detection interval in the receiver. It is shown that with this approach the guard interval can be reduced by about 50%, resulting in increased bandwidth efficiency and only moderate SNR degradation of less than 1 dB.

References

[1] S. Kaiser and K. Fazel, "A spread-spectrum multi-carrier multiple-access system for mobile communications," in *Proc. First Intern. Workshop on Multi-Carrier Spread-Spectrum*, pp. 49–56, April 1997.

[2] S. Weinstein and P. M. Ebert, "Data transmission by frequency-division multiplexing using the discrete Fourier transform," *IEEE Trans. Commun. Tech.*, vol. 19, pp. 628–634, Oct. 1971.

[3] S. Kaiser and W. A. Krzymien, "Performance effects of the uplink asynchronism in a spread spectrum multi-carrier multiple access system," accepted for publication in the *European Transactions on Telecommunications (ETT)*, 1999.

[4] P. Hoeher, S. Kaiser, and P. Robertson, "Pilot-symbol-aided channel estimation in time and frequency," in *Proc. IEEE Global Telecommun. Conf. (GLOBECOM'97), Commun. Theory Mini Conf.*, pp. 90–96, Nov. 1997.

[5] K. Pahlavan and A. H. Levesque, *Wireless Information Networks*. New York: John Wiley & Sons, 1995.

[6] W. C. Jakes, *Microwave Mobile Communications*. New York: John Wiley & Sons, 1974.

Section II

APPLICATIONS OF MULTI-CARRIER SPREAD-SPECTRUM & RELATED TOPICS

Performance of a Flexible Form of MC-CDMA in a Cellular System

Heidi Steendam and Marc Moeneclaey
Department of Telecommunications and Information Processing, University of Ghent, B-9000 GENT, BELGIUM

Key words: MC-CDMA, flexibility

Abstract: In this contribution, we investigate a variant of the traditional MC-CDMA system in the case of downlink communication. In the proposed MC-CDMA system, we can independently select the number of chips per symbol (N_{chip}), the number of carriers (N_{carr}) and the FFT length (N_{FFT}), so the available resources can be used more effectively. The bandwidth of this flexible MC-CDMA system is proportional to N_{chip}, while the spectral density of the power spectrum is inversely proportional to N_{chip}: the transmitted power is independent of N_{chip}. Furthermore, the flexible MC-CDMA system spreads the power of a smallband interferer over a large bandwidth, so the immunity of the system to smallband interferers increases for increasing N_{chip}. In the presence of a dispersive channel and for the number of users equal to N_{chip}, the powers of the useful component, the interference and noise are independent of the number of chips per symbol, while an optimal guard interval can be found that maximises the performance.

1. INTRODUCTION

As orthogonal multicarrier (MC) techniques have good bandwidth efficiency and can offer an immunity to channel dispersion, these techniques are excellent candidates for high data rate transmission over dispersive channels. To cope with the high bit error rates caused by the strong attenuation of some carriers, orthogonal MC systems have been investigated in combination with code-division multiple-access (CDMA) [1-10]. By combining CDMA with the orthogonal MC technique, coding is provided by spreading the data on the different carriers using the CDMA codes, so frequency diversity is obtained. One of the combinations of CDMA and orthogonal multicarrier is the MC-CDMA system, where the data symbols are first spread using the CDMA codes and then modulated on the orthogonal carriers. The MC-CDMA system has been proposed for downlink communication in mobile radio [3-10].

In this contribution, we consider the downlink transmission of a cellular MC-CDMA system. We investigate a variant of the traditional MC-CDMA system. In the proposed, flexible MC-CDMA system, the number of chips per symbol N_{chip} can be chosen independently of the number of carriers N_{carr}, which offers us a higher flexibility as compared to the traditional MC-CDMA system, where the number of carriers was fixed to the number of chips per symbol. Furthermore, as the carriers inside the rolloff area give rise to severe performance degradation [11], we only use the carriers outside the rolloff area, which means that the FFT length N_{FFT} exceeds the number of carriers ($N_{FFT}>N_{carr}$). In the proposed variant of the MC-CDMA system, we can independently choose N_{chip}, N_{carr} and N_{FFT}, so the available resources can be used more effectively.

2. SYSTEM DESCRIPTION

The conceptual block diagram of the considered system is shown in figure 1 for one user. The data symbols $\{a_{i,m}\}$ transmitted at a rate R_s, where $a_{i,m}$ denotes the i-th symbol transmitted to the user m, are multiplied by a higher rate chip sequence $\{c_{n+iN_{chip},m}|n=0,\ldots,N_{chip}-1\}$, $c_{n+iN_{chip},m}$ denoting the n-th chip of the sequence belonging to user m during the i-th symbol interval. The complex chip sequence corresponding to user m consists of the product of a real-valued orthogonal chip sequence of length N_{chip} (e.g. Walsh-Hadamard), corresponding to the considered user, and a complex-valued random chip sequence (e.g. a complex-valued pseudo-noise sequence of length $L>>N_{chip}$), equal for all users of the same cell and having the same rate as the orthogonal sequences. In other cells, the orthogonal sequences are reused by multiplying them with another random sequence. These hybrid sequences have better correlation properties than the pure Walsh-Hadamard sequences.

2.1 Single User Transmission

Let us first consider the transmission to one user. The spread data symbols are mapped on the carriers as shown in figure 2 and then modulated using the inverse FFT of length N_{FFT}, resulting in the time-domain samples

$s_{j,k}$, the k-th sample of the j-th FFT block, at a rate 1/T. The transmitted time-domain samples are cyclically extended with a guard interval ν to cope with the channel dispersion and applied to a unit-energy square-root Nyquist filter. The power spectrum of the complex envelope of the resulting transmitted signal is shown in figure 3. The flexible MC-CDMA system uses a bandwidth $B = (N_{carr}/N_{FFT})/T = N_{chip}R_s\ (N_{FFT}+\nu)/N_{FFT}$. The occupied bandwidth B is proportional to N_{chip} and, as the transmitted power P_s is independent of the number of chips, the spectral density of the power spectrum is inversely proportional to N_{chip} (figure 3a). A guard interval ν introduces a ripple in the power spectrum proportional to ν and at the same time it gives rise to an increase of the used bandwidth B (figure 3b). To normalise the transmitted power ($P_s=R_sE_s$, where E_s is the energy per symbol) the transmitted samples are multiplied by a factor $\mathrm{sqrt}(N_{FFT}/(N_{FFT}+\nu))$.

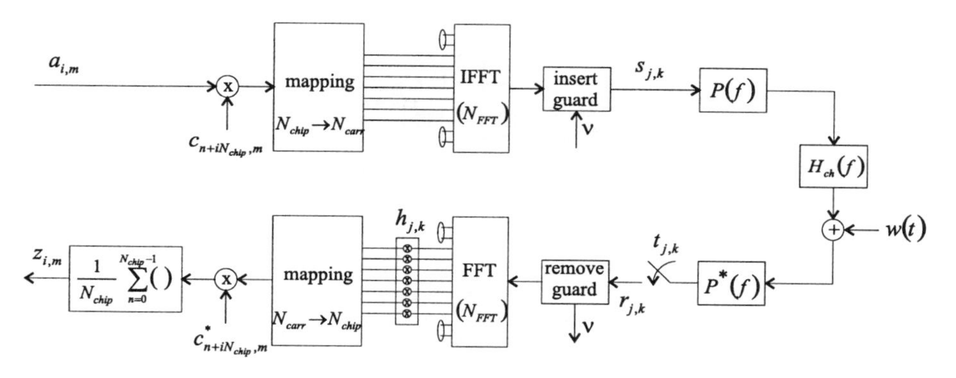

Figure 1. Conceptual block diagram of the flexible MC-CDMA system for one user

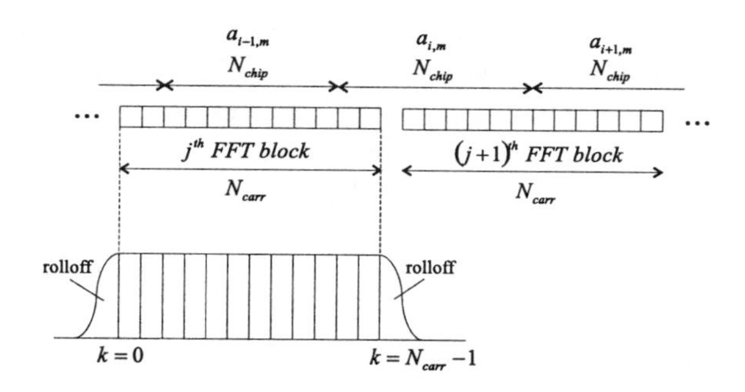

Figure 2. Mapping of the chips on the carriers

In the case of an ideal channel and when the transmitted signal is only disturbed by additive white Gaussian noise (AWGN) with a spectral density

N_0, the guard interval can be omitted. The resulting signal-to-noise ratio at the input of the decision device equals E_s/N_0.

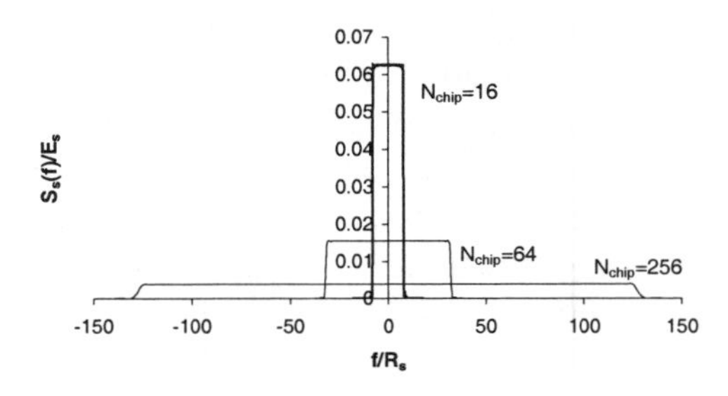

Figure 3a. Power spectrum for different N_{chip}

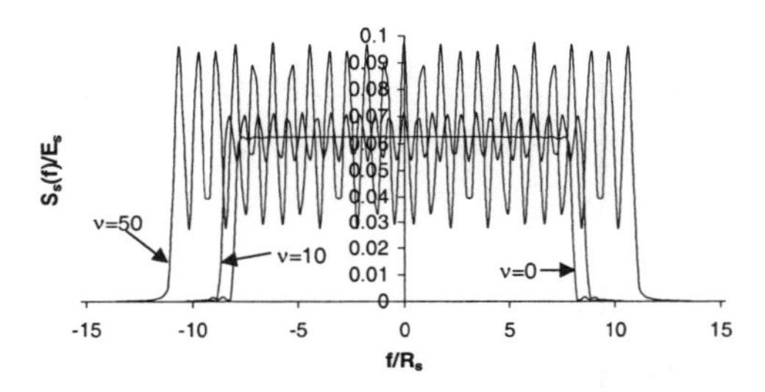

Figure 3b. Power spectrum for different guard intervals

2.2 Influence of a Narrowband Interferer

Let us consider the case of the transmitted signal disturbed by a narrowband interferer with power P_J, frequency f_J and phase θ_J. Due to the random character of the chip sequence, the interference components after multiplying with the chip sequences are uncorrelated and behave as AWGN. The interference power is equally spread over a bandwidth $N_{chip}R_s$ and has a power spectral density of $1/N_{chip}\, P_J X(f_J)$, where $X(f_J) \leq 1$ is shown in figure 4 as function of the interferer frequency. The quantity $X(f_J)$ equals 1 for interferer frequencies f_J outside the rolloff area that coincide with a carrier frequency. The signal-to-interference ratio at the input of the decision device is given by

$$SIR = N_{chip} \frac{P_s}{P_J X(f_J)} \geq N_{chip} \frac{P_s}{P_J} \tag{1}$$

The SIR is independent of N_{carr} and N_{FFT} and as the SIR increases for an increasing N_{chip}, the immunity of the flexible MC-CDMA system to narrowband interferers increases.

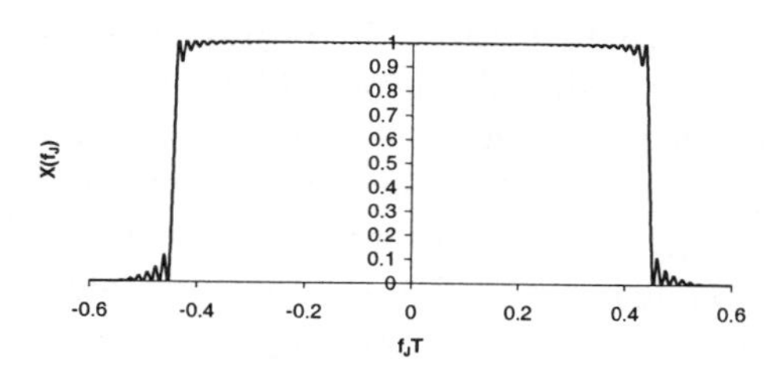

Figure 4. Transfer function smallband interferer power, $N_{FFT}=64$

3. MULTIUSER INTERFERENCE

When different users are present, multiuser interference (MUI) can occur. We can distinguish two types of MUI: intracell and intercell MUI. In the case of an ideal channel, no intracell interference is introduced, as the signals of the different users of the same cell are orthogonal. Signals of users belonging to adjacent cells are uncorrelated with the signal of the considered user, as the hybrid sequences are constructed with different random sequences, so the users of the adjacent cells will introduce MUI. The intercell MUI caused by the users of cell β is given by

$$MUI_{\beta} = \frac{N_{FFT}}{N_{FFT} + \nu} \frac{1}{N_{chip}} \sum_{\ell} E_{s,\ell}^{\beta} \approx \frac{1}{N_{chip}} \sum_{\ell} E_{s,\ell}^{\beta} \qquad (2)$$

where $E^{\beta}_{s,\ell}$ is the energy per symbol of user ℓ of cell β. When the guard length is small as compared to the FFT length, the intercell MUI is essentially independent of the guard length and the FFT length.

3.1 Dispersive Channel

In the case of a dispersive channel, intracell interference will be introduced as the orthogonality between the different users is lost. Assuming g(t) is the impulse response of the cascade of the transmit filter, the channel and the receiver filter, which is matched to the transmit filter, the samples at the input of the receiver yield

$$r_{j,k} = \sum_{j'} \sum_{k'} g\left(\left(k - k' + (j - j')(N_{FFT} + \nu)\right)T\right) s_{j',k'} + w_{j,k}$$
$$k = -\nu, \ldots, N_{FFT} - 1 \quad ; \quad j = -\infty, \ldots, +\infty \tag{3}$$

where $w_{j,k}$ is white Gaussian noise with power spectral density N_0. The channel is normalised such that the power of the signal of user m equals $P_{s,m}=E_{s,m}R_s, \forall m$. The receiver selects the N_{FFT} samples outside the guard interval for further processing. The selected samples are demodulated using the FFT. The FFT outputs outside the rolloff area are applied to a one-tap MMSE equaliser with equaliser coefficients $h_{j,k}$, to scale and rotate the FFT outputs. The resulting samples are mapped into blocks of N_{chip} samples and correlated with the chip sequence of the considered user to obtain the samples at the input of the decision device

$$z_{i,m} = \sum_{i'} \sum_{m'} \sqrt{E_{s,m'}} a_{i',m'} I_{i,i',m,m'} + W_{i,m} \tag{4}$$

where $W_{i,m}$ is a zero-mean complex-valued Gaussian noise term and $I_{i,i',m,m'}$ is the interference caused by the i'-th symbol of user m' on the i-th symbol of the considered user m. In order to eliminate the dependency of the symbol interval i, we consider the time-average of the power of the samples at the input of the decision device, given by

$$E\left[\left|z_{i,m}\right|^2\right] = \frac{N_{FFT}}{N_{FFT} + \nu}\left(P_U + P_I\right) + P_N \tag{5}$$

where P_U is the time-average of the power of the average useful component, P_I consists of the time-average of the sum of the powers of the fluctuation of the useful component, caused by the random character of the chip sequences, the intracell multiuser interference and intersymbol interference. The last contribution in (5) P_N is the time-average of the power of the additive Gaussian noise component.

In figure 5, the powers $(1-P_U)$, P_I and P_N are shown as function of the number of chips per symbol for the maximum load, i.e. the number of users equals N_{chip}. For large N_{chip}, the powers are essentially independent of N_{chip}: as for large N_{chip} the interference power is proportional to the number of users and inversely proportional to the number of chips per symbol, the total interference power is essentially independent of N_{chip} for the maximum load.

In figure 6, the powers $(1-P_U)$, P_I and P_N are shown as function of the guard interval for the maximum load. The interference power decreases for an increasing guard length, as the signals at the borders of the FFT blocks are less affected by the dispersive channel when the guard length increases, i.e. the intersymbol interference decreases. For large guard lengths, the interference power reaches a lower limit: the interference is mainly determined by the multiuser interference. The useful power and noise power only slightly vary with the guard length. However, due to the factor $N_{FFT}/(N_{FFT}+\nu)$, the performance will decrease for an increasing guard length. An optimal guard length can therefore be found at intermediate guard lengths.

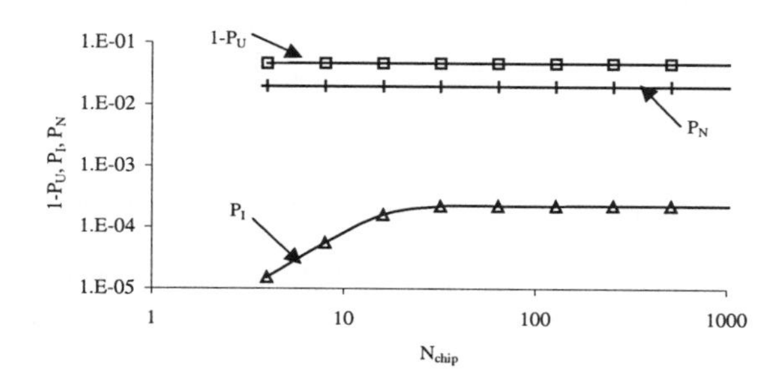

Figure 5. 1-P_U, P_I and P_N as function of the number of chips per symbol

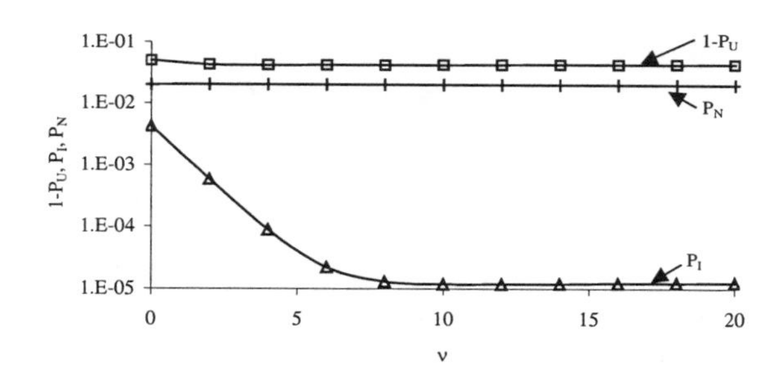

Figure 6. 1-P_U, P_I and P_N as function of the guard interval

4. CONCLUSIONS

We have investigated a flexible form of MC-CDMA in a cellular system. The power spectrum of the transmitted signal has a bandwidth that increases with the number of chips per symbol, while the power spectral density is inversely proportional to N_{chip}. Furthermore, the immunity of the flexible MC-CDMA system to narrowband interferers increases for increasing N_{chip}, as the interferer power is spread over a large bandwidth. The presence of different users gives rise to multiuser interference. However, in an ideal channel, no intracell MUI is introduced as all users in the same cell are orthogonal. Users of other cells will introduce intercell MUI, as they are uncorrelated with the users of the considered cell. The flexible MC-CDMA system is also investigated in the presence of a dispersive channel. For the maximum load, the flexible MC-CDMA system is essentially independent of the number of chips per symbol. Furthermore, an optimal guard interval can be found that maximises the performance.

REFERENCES

[1] S. Hara, R. Prasad, "Overview of Multicarrier CDMA", IEEE Comm. Mag., no. 12, vol. 35, Dec 97, pp. 126-133

[2] L. Vandendorpe, O. van de Wiel, "Decision Feedback Multi-User Detection for Multitone CDMA Systems", Proc. 7th Thyrrenian Workshop on Digital Communications, Viareggio Italy, Sep 95, pp. 39-52

[3] E.A. Sourour, M. Nakagawa, "Performance of Orthogonal Multicarrier CDMA in a Multipath Fading Channel", IEEE Trans. On Comm., vol. 44, no 3, Mar 96, pp. 356-367

[4] S. Hara, T.H. Lee, R. Prasad, "BER comparison of DS-CDMA and MC-CDMA for Frequency Selective Fading Channels", Proc. 7th Thyrrenian Workshop on Digital Communications, Viareggio Italy, Sep 95, pp. 3-14

[5] N. Yee, J.P. Linnartz, G. Fettweis, "Multicarrier CDMA in Indoor Wireless Radio Networks", Proc. PIMRC'93, Yokohama, Japan, 1993, pp. 109-113

[6] V.M. Da Silva, E.S. Sousa, "Multicarrier Orthogonal CDMA Signals for Quasi-Synchronous Communication Systems", IEEE J. on Sel. Areas in Comm., vol. 12, no 5, Jun 94, pp. 842-852

[7] Multi-Carrier Spread-Spectrum, Eds. K. Fazel and G. P. Fettweis, Kluwer Academic Publishers, 1997

[8] N. Yee, J.P. Linnartz, "Wiener Filtering of Multicarrier CDMA in a Rayleigh Fading Channel", Proc. PIMRC'94, 1994, pp. 1344-1347

[9] N. Morinaga, M. Nakagawa , R. Kohno, "New Concepts and Technologies for Achieving Highly Reliable and High-Capacity Multimedia Wireless Communications Systems", IEEE Comm. Mag., Vol. 35, no. 1, Jan 97, pp. 34-40

[10] L. Tomba and W.A. Krzymien, "Effect of Carrier Phase Noise and Frequency Offset on the Performance of Multicarrier CDMA Systems", ICC 1996, Dallas TX, Jun 96, Paper S49.5, pp. 1513-1517

[11] H. Steendam, M. Moeneclaey, "The Effect of Synchronisation Errors on MC-CDMA Performance", ICC'99, Vancouver, Canada, Jun 99, Paper S38.3, pp. 1510-1514

A Multi-Carrier Spread-Spectrum Approach to Broadband Underwater Acoustic Communications

R F Ormondroyd, W K Lam, and J J Davies

(RFO) Communications and Wireless Networks Group, RMCS, Cranfield University, Swindon, SN6 8LA, UK.

(WKL) Roke Manor Research Ltd, Roke Manor Romsey,SO51OZN, UK.

(JJD) Defence Evaluation Research Agency, Winfrith Technology Centre, Winfrith, Dorchester, DT2 8XJ, UK.

Key words: Underwater acoustic communications, channel estimation, multi-carrier spread-spectrum modulation, OFDM, synchronisation

Abstract: The paper presents a detailed Monte-Carlo simulation study of a novel MC-CDMA system operating in the medium-range shallow underwater acoustic channel. The new system uses random phase signals to multiplex the data onto orthogonal sub-carriers. The frame structure of the signal allows the random phase signal to be used for channel sounding *and* synchronisation. Carrier and frame synchronisation is achieved by exploiting the excellent auto-correlation property of the random phase signal and synchronisation is achieved in levels of noise down to -3dB SNR under conditions of limited Doppler. By adjusting the phases of the carriers, the 'crest factor' of the transmitted signal can be optimised. Spectrum spreading is used in conjunction with COFDM modulation to provide a flexible system that is able to maintain an acceptable BER in low SNR conditions by trading data rate for spread-spectrum process gain.

1. INTRODUCTION

There is a current requirement for broadband communications in the shallow underwater acoustic (UWA) channel where co-channel interference, extremely high Doppler frequency offsets and time-varying and frequency-selective deep fades due to multipath propagation make reliable data reception difficult. Because of the extremely low carrier frequencies and the nature of propagation in the underwater acoustic channel, such channels can be doubly-spread in both frequency and time. There has been some progress in broadband UWA communications in less dispersive, deep-water, channels, using robust coding and interleaving together with bandwidth efficient coherent modulation schemes such as m'PSK and QAM and time-domain equalisation [1-3]. Over short to medium ranges, it is possible to exploit spatial processing using a vertical array transducer to enable effective equalisation to be achieved with acceptable levels of complexity. However, at longer ranges the effectiveness of spatial processing is reduced and equalisation becomes a problem for long delay-spreads. An alternative approach to UWA communications that is robust in a multipath fading channel is to use a coded orthogonal frequency division multiplexed (COFDM) system.

2. COFDM SYSTEM

The COFDM technique has been well researched in the past, particularly for terrestrial digital television and digital audio broadcasting. Recently, the method has been applied to underwater acoustic communications [4,5]. A schematic diagram of the system used here is shown in figure 1. In essence, OFDM transmits a high-speed data stream as N parallel data streams on N orthogonal sub-carriers. The rate of the data on each sub-carrier is $1/N$ of the original rate. By coding and interleaving the data in time and frequency, the COFDM system possesses time- and frequency-diversity and can be very resilient in time-varying and frequency-selective channels.

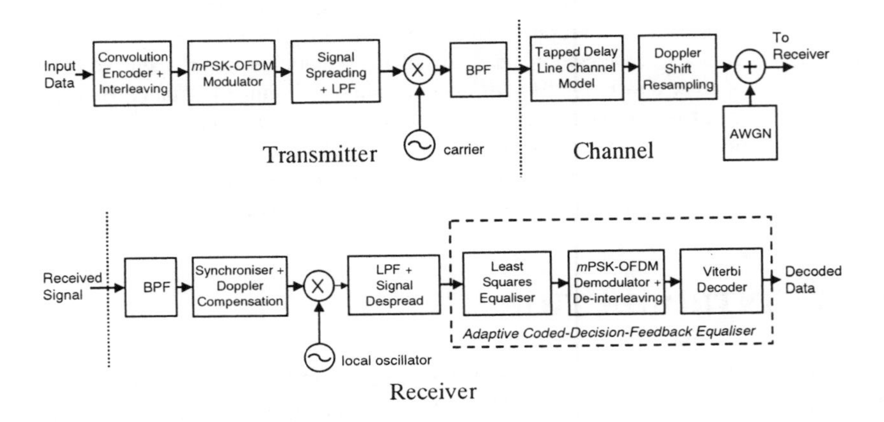

Figure 1. OFDM System configuration

A common implementation of the OFDM transmitter uses the IFFT to generate the orthogonal sub-carriers. However, we use a method whereby each orthogonal sub-carrier has a pseudo-random phase relative to the other sub-carriers. The resulting OFDM signal is often called *pseudo-random Gaussian noise* (PRGN). This has a number of advantages compared with the normal OFDM implementation. First, by appropriate choice of the phasing of the carriers [6], it is possible to optimise the *crest factor* of the transmitted signal and this has an important consequence if the amplifiers and transducers used in the UWA system have non-linear transfer characteristics. Second, the PRGN carriers randomise the mapped phase sequence of the data. Third, the transmitted signal has a similar spectrum to conventional IFFT generated signals so that it can still be used for channel sounding to estimate the channel transfer function prior to equalisation. Fourth, the transmitted signal has an auto-correlation function that has a single main lobe and small side-lobes. This property can be exploited in the receiver to allow the receiver to obtain carrier and frame synchronisation, which is vital to ensure that the received OFDM blocks are correctly de-multiplexed. The PRGN signal only randomises the data, however, it does not spread it. In order to maximise efficiency, no cyclic extension is used in this system. In order to improve performance in high levels of noise, each OFDM block is retransmitted four times in a very bandwidth efficient way [7,8].

The channel is modelled as a multi-tap transversal filter with Rayleigh fading of the signal envelope on each tap [6,7]. For the UW channel, the acoustic carrier frequency is very low relative to RF systems and Doppler, due to platform motion, modulates the width of the data pulses as well as shifting the frequencies of the carriers. The channel model incorporates Doppler re-sampling at a slightly different rate to simulate the effect of the Doppler on the width of each modulated data bit.

2.1 Frame Structure

The transmitted signal comprises a PRGN 'burst' of pilot tones that is transmitted for 256ms. This is used in the receiver to obtain both an estimate of the channel transfer function and frame synchronisation. Synchronisation is achieved by correlating the received signal against a replica PRGN sequence. However, this is made difficult by the effect of the Doppler on the data timing and the sub-carrier frequencies. In this system, Doppler compensation is added prior to OFDM demodulation [8,9]. After each sounding signal, there are 16 data blocks that are modulated by the PRGN. During this period, channel estimates are taken directly from the received data using a decision-directed feedback channel estimator system, similar in concept to the CD3-

OFDM system [10]. After the data blocks have been transmitted, a new channel estimate is obtained using the pilot tone burst. This reduces the likelihood of the breakdown of the feedback estimator due to the propagation of errors in the data estimates, and hence the channel estimates. The structure of the receiver is shown in figure 2.

This feedback system must maintain an accurate estimate of the channel transfer function between the relatively infrequent channel sounding signals, especially in high levels of noise. For the UWA communications system, the situation is made more complicated by (i) the extreme Doppler spread, which reduces the channel coherence time and (ii) multipath echoes that may be difficult to resolve. For these reasons, the traditional approaches to channel estimation are not adequate and a new method has been developed.

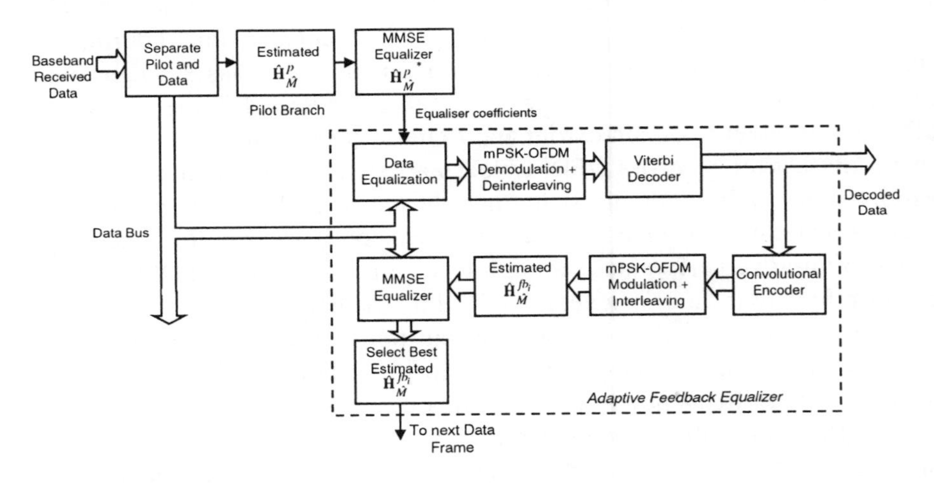

Figure 2. Schematic of adaptive feedback equaliser

2.2 Channel Estimation

Because the transmitted channel sounding signal has a flat spectrum, the amplitude and phase of the received pilots provide a noisy estimate of the channel transfer function directly. In many applications it is simply necessary to filter this estimate to arrive at an adequate estimate of the channel transfer function. That method is not appropriate for the underwater channel and we have used a parametric model of the channel to improve the channel estimate. The channel impulse response is a very sensitive measure of the UWA channel behaviour. Each echo gives rise to a sharply peaked impulse response, although several peaks may be very closely spaced. Consequently, in this work, the parametric model of the channel focuses on the impulse response. An 'all-zero' approach does not model the UWA impulse response with sufficient accuracy. A better technique is to use an 'all-pole' auto-regressive (AR) method. The PRGN provides noisy samples of the frequency characteristics of the channel, $\hat{H}_p$, from which we need to estimate the channel impulse response. One problem is to decide upon the number of poles that is needed to obtain the most accurate impulse response from the frequency domain data. There are many approaches to this problem. In our case the PRGN provides us with 128 samples of the channel impulse response, from this we need to synthesise a channel model which has considerably fewer resolvable echoes (up to 40 in this case). In our system we have used a sub-optimum technique to compute the AR coefficients that is based on the Prony method [11].

The technique uses the forward and backward linear predictor (FBLP) method [11]. This requires the computation of the inverse of the product of two matrices: (i) the windowed forward and backward sections of the channel transfer function and (ii) its Hermetian. This is inefficient. An elegant solution that can take advantage of the reduction of the order of the problem is to use singular value decomposition (SVD). However, to use this method effectively, the dimension of the signal sub-space corresponding to the largest eigenvalues is needed. In this work we have used the *minimum description length method* (MDL) [12] to estimate the order of the AR model. Once the AR coefficients have been found, the estimated channel impulse response can be generated from the AR model. This provides a 'clean' channel frequency response, $\hat{H}_M$.

In this work, QPSK and 8PSK coherent phase modulation was used. Consequently, 'phase only' equalisation was used whereby the received complex signal samples on each sub-carrier are equalised by multiplication with the complex conjugate of the channel transfer function, $\hat{H}_M^*$. The OFDM modulator and demodulator were implemented using a SHARC signal processor. The channel estimator algorithm and synchroniser were implemented in software running in real-time on a PC.

3. SIMULATION RESULTS

Detailed Monte-Carlo simulations of the system were carried out for three typical channels models: (i) an AWGN channel with no multipath propagation, (ii) a 2-ray time-varying frequency-selective (2-TSFS) channel and (iii) a 4-ray time-varying frequency-selective channel (4-TSFS). In each case, Doppler was included in the channel model, as described earlier. The model parameters were chosen to generate a time-varying, frequency-selective channel path loss that closely matched experimental propagation studies. The main system parameters are detailed in Table 1. In the interest of space, only the rate 1/2 QPSK system is considered here. The performance of the OFDM frame synchroniser of the system in these types of channel and modulation scheme has been detailed in [6] and will not be considered here.

Table 1. Table of Parameters

Acoustic carrier frequency	1kHz
OFDM bandwidth	1kHz
Modulation scheme	QPSK and PRGN modulation
Coding	Rate ½ K=7 convolutional coding
COFDM data rate/MC-CDMA chip rate	1kb/s, 1kchip/s
No of sub-carriers	128
OFDM Block length	256 ms

3.1 Bit Error Probability

The bit-error-probability of the rate-1/2 QPSK-OFDM system was tested for the three channel types and typical results are presented in figure 3.

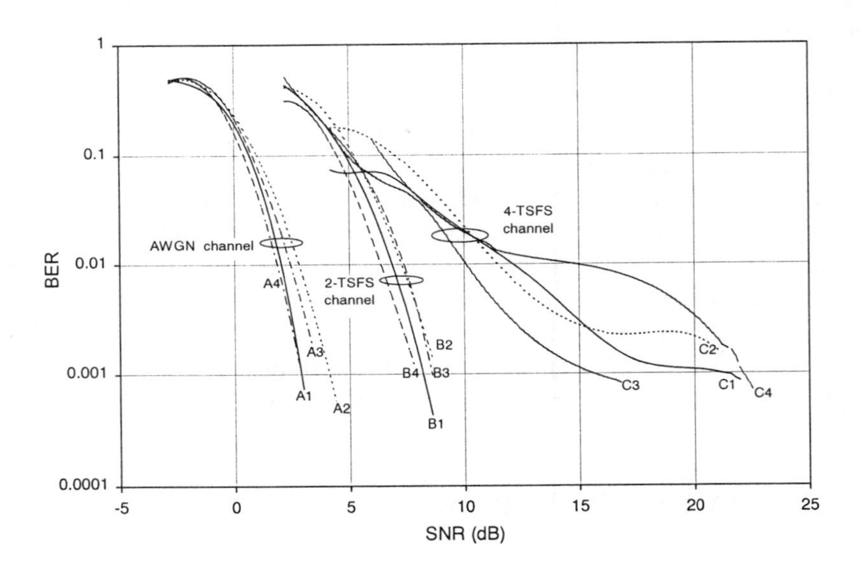

Figure 3. Bit error rate performance of rate1/2 QPSK-OFDM system

(AWGN Channel: A1=0 knots, A2=2 knots, A3=4 knots, A4=6 knots
2-TSFS Channel: B1=0 knots, B2=2 knots, B3=4 knots B4=6 knots
4-TSFS Channel C1=0 knots, C2=2 knots, C3=4 knots C4=6 knots)

For the AWGN channel (curves A1-A4), the system compensates for the Doppler shift extremely well and it was possible to correctly decode the data even at an SNR of approx. 4dB. This is due, in part, to the coding and interleaving, but the performance is improved by 6dB because the OFDM blocks are repeated in the Tx by the factor of four repetition code and these are then combined in the Rx. For the 2-TSFS channel (curves B1-B4), the BER performance is generally 4 to 5dBs worse than the AWGN case. However, there is no error-rate floor due to either burst errors in the fading channel or failure of the equaliser. There is a spread of only 3dB in performance as the platform velocity is increased, indicating that the receiver works well to compensate the Doppler shift and equalise the multipath channel.

For the 4-TSFS channel (curves C1-C4s), there is some degradation in performance as the Doppler shift is increased and there is now a floor to the BER. This is mainly due to incorrect Doppler correction in the receiver.

4. SPREAD-SPECTRUM OVERLAY

The system above is generally robust to multipath propagation and Doppler spread. In order to provide the system with additional robustness to both AWGN and frequency-selective fading, the system configuration includes a direct-sequence spread-spectrum overlay. This converts the system to an MC-CDMA system with minimal modification to either the transmitter or receiver. The COFDM system described above has been carefully optimised in terms of its bandwidth and OFDM block length in relation to the coherence bandwidth and coherence time of the channel. It was important to ensure that the original OFDM block structure was maintained to preserve the good multipath performance. For the MC-CDMA operating mode, the data-rate was reduced by the code spreading factor, P, in order to ensure that the MC-CDMA system had the same OFDM block rate and block length as the original system and the total transmission bandwidth remains unchanged.

The chips of each data bit were interleaved prior to transmission on different sub-carriers, thereby providing frequency diversity to the chips. A schematic diagram of the transmitter overlay is shown in figure 4a. In the receiver, the FFT demodulates the signals on the sub-carriers, which are then frequency domain equalised in the usual way. (Note that in this system, because pseudo random phase signals are used in the transmitter, the phase of each sub-carrier is corrected from prior knowledge of the phases used on each sub-carrier in the transmitter). Groups of P symbols representing the chips of each data bit are then de-interleaved and averaged to recover the data bits. The data bits are de-interleaved in time, and decoded using the

Viterbi decoder. The schematic diagram of the modified receiver is shown in figure 4b. This approach provides improved robustness in AWGN as well as frequency diversity.

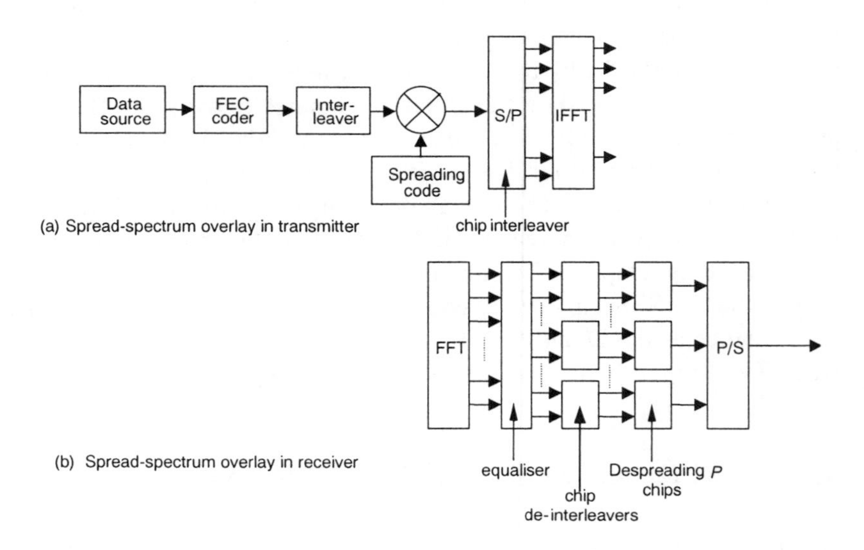

Figure 4. Schematic diagram of modified receiver

Figure 5, overleaf, shows typical bit error probability curves for the modified system for the AWGN, 2-TSFS and 4-TSFS channel models at four platform velocities (0 knots, 2 knots, 4 knots and 6 knots). To illustrate the method, QPSK modulation was used with a ×4 spreading ratio to achieve approx. 6dB process gain in addition to that provided by the repetition code. Correspondingly, the data rate was reduced from 1kb/s to 250b/s. Comparing these results with those of figure 3, it is seen that spreading allows the system to operate in at least 6 dB more noise for the AWGN and the 2-TSFS channels. However, there is a significant improvement in performance for the 4-ray TSFS channel, showing that the MC-CDMA approach has provided significant frequency diversity in addition to the AWGN process gain to combat the frequency selective fading.

5. CONCLUSIONS

A coherent COFDM modulation system has been applied to the problem of broadband underwater acoustic communications. The new system uses PRGN signals to provide robust OFDM frame synchronisation and periodic channel sounding. Decision-directed channel estimation reduces the frequency of the channel sounding signal. The channel estimator is based on the Prony method and uses a least-squares AR model of the channel. A spread-spectrum overlay provides improved noise performance without compromising the performance of the channel estimator/equaliser. The spread-spectrum overlay offers a flexible solution to the problem of using OFDM systems in poor SNR conditions and provides additional frequency diversity.

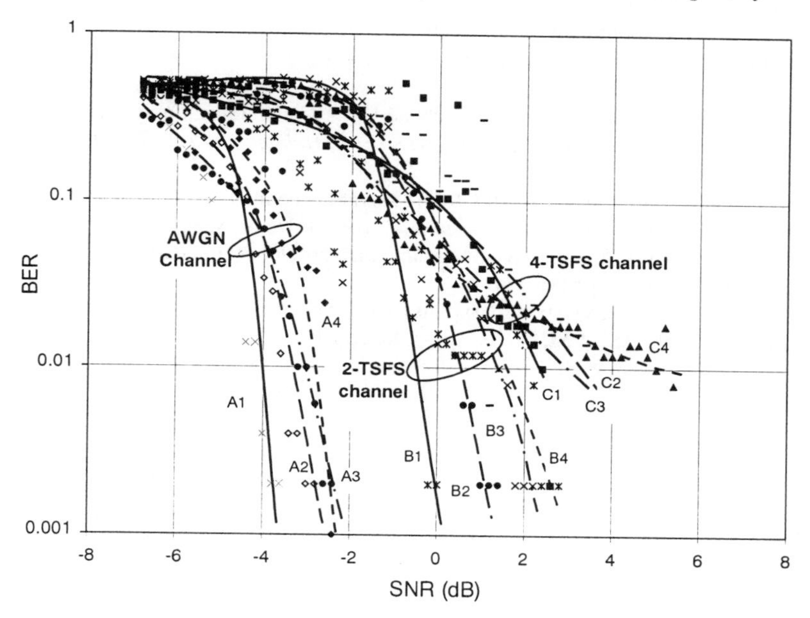

Figure 5. Bit error rate performance of the rate 1/2 QPSK-OFDM system with spread-spectrum overlay
(AWGN Channel: A1=0 knots, A2=2 knots, A3=4 knots, A4=6 knots
2-TSFS Channel: B1=0 knots, B2=2 knots, B3=4 knots B4=6 knots
4-TSFS Channel C1=0 knots, C2=2 knots, C3=4 knots C4=6 knots)

6. REFERENCES

1. M. Stojanovic et. al, "Phase-Coherent Digital Communications for Underwater Acoustic Channels", *IEEE Journal of Oceanic Eng*., vol. 19, No.1, Jan. 1994., pp100-111.
2. G. Ayela and J. M. Coudeville, "TIVA: A long range, high baud rate image/data acoustic transmission system for underwater applications," Proc. Underwater Defence Technology Conference, Paris, France,1991.
3. M. Stojanovic, "Recent advances in high-speed underwater acoustic communications", IEEE J. Oceanic Engineering, vol. 21, pp. 125-136, April 1996.
4. A.G. Bessios and F.M. Caimi, "Fast Underwater Acoustic Data Link Design via Multicarrier Modulation and Higher-order Statistics Equalization", *Proc. OCEANS'95, MTS/IEEE, Oct. 95, San Diego California*, pp594-599.

5. E Bejjani and J-C Belfiore, "Multi-carrier Coherent Communications for the Underwater Acoustic Channel", *Proc. OCEANS'96, MTS/IEEE, Florida*, Sep. 1996, pp1125-1130.
6. W.K. Lam and R.F. Ormondroyd: "A coherent COFDM modulation system for a time-varying, frequency selective underwater acoustic channel", 7th IEE International Conference on Electronic Engineering in Oceanography, Southampton Oceanography Centre, pp 198-203, June 1997
7. W.K. Lam and R.F. Ormondroyd, "A Broadband UWA Communication System Based on COFDM Modulation", Proc. OCEANS'97, MTS/IEEE, Halifax, Canada, Oct. 1997, pp862-869.
8. W K Lam and R F Ormondroyd, "A novel broadband COFDM modulation scheme for robust communication over the underwater acoustic channel", MILCOM'98, Bedford, Massachusetts, Oct 1998, pp 128-133
9. W.K. Lam and R.F. Ormondroyd, "A Broadband UWA Communication System Based on COFDM Modulation", *Proc. OCEANS'97, MTS/IEEE, Halifax, Nova Scotia*, Oct. 1997, pp 862-869.
10. V. Mignone and A. Morello, "CD3-OFDM: A Novel Demodulation Scheme for Fixed and Mobile Receivers", IEEE Trans. On comm., vol. 44, No. 9, Sep. 1996, pp1144-1151.
11. S.L. Jr. Marple, "New Autoregressive Spectrum Analysis algorithm" IEEE Trans. ASSP, vol. ASSP-28, Aug. 1980, pp441-454.
12. M. Wax and T. Kailath, "Detection of Signals by Information Theoretic Criteria", IEEE Trans. on ASSP, vol. ASSP-33, No. 2, Apr. 1985, pp387-392

A New Combined OFDM-CDMA Approach to Cellular Mobile Communications

Li Ping
Department of EE, City University of Hong Kong, email: eeliping@cityu.edu.hk

Key words: *Multiple Access, CDMA, OFDM, Mobile Communications, Indoor Communications*

Abstract

We present a combined OFDM and CDMA cellular system derived from the OFDM model by introducing an extra CDMA layer. This mitigates inter-cell interference without affecting intra-cell orthogonality and such property can be maintained in multi-path environments. The new scheme is suitable for both up and down links. It retains the low receiver complexity property of the OFDM system.

1. Introduction

The capacity of a CDMA (Code Division Multiple Access) system can be enhanced by suppressing multiple access interference (MAI). Maximum likelihood (ML) estimation technique can be used for this purpose but the complexity involved is generally high [1-3].

In this paper we present a new CDMA scheme free from intra-cell MAI. Starting from an OFDM (Orthogonal Frequency Division Multiplexing) model [4], we derive the proposed scheme by introducing an extra CsDMA (Code-shift Division Multiple Access) layer [5]. The new layer does not affect the normal transmission of the original system but it provides randomization effect to alleviate the inter-cell MAI problem. The receiver is realized by a simple quasi-coherent detection technique without the necessity of high cost ML estimation. The orthogonality among same-cell users is maintained in a multi-path environment. This is compared with some CDMA systems (such as the down-link of IS-95) with intra-cell orthogonality in ideal circumstances but this property is lost when multi-path reflection is present.

II. The cyclic prefix technique

Let $x=\{x_n\}$ and $y=\{y_n\}$ be two discrete sequences and $<n-n'>=n-n'$ modulo N. Define,

sliding convolution $z=x*y$:
$$z_n = \sum_{n'=-\infty}^{\infty} x_{n'} y_{n-n'} \qquad (1a)$$

N-point cyclic convolution $z=x\otimes y$:
$$z_n = \sum_{n'=0}^{N-1} x_{n'} y_{<n-n'>} \qquad 0 \le n \le N-1 \qquad (1b)$$

where $\bar{y}$ is the conjugate of y. The multi-path effect in a digital system is usually modeled by a sliding convolution $\boldsymbol{c}*\boldsymbol{w}$, where $\boldsymbol{w}$ is the transmitted sequence and $\boldsymbol{c}$ a vector of the path reflection coefficients. Assume that $\boldsymbol{c}$ has a limited nonzero span in $(0, D)$, i.e., $c_n = 0$ for $n<0$ and $n>D$, see Fig.1. Given $\boldsymbol{w}=\{w_n \mid n=0,1, ...N-1\}$, the prefix of $\boldsymbol{w}$ is its cyclic extension defined by $w_n=w_{n+N}$, $n\in[-D,-1]$. Assuming $D<N$, the prefix technique transforms a sliding convolution into a circular one as

$$\sum_{n'=-\infty}^{\infty} c_{n'} w_{n-n'} = \sum_{n'=0}^{D} c_{n'} w_{n-n'} = \sum_{n'=0}^{N-1} c_{n'} w_{<n-n'>} \qquad (2)$$

or simply, $\boldsymbol{c*w=c\otimes w}$, with the understanding that (2) is defined only in $[0, N-1]$. This relationship is the basis for OFDM [4] as well as for the new system introduced below.

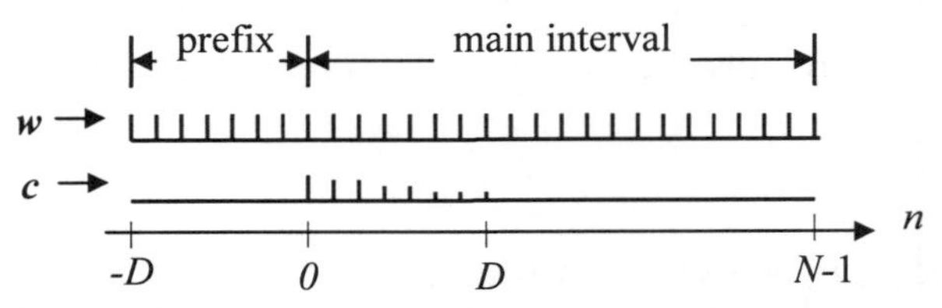

Fig.1. The cyclic prefix transforms a sliding convolution into a circular one.

III. The system model

A. The OFDM system

Consider the system in Fig.2 where F and F^{-1} are a pair of DFT (Discrete Fourier Transformation) and IDFT (inverse DFT) operators. We first ignore the effect of sequence $\boldsymbol{p}$ by setting $\boldsymbol{p}=\delta=\{1, 0, ... 0\}$. We also ignore the block labeled by "multiplexing". Then $\boldsymbol{w=v\otimes\delta=v}$, $\hat{\boldsymbol{v}}=\hat{\boldsymbol{w}}\otimes\delta=\hat{\boldsymbol{w}}$ and the system reduces to an common OFDM [4]. A prefix is padded to $\boldsymbol{w}$ before transmitting. Let the underlying QAM layer be described by the well-known model $\hat{\boldsymbol{w}}=\boldsymbol{c}*\boldsymbol{w}+\eta$ [6], where $\boldsymbol{c}=\{c_n\}$ and $\eta=\{\eta_n\}$ represent the intersymbol interference and additive noise respectively. Provided that the prefix length is longer than the maximum delay dispersion, from (2)

$$\hat{\boldsymbol{w}} = \boldsymbol{c} \otimes \boldsymbol{w} + \eta \tag{3}$$

or

$$\hat{\boldsymbol{v}} = \boldsymbol{c} \otimes \boldsymbol{v} + \eta \tag{4}$$

when $\boldsymbol{w}$=$\boldsymbol{v}$ and $\hat{\boldsymbol{v}} = \hat{\boldsymbol{w}}$. Applying DFT to $\hat{\boldsymbol{v}}$ in (4), we have the end-to-end relationship

$$\boldsymbol{u} = F(\hat{\boldsymbol{v}}) = \{C_k u_k + \zeta_k\} \tag{5}$$

where $\{C_k\}$=$F(\boldsymbol{c})$ and $\{\zeta_k\}$= $F(\eta)$. Eqn. (5) implies that the orthogonality among the OFDM frequency carriers is maintained in a multi-path environment [4].

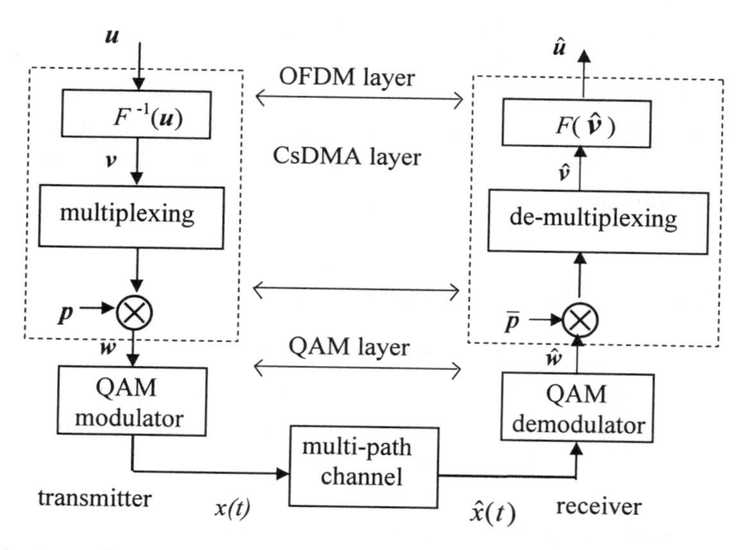

Fig.2. Overview of the proposed OFDM-CsDMA system. The QAM layer is based on the well known quadrature amplitude modulation principle [6].

B. The OFDM-CsDMA system

We now derive the proposed system by relaxing the constraint on $\boldsymbol{p}$ as,

$$\boldsymbol{p} \otimes \overline{\boldsymbol{p}} = \delta \tag{6}$$

Now $\boldsymbol{w} = \boldsymbol{v} \otimes \boldsymbol{p}$ in Fig.2. Based on (1b),

$$\boldsymbol{w} = \boldsymbol{v} \otimes \boldsymbol{p} = \sum_{n'=0}^{N-1} v_{n'} \boldsymbol{p}^{(n')} \tag{7}$$

with $\boldsymbol{p}^{(n')}$ the cyclic shift of $\boldsymbol{p}$ by n' positions. Eqn.(7) can be regarded as spreading $v_{n'}$ by $\boldsymbol{p}^{(n')}$, which is equivalent to the CsDMA principle in [5] and hence the name for the middle layer.

Suppose that a cyclic prefix is padded to $\boldsymbol{w}$ so that (3) still holds. Substituting the relationships $\boldsymbol{w}=\boldsymbol{v}\otimes\boldsymbol{p}$, $\hat{\boldsymbol{v}}=\hat{\boldsymbol{w}}\otimes\overline{\boldsymbol{p}}$ and $\boldsymbol{p}\otimes\overline{\boldsymbol{p}}=\delta$ into (3), we have

$$\hat{\boldsymbol{v}} = \boldsymbol{c} \otimes \boldsymbol{v} + \eta' \quad (8)$$

with $\eta' = \eta \otimes \overline{\boldsymbol{p}}$. The similarity between (4) and (8) indicates that the upper layer transfer function of the wanted signal is not affected by the CsDMA layer. In particular, the orthogonality relationship implied by (5) still holds.

The functional block labeled by "multiplexing" is employed to produces a more general structure for $\boldsymbol{v}$ as shown in Fig.3. Refer to the signal in the form of $F^{-1}(\boldsymbol{u})$, i.e., the signal before the multiplexing block in Fig.2, as an OFDM frame. In Fig.3, F OFDM frames together with their prefixes are multiplexed in $\boldsymbol{v}$, which are convoluted with $\boldsymbol{p}$ together. An extra prefix is padded to the resultant $\boldsymbol{w}$. In this way, it can be verified that (5) is again valid. Let the OFDM frame length be M. Each frame contains M orthogonal carriers distinguished by their discrete frequencies [4]. A total of $F\times M$ orthogonal carriers can be established within $\boldsymbol{v}$ in this way and they can carry $F\times M$ complex symbols.

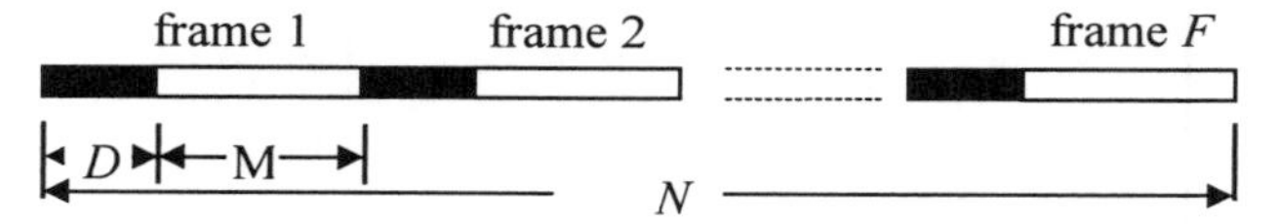

Fig.3. A more general signal structure of $\boldsymbol{v}$. The dark signals represent prefixes. This vector is convoluted with $\boldsymbol{p}$ to give $\boldsymbol{w}$ and then an extra prefix is padded.

Consider to share these $F\times M$ carriers among users in a cell for the up-link. Assume that the system is approximately synchronized with certain tolerance for frame synchronization error among different users. Such error has essentially the same effect as multi-path delay. Provided that the combined effect of frame synchronization error and multi-path delay is covered by the prefix length, the orthogonality among all the users can be maintained.

C. Inter-cell MAI mitigation

We assign a set of different $\boldsymbol{p}$ sequences, referred to as master codes, to the neighboring cells. Suppose that they are properly designed to be approximately random to each other. The convolution $\hat{\boldsymbol{v}}=\hat{\boldsymbol{w}}\otimes\overline{\boldsymbol{p}}$ at the receiver then randomly distributes the inter-cell MAI among all the users, which mitigates the worst case MAI. This property distinguishes the proposed system with a common OFDM, which will be demonstrated by the simulation results later.

Notice that when F=1 in Fig.3, we have $v \otimes p = F^{-1}(u) \otimes p = F^{-1}(u \circ P)$ where $P=F(p)$ and "$\circ$" stands for symbol-by-symbol multiplication between two vectors. This actually implies no processing gain from the scheme. However, it is not the case for $F>1$, for which it can be shown that a process gain can be achieved.

D. Transceiver complexity

So far as transceiver is concerned, the main difference between OFDM and OFDM-CsDMA is the convolutions involving $\boldsymbol{p}$ and $\overline{\boldsymbol{p}}$. These operations incur very modest costs if $\boldsymbol{p}$ is carefully chosen. For example, let $\boldsymbol{s}$ be a length-N m-sequence, $\boldsymbol{e}$ an all-1 sequence, $\alpha=1/\sqrt{N+1}$ and $\beta=(1\pm\sqrt{N+1})/N$. Then $\boldsymbol{p}=\alpha(\boldsymbol{s}+\beta\boldsymbol{e})$ satisfies (6) [7]. Now $\boldsymbol{v}\otimes\boldsymbol{p}=\alpha(\boldsymbol{v}\otimes\boldsymbol{s}+\beta\boldsymbol{v}\otimes\boldsymbol{e})$. Since $\boldsymbol{v}\otimes\boldsymbol{e}$ is trivial, we will concentrate on $\boldsymbol{v}\otimes\boldsymbol{s}$, which can be rewritten in a matrix form as,

$$\boldsymbol{v}\otimes\boldsymbol{s} = \begin{bmatrix} v_0 & \cdots & v_{N-1} \end{bmatrix} \begin{bmatrix} s_0 & s_1 & \cdots & s_{N-1} \\ s_{N-1} & s_0 & \ddots & \\ & \ddots & \ddots & s_1 \\ s_1 & \cdots & s_{N-1} & s_0 \end{bmatrix} = \boldsymbol{v}\boldsymbol{S} \tag{9}$$

Augment $\boldsymbol{S}$ by one zero row and one zero column to

$$\boldsymbol{S}' = \begin{pmatrix} \boldsymbol{0} & \boldsymbol{0} \\ \boldsymbol{0} & \boldsymbol{S} \end{pmatrix} \tag{10}$$

Then $[\boldsymbol{0}, \boldsymbol{v}\otimes\boldsymbol{s}] = [\boldsymbol{0}, \boldsymbol{v}]\boldsymbol{S}'$. The columns of $\boldsymbol{S}'$ form the codeword set of an length $N'=N+1$ augmented m-sequence, which is equivalent to a Hadamard code up to an interleaving. There exist permutation matrixes $\boldsymbol{A}$ and $\boldsymbol{B}$ such that $\boldsymbol{S}' = \boldsymbol{AHB}$, where $\boldsymbol{H}$ is a Hadamard matrix. Hence

$$\begin{bmatrix} 0 & \boldsymbol{v}\otimes\boldsymbol{s} \end{bmatrix} = \begin{bmatrix} 0 & \boldsymbol{v} \end{bmatrix} \boldsymbol{AHB} \tag{11}$$

Eqn. (11) can be implemented very efficiently by an FHT (Fast Hadamard Transform), costing $N'\log_2 N'$ additions, and two interleavers. For detailed discussion, see [5]. The corresponding transmitter and receiver and structures are shown in Fig.4 (QAM layer omitted).

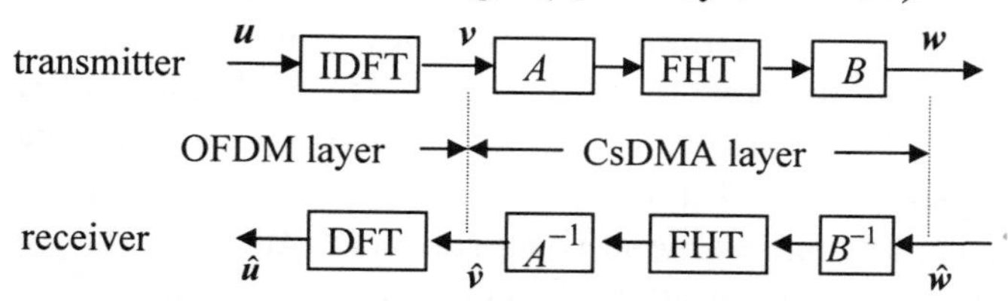

Fig. 4. Transceiver structures of the proposed system. Blocks labeled by A, A^{-1}, B and B^{-1} are interleavers realizing the transformation between the m-sequence and the Walsh sequence. Both DFT and IDFT can be implemented by FFT.

IV. Simulation study

The underlying propagation is based on COST-207 TU (Typical Urbane) models [8] with maximum delay spread = 5μs, maximum frame alignment error = 5μs, chip duration = 0.9977μs, bandwidth=1.002 MHz and carrier frequency =1GHz. The prefix length is D = 11. Rectangular pulse shaping and simple mixing-integration demodulator are used for the QAM layer.

Input information bits are encoded by a rate 1/2, constraint length 9 convolutional code (IS-95 rate-1/2 code) with 192 bits in a frame, producing a length 400 binary sequence (including 16 tailing bits), from which a length-200 complex sequence is formed in a QPSK manner. It is convolutionaly interleaved to generate the input vector $\boldsymbol{u}$ in Fig.2.

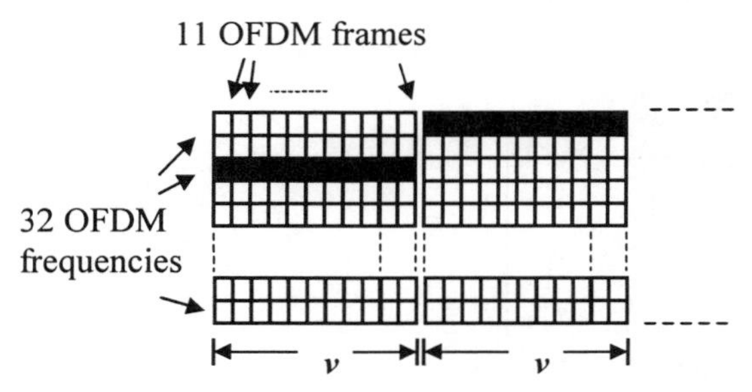

Fig.5. The structure of two consecutive $\boldsymbol{\nu}$-frames for a particular user. Here each ν-frame has the structure as shown in Fig.3 with F=11. A user occupies all 11 carriers of the same frequency in a $\boldsymbol{\nu}$-frame. This frequency hops for different $\boldsymbol{\nu}$-frame, as indicated by the dark areas, to average out the fading effect of the channel. The first OFDM frame in each $\boldsymbol{\nu}$-frame is used as reference.

The master codes $\{\boldsymbol{p}\}$ are constructed from length 511 m-sequences (different cells using different m-sequences). The parameters for $\boldsymbol{\nu}$ (see Fig.3) are N=511, D=11, F=11 and M=32. One OFDM frame is used to estimate phase reference, which avoids the complicated pilot scheme. The remaining 10 frames are used to carry information. We observed that the performance can be improved by setting the power level of the reference frame to about 5dB above the information frames, which is used in all of the results presented below. The information symbols, prefixes and reference occupy a total of 473 positions in $\boldsymbol{\nu}$, leaving 38 positions unused. Within one frame of $\boldsymbol{\nu}$, every user is assigned with one fixed carrier, distinguished by a unique discrete OFDM frequency, in all 11 OFDM frames, see Fig.5. This carrier changes for every frame of $\boldsymbol{\nu}$ to average out fading effect in a frequency hopping manner. Leaving one carrier as reference, the remaining 10 carriers carry a total of 20 coded bits. With a gating factor of 1/2, we obtain user rate R=9.6kbps.

Using dual antenna diversity and equal gain combining, BER=10^{-3} can be achieved at E_b/N_0 ≈7dB. Leaving 1dB implementation margin, we set required E_b/N_0=8dB. Substitute this into the formula of [9] and consider the elimination of the intra-cell MAI, the proposed system can achieve a up-link capacity of 54 user/cell. This is compared with about 24 user/cell for a random waveform CDMA system with similar spreading ratio [9].

We observed that the proposed system is not very sensitive to vehicle speed.. Moderately higher speeds result in better performances due to better interleaving effect. Performance loss start to takes place at extremely high speed above 400 km/h.

An inherent problem of the OFDM technique is the RF amplifier efficiency due to the multi-level modulation involved. This is also the case for the proposed scheme. A common treatment is to clip the signal (in this case $\boldsymbol{w}$). Clipping may affect the orthogonality between the same-cell users. When peak-to-mean power ratio is limited to 0dB (Here mean power refers to that before clipping), we observed that a performance loss of about 0.8dB occurs.

A distinguished feature of CDMA is its robustness against worst-case inter-cell MAI. This is demonstrated for the proposed scheme in Fig. 7 (vehicle speed = 50km/h).

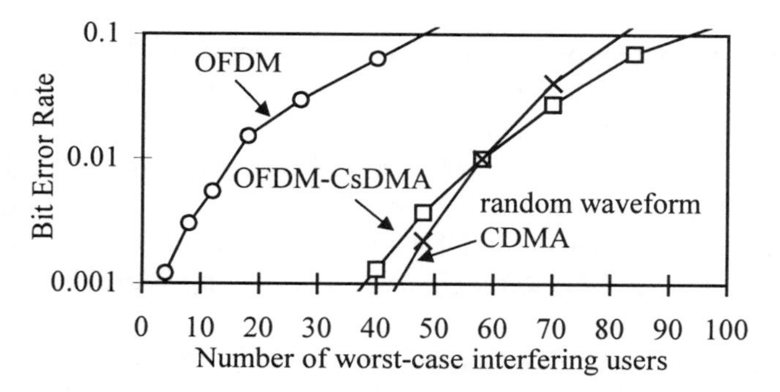

Fig.7. Simulated up-link performance comparison with worst-case interference. Every interfering signal has the same mean arrival power as the wanted signal. The comparable random waveform CDMA employs spreading ratio≈105, a rate 1/3, constraint length 9 convolutional code, dual receiving antenna and a four-finger rake receiver. The OFDM system is obtained by setting $\boldsymbol{p}=\boldsymbol{\delta}$ and keeping everything else unchanged.

For Fig.7, perfect power control is used. Every user is allocated to the cell that results in minimum transmission power. Consequently, the maximum arrival power level (averaged over fast fading) of an interfering signal cannot exceed that of the wanted signal. Let the service quality be

BER<10^{-3} and N_{worst} be the maximum number of worst-case interferers that can be tolerated. It is seen that the values of N_{worst} are, respectively, 3 for OFDM, 38 for OFDM-CsDMA and 44 for a comparable random waveform CDMA. Recall that N_{worst} for a random-waveform CDMA includes all the same-cell users. As a fair comparison, a fully loaded random waveform CDMA system with 24 user/cell can only tolerate about 20 worst-case other-cell interferers, which is considerably lower than OFDM-CsDMA. This clearly demonstrates the advantage of the proposed scheme.

V. Discussions and conclusions

Eliminating intra-cell MAI may potentially leads to 2.8 times of CDMA capacity increase [9] relative to a random waveform CDMA. The example in Section IV achieves a capacity increase of about 2.2 times. The difference is due to the overheads (prefix and reference) involved. The proposed scheme is most suitable for applications with relatively low delay dispersion, such as indoor systems. The low receiver complexity of the proposed scheme makes it suitable to handle relatively high data rates typically required in such applications.

References

[1] S. Moshavi, "Multi-user detection for DS-CDMA communications", *IEEE Communication Magazine*, vol. 34, no. 10, pp.124-136, Oct. 1996.
[2] J. Ruprecht, F.D. Neeser and M. Hufschmid, "Code time division multiple access: An indoor cellular system", *Proc. IEEE Vehicle Technology Conference*, pp.736-739, 1992.
[3] K. Fazel, "Performance of CDMA-OFDM for mobile communications" *IEEE. Int. Conference on Universal Personal Communications*, pp.975-979, 1993.
[4] M. Alard and R. Lassalle, "Principle of modulation and channel coding for digital broadcasting for mobile receivers", *EBU Review*, no. 224, pp. 168-190, August 1987.
[5] Li Ping, "Code-shift division multiple access for cellular mobile", *Proc. IEEE Vehicle Technology Conference*, pp.377-381, 1997.
[6] J.G. Proakis, *Digital Communications*, McGraw Hill, Inc., 1995.
[7] H. Harada, G. Wu, K. Taira, Y. Hase and H. Sasaoka, "A new multi-code high speed mobile radio transmission scheme using cyclic modified m-sequence", *Proc. IEEE VTC'97*, pp.1709-1713, 1997.
[8] ETSI/TC GSM Recommendation 05.05.
[9] A. Viterbi, "The orthogonal-random waveform dichotomy for digital mobile personal communication", *IEEE Personal Communications Magazine*, pp.18-24, no.1, 1994.

A Digital Microwave Point-to-Multi-Point (PMP) System Based on Multi-Carrier FDMA Transmission

Khaled Fazel, Volker Engels
Digital Microwave Systems, Bosch Telecom GmbH, D-71522 Backnang, Germany

Key words: Microwave-PMP, FDMA, DBA, Adaptive Coding & Modulation, Cellular

Abstract: In this paper the concept of a PMP system based on multi-carrier transmission, developed recently within Bosch-Telecom has been presented. The use of advanced microwave RF-technologies, dynamic bandwidth allocation (DBA), adaptive channel coding & modulation and efficient cell planning tools allows the service providers to build efficient local distribution networks and to offer circuit switched (e.g. POTS, ISDN, E1) and packet data (e.g. IP) services.

1. INTRODUCTION

Fixed microwave Point-to-Multi-Point (PMP) digital transmission systems will become an important part of the local distribution and feeder networks. The currently allocated microwave frequency ranges for PMP applications are from 3.5- up to 28 GHz. In the *Channel-allocations* of the CEPT, the available spectrum is split into smaller channels, e.g. 14-28 MHz channels using Frequency Division Duplex (FDD) (see *Table 1*) [1], where several service providers with a moderate number of channels are licensed to serve the same area. Channel-allocations form the basic criteria for the design of a suitable cellular system to achieve i) high data rates, ii) full coverage and iii) high inter-cell interference immunity.

PMP-Frequencies	3.5 GHz	10.5 GHz	26 GHz
Channelization, *B*	3.5 / 7 / **14**	3.5 / 7 / 14 / 28 / **30**	3.5 / 7 / 14 / **28** / 56
Duplex in MHz	100	350	1008
N° of FDD channels	6 x 14 MHz	5 x 30 MHz	18 x 28 MHz

Table 1: Microwave PMP-Frequency channelization

For microwave PMP systems usually the presence of a line of sight (LOS) between the base station (BS) and the terminal station (TS) is guaranteed. Indeed, the microwave radio link is subject to high rain-attenuation and high amount of interference. Furthermore, the conventional microwave RF-technology suffers from high phase noise.

In addition, the microwave PMP-systems could compete with wired technologies (e.g. xDSL) by providing the following additional features: i) wide range of data rates, e.g. applications (switched and packet), ii) lower latency, iii) higher coverage, iv) lower cost, and v) higher spectral efficiency.

Hence, the design of an efficient digital microwave transmission system for PMP applications will be a technological challenge. An attractive approach would be based on the combination of multi-carrier transmission employing advanced microwave RF-technologies (with very low phase noise), dynamic bandwidth allocation (DBA), adaptive channel coding & modulation, and efficient frequency planning tools; where, this approach has been adopted by Bosch-Telecom in his PMP product portfolio called *Digital Multi-point System* (DMS).

2. TRANSMISSION SYSTEM OVERVIEW

A digital *point-to-multi-point* cellular access network, called *Digital Multi-point System DMS*, is made out of a Base-Station (BS) and several fixed Terminal Stations (TS). Each cell is divided into different sectors (e.g. 15-90 degree sector angle), where for each sector small beam antennae with Frequency Division Duplex (FDD) are used. The BS will be connected through the Service Network Interface (SNI) to different core networks; where the TS will be connected for instance to a PABX or to different local networks by the User Network Interface (UNI). *Figure 1* illustrates the system under study which is based on a FDMA concept employing dynamic bandwidth allocation (DBA) [2]. The system is quite flexible and offers services with basic data rate between 64 kbit/s and 2Mbit/s per link with 64kbit/s granularity (e.g. POTS, ISDN, E1/T1, IP applications). The transmitter (BS or TS) comprises RF-, IF-Units and a digital modem (modulation, channel coding, encryption). In addition to a digital IF combiner/splitter, the BS is equipped with a Radio System Controller to supervise the traffic and the bandwidth assignments within a sector.

The channel coding and modulation can be adapted to the TS-reception conditions. The adaptive channel coding and modulation is based on a combination of punctured convolutional coding (mother code rate ½) and MPSK modulation (M=4, 8, 16). Each TS transmits/receives a narrow band digital signal with a given roll-off factor within the allocated bandwidth, where the spectra of the transmit signals are not orthogonal to each other. This small loss in spectral efficiency can be acceptable comparing to the simplicity of all synchronization's mechanisms.

For each TS, the amount of allocated bandwidth can be adapted dynamically, depending on the traffic load; where this combination of DBA with adaptive coding & modulation results in an efficient sharing of the available capacity.
Furthermore, comparing to a TDM/TDMA solution, this strategy provides higher immunity against multi-path propagation & interference, simpler frequency planning, and smaller TS transmit power [3-5].

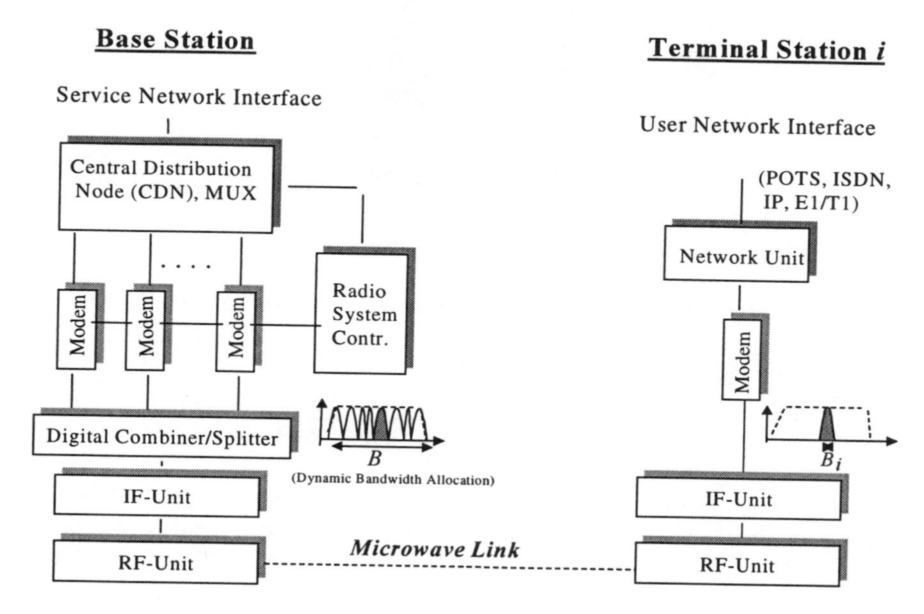

Figure 1. Digital microwave PMP-transmission system based on FDMA

3. CONCEPT OF MULTI-CARRIER TRANSMISSION

In DMS, using advanced microwave technologies the multi-carrier transmission with dynamic bandwidth allocation is chosen that offers following main advantages:

- low power Terminal Station,
- support of variable data rates: from 64 kbit/s up to 2 Mbit/s per carrier,
- simpler frequency planning,
- high spectral efficiency,
- and high flexibility by providing packet and switched traffic.

The available bandwidth B (e.g. 28 MHz @ 26 GHz) is shared among all users within a sector. It will be divided into several non-equal sub-bands B_i $(i=1, .. N)$ according to the requested bandwidths of all active N Terminal Stations. The BS and the TS will use a given sub-band during the traffic time (see *Figure 2*). If there is no more traffic, the used sub-band will be assigned to the next active users. Here, the sophisticated dynamic bandwidth allocation (DBA) algorithm will be used [2]. The time to jump from a sub-band B_i to another B_j by DBA is quite fast, about 100 ms. Furthermore, for each sub-band the modulation and the coding could be adapted independently. The power/frequency control is done independently for each link.

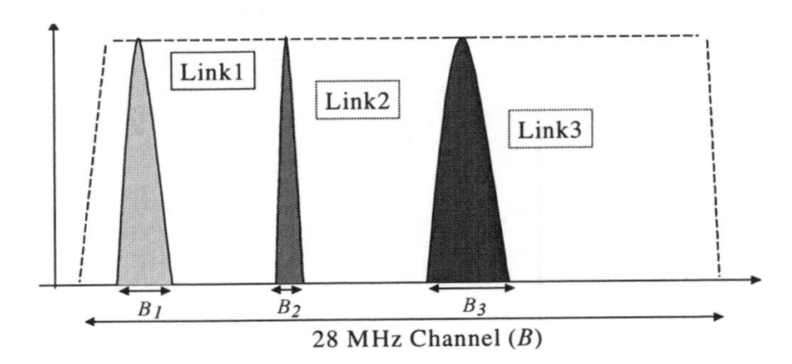

Figure 2. Dynamic Bandwidth Allocation (Frequency allocation at a given time)

Mod. Constellation	QPSK1/2	QPSK3/4	QPSK7/8	8TCM	16TCM
No of 2 Mbit/s links	10	15	18	21	31

Table 2: Number of 2 Mbit/s links in 28 MHz channel

According to the data rate, in TS different kind of radio network units can be used. Higher data rate (> 2 Mbit/s) can be achieved by using several modems, i.e. several sub-bands for a given terminal. This concept allows a very flexible use of the spectrum and provides a high multiplexing gain in the air. Furthermore, if the traffic load is increased/decreased during a call, the bandwidth adaptation to the actual load is done without link interruption. In *Table 2* the maximum achievable number of 2 Mbit/s links in a 28 MHz channel with a filter roll-off factor of 0.3 is given. By using 16TCM up to 31x 2 Mbit/s can be simultaneously transmitted from the BS.

4. CELLULAR ASPECTS

In cellular environments cell sectorization together with directive antennae could be deployed for increasing the system capacity. However, in a dense cellular system, only cell sectorization will not be sufficient to de-couple the inter-cell (or inter-sector) interference. In DMS, the following methods are used to achieve the highest capacity in a cellular environment:

- space division, i.e. cell sectorization,
- antennae polarisation,
- employing highly directive TS antennae,
- frequency division (frequency-decoupling) with fine granularity,
- and individual link optimisation.

Different cell-sectorization is possible: from four (90° BS-Antenna) up to 24 (15° BS-Antennae) sectors could be envisaged. Vertical or horizontal polarisation will be used for each sector (see *Figure 3*, case four sectors). The TS could employ high gain directive antenna (e.g. 5° planar or parabolic antenna). Finally, if the cell sectorization and polarisation is not sufficient to combat the interference, the frequency division could be used to de-couple the remaining other-cell interference.

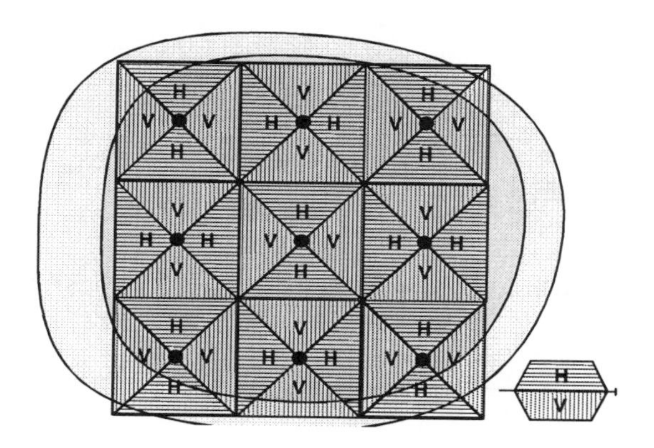

Figure 3. DMS in Multi-cell environment

In a cellular environment with high sectorization the use of high order modulation (e.g., M=32, 64) is quite difficult due to the presence of high amount of interference. Indeed, the amount of interference depends strongly on the position of the TS within a sector. Areas close to the regular grid formed by the base stations suffer from higher interference, since the directional antenna of a TS at this location shows both to the wanted and to interfering base stations. Therefore, the size of areas sensitive to high interference depends on the beamwidth of the TS antenna.

The carrier to interference power ratio C/I is given by the ratio of distances between the TS to the wanted BS, and the TS to the interfered BS. The C/I becomes worse with larger distance and increasing cell overlap. In *Figure 4* the interference scenario corresponding to *Figure 3* is calculated for equal wanted power densities levels at all base stations under clear sky conditions, which is an optimistic assumption [5]. Due to different modulation schemes or increased power levels for compensating locally limited rain attenuation, the power densities might vary between base stations, which can impair the interference conditions.

The interference as calculated in *Figure 4* is only emanated from other cells. Indeed, the interference from neighbor sectors within a cell (see *Figure 5*) has to be further considered. For a well balanced system, the interference contributions should not exceed the noise contributions, meaning that the demodulator has to cope with a C/(N+I)-ratio, which is by 3 dB worse than the pure C/I-ratio indicated in *Figure 4*.

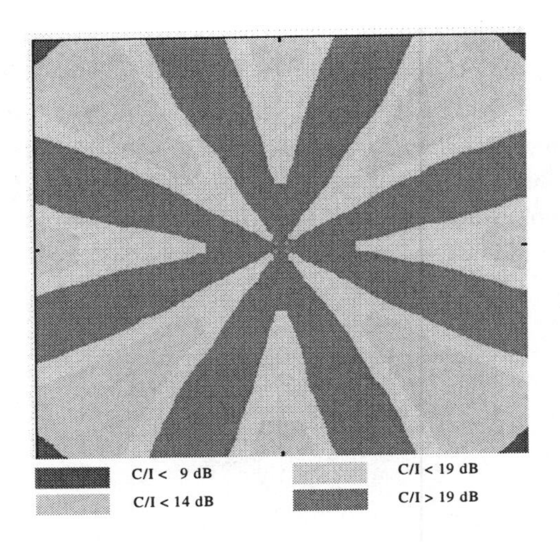

Figure 4. C/I in different cell regions (Only inter-cell interference contributions)

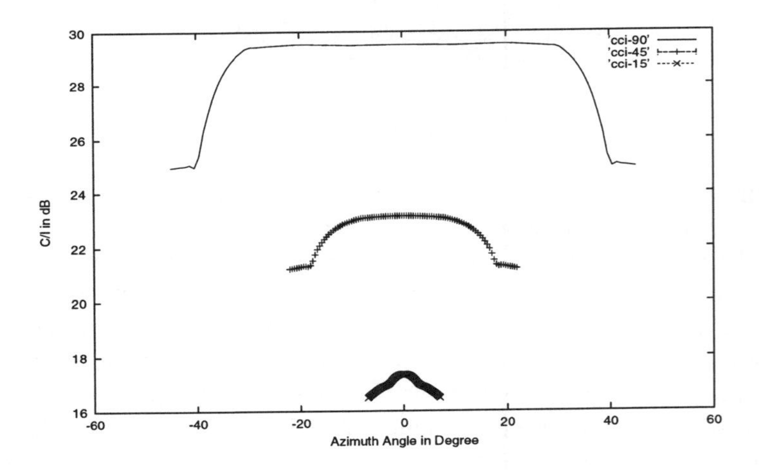

Figure 5. C/I in TS with planar antenna vs. Azimuth-Angle for different sectorized antennae (Only inter-sector interference contributions)

As shown in *Figure 4* the C/I-values for some critical areas could be lower than 9dB, where only for links in this area, the most robust modulation schemes such as QPSK ½ has to be used. Therefore, by exploiting this fact, in DMS several kind of modulation constellations and code rates could be used for different TS located in different region of a sector. This allows to individually optimise each link and therefore double the spectral efficiency comparing to the use of the same modulation (e.g. QPSK, r=1/2) for all TS, as it would be the case for a TDM/TDMA scheme [5].

5. SYSTEM PERFORMANCE

In *Table-3* the required C/(N+I) values for different coding & modulation schemes are given. In the same *Table*, for a 45° cell-sectorization (8 sectors per cell) the distance between the BS and TS is given for DMS-26 carrying 2 Mbit/s per link with parabolic antenna in rain zone H (e.g. Germany) by guaranteeing 99.99% link availability. The transmit TS power is quite low: Only 8.5 dBm. It should be noticed that for 5.2 km, the rain attenuation in zone H can be more than 20 dB!

In *Table-4* by taking into account a roll off factor of 0.3, the spectral efficiency s_{eff} of a FDMA/adaptive coding & modulation system in a cellular environment is given [4]. The results show that by using adaptive coding & modulation the spectral efficiency can be doubled comparing to a FDMA/QPSK1/2 modulation with spectral efficiency of 0.77 bit/s/Hz per sector.

Modulation	Code rate	C/(N+I) in dB	BS-TS Distance in km
QPSK	1/2	7	5.2
8-TCM	2/3	13	4.5
16-TCM	3/4	20	3.4

Table 3: Estimated C/(N+I), BER=10E-7, 26 GHz, Rain Zone H, 99.99% Avail.

No of Adj. Cells	No of sect. per cell	Average s_{eff} per sector (in bit/s/Hz)	Average s_{eff} per cell (in bit/s/Hz)
1	4	1.54	6.16
1	8	1.50	12
2	4	1.54	6.16
2	8	1.29	11.76
9	4	1.36	5.44
9	8	1.29	10.32

Table 4: Estimated Spectral Efficiency of DMS in a Cellular Environment using Adaptive Channel Coding and Modulation

It can be noticed that even in the case of dense cellular system (9 cells), the system offers a high average spectral efficiency per cell of 10.32 bit/s/Hz.

6. APPLICATIONS

Multi-carrier FDMA with DBA can be seen as a next step for higher flexibility, providing «multiplexing in the air» capability. It combines the flexibility of the switched line traffic and the capability to react on dynamic changes of the reception condition. Hence, it is best suited for leased and switched circuit applications, where dial tone and transmission delays fulfill the specified requirements. A large variety of end user equipment can be connected at the standard interfaces, providing economical solution and a perfect integration into switched circuit networks.

As an example, in Table 5, the number of simultaneously active telephone lines (each with 64 kbit/s) offered by DMS in a single cell is compared with FDMA/QPSK1/2. If we further consider the multiplexing gain offered by DBA, the real number of telephone lines served by DMS will be much higher.

Modulation	**FDMA/Adaptive Modulation**	**FDMA/QPSK**
No of active Tel. lines per cell	4920	2526

Table 5: Estimated No of simult. active Tel. Lines in 28 MHz , 8-sectors/cell

Interactive packet data applications require both minimum reaction delays and high data rates. Fast reaction times are guaranteed by the permanent interconnection between base and TS. In one sector, all carriers from/to the TSs are permanently present at a minimum data rate to carry at least the control and monitoring overhead, and they have to share the same spectrum. A single TS therefore can at maximum use the remaining spectrum; thus making it well suited for IP or file transfers.

7. CONCLUSIONS

In this article we have analysed the concept of a digital microwave point to multi-point transmission system based on multi-carrier transmission with dynamic bandwidth allocation and adaptive channel coding & modulation. The combination of FDMA with DBA using advanced microwave technologies can be seen as a next step for higher flexibility and providing «multiplexing in the air» capability. It allows service providers to offer both circuit switched and packet services with high spectral efficiency in a cellular environment.

8. REFERENCES

[1] Draft EN 301 080 V1.1.1 (1997-08) and Draft EN 301 213 (1998-06) Transmission and Multiplexing (TM); Digital Radio Relay Systems (DRRS); Frequency Division Multiple Access (FDMA) point to multipoint DRRS in the band 3/24,2- to 11/29,5 GHz

[2] E. Auer, "Dynamic Bandwidth Allocation for FDMA Systems", 5th ECRR Conf. Proceedings, Bologna, Italy, May 14-17, 1996

[3] H-P. Petry," Multiservice Wireless Access Systems: Key Technologies and Boundary Conditions for Successful Deployment ", EUMC, 1999

[4] M Glauner, "Cell Planning and Transmission Capacity in Urban Fixed Networks Using the Digital Multipoint System (DMS) from Bosch" , Internal report, April 99

[5] Wolfgang Rümmer, "Point-to-Multi-Point - Is there an Optimum Access Scheme?" ICT-98-Conference Proceeding, 22-25 June, Greece 1998

A HYBRID TDMA/CDMA SYSTEM BASED ON FILTERED MULTITONE MODULATION FOR UPSTREAM TRANSMISSION IN HFC NETWORKS

Giovanni Cherubini
IBM Research, Zurich Research Laboratory
CH-8803 Rüschlikon, Switzerland
cbi@zurich.ibm.com

Abstract We present a novel hybrid TDMA/CDMA system for upstream transmission in multiple-access networks. The hybrid multiple-access scheme is based on a modulation technique related to orthogonal frequency-division multiplexing, named filtered multitone modulation (FMT). After introducing the principles of FMT modulation, we describe its application to upstream transmission in hybrid fiber/coax networks, and discuss the characteristics of the proposed scheme.

1. INTRODUCTION

We consider a multiple-access system, in which a head-end controller (HC) broadcasts data and medium-access control (MAC) information over a set of downstream channels to several stations, and these stations send information to the HC over a set of shared upstream channels. Examples of systems exhibiting these characteristics are the emerging two-way hybrid fiber/coax (HFC) systems [1, 2], and their wireless counterparts, multichannel multipoint distribution service (MMDS) and local multipoint distribution service (LMDS) [3].

An HFC system is a point-to-multipoint, tree and branch access network in the downlink, with downstream frequencies in the 50–860 MHz band, and a multipoint-to-point, bus access network in the uplink, with upstream frequencies in the 5–42 MHz band. The maximum round-trip delay between the HC and a station is of the order of 1 ms. The design of an HC modem transmitter and station modem receivers for downstream transmission presents some technical challenges due to transmission rates on the order of 30 to 45 Mb/s per downstream channel. However, owing to the continuous broadcast mode of downstream transmission over a channel with low distortion and a high signal-to-noise ratio (typically ≥ 42 dB by regulation), well-known signal-processing techniques can be applied. In the uplink, implementation of physical (PHY) layer transmission and MAC layer functions pose considerable technical challenges. First, because signals may be transmitted in bursts, HC receivers with fast synchronization capabilities are essential. Second, individual station signals must be received at the HC at defined arrival times and power levels. Third, the upstream channel is generally much more noisy and subject to distortion than the downstream channel.

Transmission schemes based on single-carrier quadrature-amplitude modulation (QAM) may represent a solution for upstream transmission. These schemes, however, are generally less efficient and robust in the presence of impulse noise and narrowband interference than multicarrier modulation techniques, also known as orthogonal frequency division multiplexing (OFDM) [4]. For example, versions of OFDM known as discrete multitone (DMT) modulation and discrete wavelet multitone (DWMT) modulation are considered as transmission schemes for the digital subscriber line (DSL) [5, 6]. Here we propose for upstream transmission in HFC networks a technique related to OFDM, named filtered multitone (FMT) modulation [7]. FMT modulation exhibits significantly lower spectral overlapping between adjacent subchannels and provides higher transmission efficiency than DMT, and is better suited for passband transmission than DWMT. A general description of FMT modulation with application to very high-speed DSL transmission is given in [7].

A multiple access scheme for upstream transmission in HFC networks would have to be suitable for transmission of short messages in burst mode, and also would have to ensure reliable communications in the presence of impulse noise and narrowband interference. To accommodate these requirements, we propose a novel hybrid time-division multiple access (TDMA)/code-division multiple access (CDMA) scheme based on FMT modulation.

2. FMT MODULATION

The equivalent baseband signals of an OFDM system are defined in the fundamental frequency band $(-M/2T, M/2T)$, where M denotes the number of subchannels and T denotes the modulation interval. The vector $\mathbf{A}_n = \{A_n^{(i)}, i = 0, ..., M-1\}$ denotes the block of complex symbols transmitted in the n-th modulation interval. The equivalent baseband transmitted signal x_k is expressed by

$$x_k = \sum_{n=-\infty}^{\infty} \sum_{m=0}^{M-1} A_n^{(m)} h_{k-nM}(m) , \tag{1}$$

where $h_k(i)$, $i = 0, ..., M-1$, are the impulse responses of M filters with frequency responses given by $H_i(f) = \sum e^{-j2\pi fkT/M} h_k(i), i = 0, ..., M-1$. For upstream transmission in an HFC network, the baseband signal is translated to a carrier frequency f_c. The baseband received signal is filtered by a bank of M filters with impulse responses $g_k(i)$, $i = 0, ..., M-1$. The filter output signals are sampled at the modulation rate $1/T$, and the samples are used to determine an estimate of the sequence of transmitted symbols. To ensure that transmission is free of intersymbol interference (ISI) within a subchannel, as well as free of interchannel interference (ICI) between subchannels, certain orthogonality conditions must hold [8].

The complexity of an OFDM system can be substantially reduced if the filtering operations are performed by frequency-shifted versions of a baseband prototype filter, with carrier frequencies given by $f_i = i/T, i = 0, ..., M-1$. In this case a uniform filter bank is obtained, which can be efficiently implemented by a discrete Fourier transform (DFT) and an M-branch polyphase network [9]. Here we consider causal FIR prototype filters having length γM.

DWMT and DMT represent variants of OFDM that are considered for applications. In practice, equalization must be employed in both schemes to cope with nonideal channel characteristics. Let us assume upstream transmission in an HFC network is based on DWMT. A DWMT signal is generated by real-valued input symbols and real-valued filter impulse responses, so that the signal spectrum has Hermitian symmetry around the frequency $f = 0$. The passband signal can be obtained by single side-band (SSB) modulation. Pilot tones are usually employed to provide carrier-phase information [10]. Transmission of pilot tones, however, would not be practical in the multiple-access environment we are considering. Assume now upstream transmission is based on DMT. In this case the passband signal can be obtained by double side-band amplitude and phase modulation (DSB-AM/PM) with zero excess bandwidth. Carrier-phase recovery does not represent a problem. Recall, however, that in DMT systems orthogonality holds only if the individual subchannel signals are received in proper synchronism. Because of the large amount of spectral overlap between contiguous subchannels, reception of a signal with improper timing phase results in ICI, i.e., the signal will disturb several other subchannel signals and vice versa. This situation cannot be avoided when a cable modem sends a request for registration in a subchannel specified by the HC with no prior knowledge of the correct timing phase and transmit power level.

To solve the dilemma posed by passband OFDM transmission in a multiple-access environment we propose FMT modulation, i.e., a filter-bank modulation technique where the filters are frequency-shifted versions of a prototype filter yielding a high level of subchannel spectral containment, such that the ICI is negligible as compared to the level of other noise signals. The design of filtering elements for FMT systems turns out to be easier if ISI is allowed within a subchannel. In this paper, the amplitude characteristic of the prototype filter approximates the frequency response of an ideal filter given by

$$H_{\text{ideal}}(f) = \begin{cases} \left|\frac{1+e^{-j2\pi fT}}{1+\rho e^{-j2\pi fT}}\right| & \text{if } -1/2T \leq f \leq 1/2T \\ 0 & \text{otherwise}, \end{cases} \tag{2}$$

where the parameter $0 \leq \rho \leq 1$ controls the spectral roll-off of the filter. Note that the frequency response (2) exhibits spectral nulls at the band edges. The general case of FMT modulation with excess bandwidth within a subchannel is addressed in [7]. Figure 1 illustrates the level of subchannel spectral containment achieved by a system with $M = 64$ subchannels, and prototype filter parameters $\rho = 0.1$ and $\gamma = 10$.

The frequency responses of FMT subchannels are characterized by steep roll-off towards the band-edge frequencies, where they exhibit near spectral nulls. This suggests that per-subchannel decision-feedback equalization be performed to recover the transmitted symbols. In transmission systems with trellis coding, the function of decision-feedback filtering is preferably performed at the transmitter by employing precoding techniques [11]. Optimal detection is achieved by implementing at the receiver only the equalizer forward section, which approximates the whitened matched filter, and by implementing at the transmitter the feedback section as a precoder. We consider uncoded FMT transmission employing per-subchannel Tomlinson–Harashima (TH) precoding [12, 13].

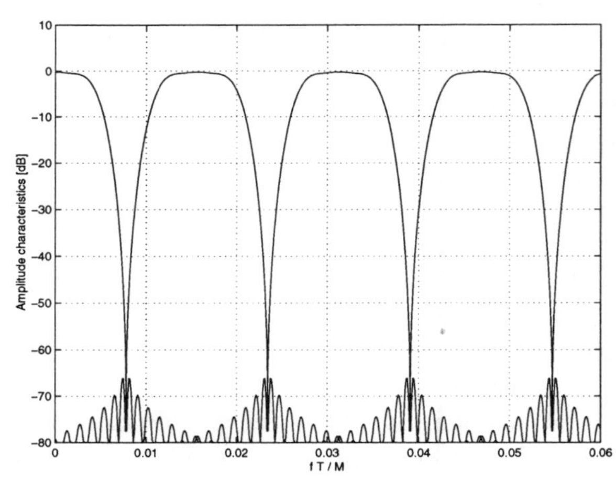

Figure 1 Frequency responses for $f \in (0, 0.06\ M/T)$ of subchannel filters in an FMT system with $M = 64$ and prototype filter designed for $\rho = 0.1$ and $\gamma = 10$.

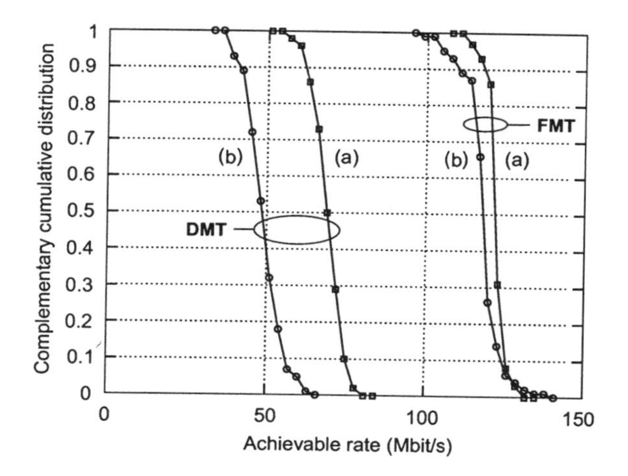

Figure 2 Complementary cumulative distributions of achievable rates for FMT and DMT upstream transmission systems with $M = 64$ subchannels.

In Figure 2, the performance of FMT modulation is compared to that of DMT in terms of complementary cumulative distributions of achievable rates, assuming an average signal-to-noise ratio at the HC input of 30 dB, and an error probability of 10^{-5} for symbol detection on each subchannel, achieved with 3 dB of additional margin. Upstream transmission with $M = 64$ subchannels over the 5–37 MHz band is considered, with channel distortion introduced by four bi-directional amplifiers as well as (a) strong or (b) weak reflections at random points in the network. The distributions are obtained with 100 upstream channel realizations. In the considered FMT system, the same linear-phase prototype filter designed for parameter values $\rho = 0.1$ and $\gamma = 10$ is used for the realization of the transmit and receive filter banks. Per-subchannel equalization is performed by employing a Tomlinson-Harashima precoder with 8 taps at the transmitter and a linear equalizer with 16 taps at the receiver. In the DMT system, a cyclic prefix of 16 samples, no time-domain equalizer and one-tap frequency equalizers are employed. The coefficients of the equivalent minimum mean-square error decision-feedback equalizers for FMT and of the one-tap frequency equalizers for DMT are computed assuming perfect knowledge of the channel characteristics.

3. A HYBRID TDMA/CDMA SYSTEM

CDMA is well suited for upstream transmission in HFC networks because it represents a robust transmission technique in the presence of narrowband interference [14, 15]. Furthermore, adjustment of the transmitted signal power during the initial registration process avoids the near-far problem. On the other hand, some or all of the stations might only occasionally access the channel and transmit in burst mode. To accommodate this type of traffic, which requires fast synchronization capabilities at the HC receiver, TDMA is preferable. The above observations motivate the proposal of a hybrid TDMA/CDMA scheme based on FMT modulation for upstream transmission in HFC networks. According to the time-varying characteristics of the disturbances that are present in the channel, and to the requests for resource allocation of each individual station, the HC dynamically assigns to each station a subchannel or a set of subchannels for upstream transmission. The HC also specifies whether during a mini-slot a subchannel is entirely dedicated to a single station for upstream transmission (TDMA), or whether it may be shared by several stations using different signature codes (CDMA).

A hybrid TDMA/CDMA scheme based on FMT modulation allows the HC to allocate upstream channel resources efficiently to each individual station for a wide range of data rates. FMT transmission by a single station was described in Section 2. Now we will describe a synchronous CDMA scheme in conjunction with FMT modulation. In a synchronous CDMA environment, the signals transmitted over the same subchannel as the signal to be detected can be modeled as cyclostationary interference, which is synchronous with the disturbed signal. In an HFC network, synchronism is due to common timing information provided by the HC to all stations in the network. It has been shown in [16] that $K - 1$ synchronous interferers having modulation interval equal to T can be suppressed by expanding by a factor K the bandwidth of the signals, and by employing at the receiver adaptive equalization with T/K-spaced taps. Interference suppression achieved in this manner can be interpreted as a frequency diversity technique. Here we consider the presence of narrowband asynchronous interferers as well.

In general, depending on the desired transmission rates and the interference characteristics, different CDMA parameters may be chosen by the HC over each subchannel. We consider a system where K_i stations transmit over the i-th subchannel. Each input symbol sequence is transmitted at a modulation rate of $1/KT$, with $K_i \leq K$. At the HC receiver, after demodulation by the DFT, assuming only narrowband interferers having spectral content within the i-th subchannel represent nonnegligible noise signals, the i-th subchannel output is given by

$$V_n^{(i)} = \sum_{k=0}^{K_i} \sum_{m=-\infty}^{\infty} \left[\sum_{\ell=0}^{K-1} s_\ell^{(k)} \tilde{A}_{m-\ell}^{(i,k)} \right] h_{n-m}^{(i,k)} + \sum_{j=0}^{J-1} U_n^{(i,j)} , \tag{3}$$

where $\{\tilde{A}_n^{(i,k)}\}$ denotes the interpolated sequence of k-th user symbols input to the i-th subchannel, $\{s_n^{(k)}\}$ is the k-th user signature code sequence with length K, $h_n^{(i,j)}$ denotes the overall i-th subchannel impulse response for the k-th station, and $U_n^{(i,j)}$, $j = 0, ..., J-1$, are narrowband interferers.

To recover the K_i sequences of transmitted symbols, the signal $V_n^{(i)}$ is filtered by a bank of K_i fractionally-spaced decision-feedback adaptive equalizers. Figure 3 illustrates the performance of a CDMA scheme employing Walsh-Hadamard codes. The curves show the achievable aggregate rates versus interference-to-signal ratio for various numbers of users, assuming $K = 8$, $J = 1$, an average signal-to-noise ratio at the subchannel output of 21.5 dB, and an error probability for symbol detection for each active station of 10^{-5}, achieved with 3 dB of additional margin. Upstream FMT transmission with $M = 64$ subchannels in the 5–37 MHz band is considered. Each station transmits CDMA signals in the subchannel centered at 21 MHz. A narrowband interferer with 62.5 kHz bandwidth and center frequency also at 21 MHz is added to the received signals. Symbol detection for each user is performed by a 24-tap fractionally $T/8$-spaced equalizer and K_i 2-tap feedback filters, with coefficients computed assuming perfect knowledge of the subchannel characteristics.

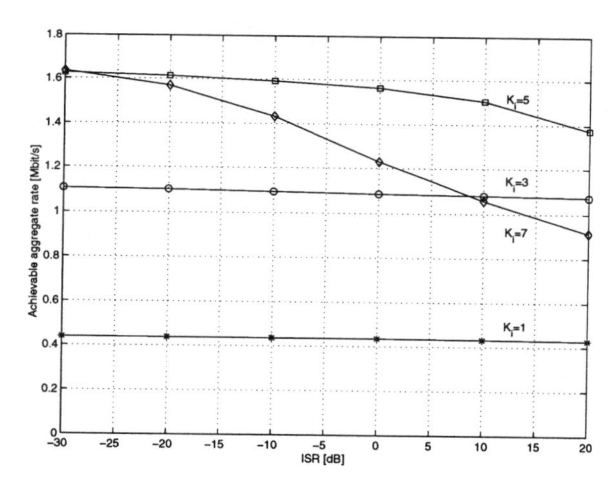

Figure 3 Aggregate achievable rates versus interference-to-signal ratio for various numbers of users sharing a subchannel in an FMT-based CDMA upstream transmission system.

4. CONCLUSIONS

The proposed hybrid TDMA/CDMA scheme based on FMT modulation is well suited for upstream transmission in HFC networks in the presence of narrow-band interference signals with time-varying spectral characteristics. TDMA is adopted over subchannels that are free of interference to accomodate the traffic of stations transmitting in burst mode. CDMA is employed over noisy subchannels for stations that require access for extended periods and transmit at low rate. Furthermore, the high level of spectral containment of individual subchannels achieved with FMT modulation allows ranging and power adjustment of unregistered stations to be performed without disturbing upstream transmission on adjacent subchannels.

Acknowledgment. The author thanks Dr. Jeyhan Karaoguz for developing the upstream channel model used to obtain the system performance results presented here.

References

[1] MCNS Interim Specification, "Data Over Cable Interface Specifications – Radio Frequency Interface Specification," MCNS Holdings, L.P., March 26, 1997.

[2] Eldering, C. A., Himayat, N. and Gardner, F. M. "CATV Return Path Characterization for Reliable Communications," IEEE Commun. Mag., vol. 33, pp. 62-69, Aug. 1995.

[3] Honcharenko, W., Kruys, J. P., Lee, D. Y. and Shah, N. J. "Broadband wireless access," IEEE Commun. Mag., vol. 35, pp. 20-27, Jan. 1997.

[4] Bingham, J. A. C. "Multicarrier Modulation for Data Transmission: An Idea Whose Time Has Come," IEEE Commun. Mag., vol. 28, pp. 5-14, May 1990.

[5] Chow, J. S., Tu, J. C. and Cioffi, J. M. "A Discrete Multitone Transceiver System for HDSL Applications," IEEE J. Sel. Areas Commun., vol. 9, pp. 895-908, Aug. 1991.

[6] Sandberg, S. D. and Tzannes, M. A. "Overlapped Discrete Multitone Modulation for High Speed Copper Wire Communications," IEEE J. Select. Areas Commun., vol. 13, pp. 1571-1585, Dec. 1995.

[7] Cherubini, G., Eleftheriou, E. and Ölçer, S. "Advanced Multicarrier Modulation Techniques for xDSL," IEEE Circuits and Systems and Communications Societies Workshop on *High-Speed Data over Local Loops and Cables*, July 26-28, 1999, Princeton University, Princeton, New Jersey.

[8] Vaidyanathan, P. P. (1992) *Multirate Systems and Filter Banks*. Englewood Cliffs, NJ: Prentice-Hall.

[9] Bellanger, M. G., Bonnerot, G. and Codreuse, M. "Digital Filtering by Polyphase Network: Application to Sample-Rate Alteration and Filter Banks," IEEE Trans. Acoust. Speech and Signal Proc., vol. ASSP-24, pp. 109-114, Apr. 1976.

[10] Kerpez, K. J. "A Comparison of QAM and VSB for Hybrid Fiber/Coax Digital Transmission," IEEE Trans. Broadcast., vol. 41, pp. 9-16, March 1995.

[11] Eyuboglu M. W. and Forney, G. D. Jr., "Trellis Precoding: Combined Coding, Precoding and Shaping for Intersymbol Interference Channels," IEEE Trans. Inform. Theory, vol. 38, pp. 301-314, March 1992.

[12] Tomlinson, M. "New Automatic Equalizer Employing Modulo Arithmetic," Electron. Lett., vol. 7, pp. 138-139, March 1971.

[13] Harashima, H. and Miyakawa, H. "Matched Transmission Technique for Channels with Intersymbol Interference," IEEE Trans. Commun., vol. COM-20, pp. 774-780, Aug. 1972.

[14] Varanasi, M. and Aazhang, B. "Near-Optimum Detector in Synchronous Code Division Multiple Access Communications," IEEE Trans. Commun., vol. 39, pp. 725-736, May 1991.

[15] Sivesky, Z., Bar-Ness, Y. and Chen, D. "Error Performance of Synchronous Multiuser Code Division Multiple Access Detector with Multidimensional Adaptive Canceller," European Trans. Commun. & Rel. Technol., vol. 5, pp. 719-724, Nov.-Dec. 1994.

[16] Petersen, B. R. and D. D. Falconer, D. D. "Minimum Mean Square Equalization in Cyclostationary and Stationary Interference – Analysis and Subscriber Line Calculations," IEEE J. Select. Areas Commun., vol. 9, pp. 931-940, Aug. 1991.

Section III

CODING AND MODULATION

ON CODING AND SPREADING FOR MC-CDMA

J. Lindner
Department of Information Technology, University of Ulm, Germany
juergen.lindner@e-technik.uni-ulm.de

1. INTRODUCTION

For future mobile communication systems, multicarrier code division multiple access (MC-CDMA) and its variants are interesting alternatives to methods based on singlecarrier (SC) on one hand and orthogonal frequency division multiplexing (OFDM) on the other. In general, the physical mobile radio channel is characterized by time-varying multipath propagation corresponding to time-varying frequency-selective transfer functions. As a consequence one has to take into account, that parts of a received signal might be faded out in frequency and/or time. Frequency and time diversity can help in this situation and there are two different ways to achieve this: Spreading and coding. Also combinations of these two basic approaches are possible.

MC-CDMA is based on OFDM and the spreading given by the CDMA part means that symbols to be transmitted are spread over some subcarriers of the OFDM scheme. A spreading factor of zero corresponds to pure OFDM and maximum spreading over all subchannels leads to a frequency diversity comparable with a SC transmission. In case of pure OFDM, i.e. without spreading, coding is needed to give a similar type of diversity, but not with code rate one as in case of spreading. Time diversity can also be achieved by spreading and coding and combinations thereof. While coding in time is the common approach, spreading in time was compared with coding in [1]. A

MC-CDMA system with spreading in time and frequency was considered first in [2] and it was given the name "Extended MC-CDMA".

It is the goal of this paper to look at the relation between coding and spreading in case of MC-CDMA in a basic way and to explain differences and commonalities. Because no spreading corresponds to OFDM, two candidates will be taken for discussion: coded MC-CDMA and coded OFDM (COFDM) or its multiuser counterpart coded OFDM-OFDMA, respectively.

Further work has been carried out in this field already. In [3], [4] frequency non-selective fading channels (Rayleigh) were considered with spreading and coding in time direction, while in [5] and [6] frequency-selective time-invariant channels with spreading and coding in frequency direction were taken. In [7] and [8] especially MC-CDMA was assumed.

2. TRANSMISSION MODEL

A transmission model with vectors and matrices will be taken as a basis, – see Fig. 1. This model is a special case of a more general model which has been derived and described before [9], [10].

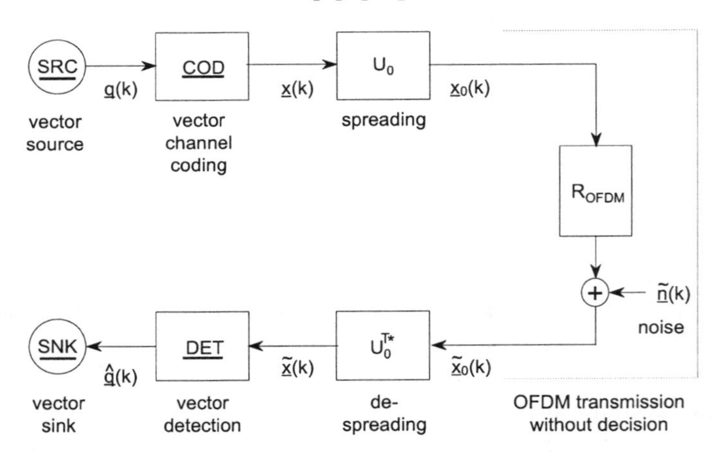

Figure 1 Model for a coded MC-CDMA transmission

The vector source in this figure generates a sequence $\underline{q}(k)$ of source symbol vectors which is mapped by vector coding (COD) into a sequence $\underline{x}(k)$ of transmit symbol vectors. The components of $\underline{q}(k)$ and $\underline{x}(k)$ are scalar sequences belonging to different users or different subchannels of one user. The transmit symbols $x_i(k)$ may be complex-valued, e.g. $x_i(k)\varepsilon\{\pm 1 \pm j\}$ in case of 4PSK. Therefore, coded modulation is included in COD.

The matrix U_0 is the *spreading matrix,* which will be the key in the discussion of the next section. It maps the vector sequence $\underline{x}(k)$ into another sequence

$\underline{x}_0(k)$ of vectors (or blocks) and defines the spreading part of MC-CDMA. If U_0 is an identity matrix, then there is no spreading at all and a coded OFDM (or OFDM-OFDMA) transmission results. If properly arranged WH (Walsh Hadamard) submatrices are taken to for U_0, a common form of MC-CDMA results. k in this model is the index of the blocks $\underline{q}(k)$, $\underline{x}(k)$ and $\underline{x}_0(k)$ with scalar components $q_i(k)$, $x_i(k)$ and $x_{0i}(k)$, respectively. The component sequences $x_{0i}(k)$ belong to the different OFDM-OFDMA subchannels and/or users. So the number M of components in $\underline{x}_0(k)$ is identical with the total number of subchannels available in the OFDM-OFDMA scheme. Because the $x_i(k)$ will be complex-valued in general, the same holds for the $x_{0i}(k)$ after the transformation with U_0, but the alphabet of the $x_{0i}(k)$ will be different in general.

At the receiving side there is a corresponding vector sequence $\tilde{\underline{x}}_0(k)$ of estimates for $\underline{x}_0(k)$:

$$\tilde{\underline{x}}_0(k) = R_{OFDM} \cdot \underline{x}_0(k) + \tilde{\underline{n}}(k) \tag{1}$$

The matrix R_{OFDM} describes the OFDM-OFDMA transmission. Due to the guard time between OFDM blocks, the cyclic repetition, the use of only one period at the receiving side and the synchronization between different users it is a diagonal matrix. Its entries are squared magnitudes of the periodically repeated transfer function of individual physical user channels at those frequencies, which are used in the OFDM scheme.

After multiplication with the conjugate complex transpose of U_0 the detection (DET) gives a sequence $\hat{\underline{q}}(k)$, which is the detected vector sequence corresponding to $\underline{q}(k)$. In general, DET is *not* the counterpart of COD at the transmitting side. For multipath propagation channels like those of interest here, it includes equalization and decoding connected in a complicated manner. If COD and U_0 are defined well, a DET algorithm at the receiving side may have the potential to produce an overall best result, i.e. to reach the capacity bound given for the physical channel. On the other hand, if U_0 is not defined well, an optimum DET algorithm has no chance to produce a good performance, because it cannot correct things which have been done wrong at the transmit side. But the definition of COD and U_0 has also an influence on the complexity of the algorithms used in DET at the receiving side.

In this model the channel was assumed to be time-invariant. General time-varying models were considered in [11]. One common assumption in this context is, that the channel does not vary during the transmission of one block (or vector) $\underline{x}_0(k)$. But over time intervals corresponding to many blocks a variation might be observed in practical applications. In Fig. 1 and eq. (1) this means that R_{OFDM} may vary slowly from block to block. Eq. (1) then becomes

$$\underline{\tilde{x}}_0(k) = R_{OFDM,k} \cdot \underline{x}_0(k) + \underline{\tilde{n}}(k) \quad (2)$$

Spreading and coding in time direction must take this into account.

It will be assumed in the following that no knowledge about the actual channel is available at the transmit side. This assumption can be justified by the fact, that there is only a marginal improvement in performance to be expected if a transmitter can use this knowledge [12].

3. NON-MULTIUSER CASE

The non-multiuser case, i.e. a transmission from one transmitter to one receiver without any interference from other users, is taken now to explain the basic differences between spreading and coding. The multiuser case will be considered after that.

3.1 SPREADING

Fig. 2 tries to illustrate the effect of spreading for a 2PSK transmission with two OFDM subchannels, where one subchannel is faded out. No time-variation is assumed in this simple example, so there is no need to consider spreading in time direction.

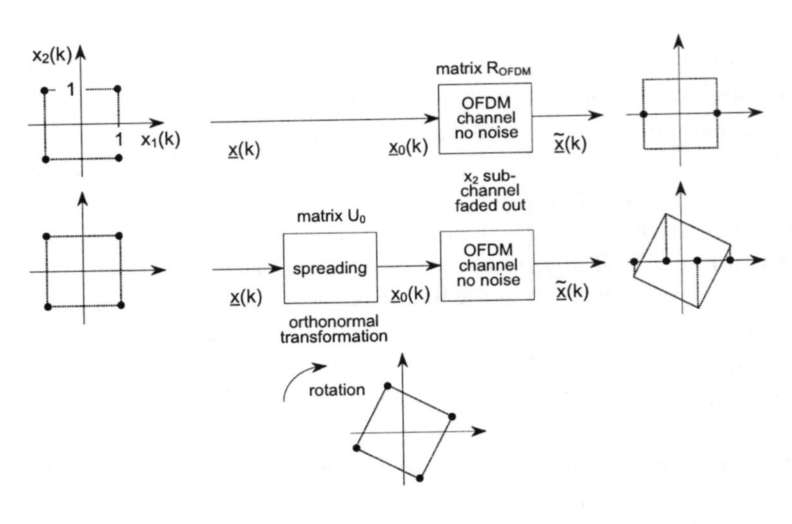

Figure 2 The Effect of Spreading

Spreading with orthogonal matrices U_0 mean a rotation of the constellation in the two-dimensional space in this example. Without this rotation the four points collapse to two, and because this means that two euclidean distances become zero, no error-free transmission is possible without spreading. With rotation (i.e. with spreading) the 4 points at the receiving side have a remaining euclidean distance, so error-free transmission is possible. In a different way one can say, that with spreading diversity for each individual symbol – or *signal space diversity* [4] – is introduced, which does not exist in the unspread OFDM case.

3.2 CODING AND SPREADING

COFDM without any interleaving means that the spreading matrix U_0 in Fig. 1 is an identity matrix. Of course, for channels with fades in time and frequency, interleaving is usually taken to distribute the code symbol errors caused by this behavior. The code must be capable to cope with errors produced by the channel behavior. If a certain percentage of symbols is faded out all time, error rates tend to become irreducible for code rates close to 1, i.e. for vanishing additive noise.

Compared with spreading, interleaving and coding also rotate the symbol constellations like in Fig. 2. But the important difference to the spreading discussed here is, that this rotation is "discrete" on *symbol basis*. No signal space diversity is possible. Signal space dimensions which are faded out, lead directly to symbol errors, erasures or soft decision values being zero. In contrast to that, spreading means to transmit each individual symbol with a certain part of its total symbol energy on many axes of the signal space, see Fig. 2. The resulting rotation is "continuous" and the transmission can be error-free in case of no noise, as long as a non-zero part of the symbol energy is received.

Coding with rates less than 1 does not change these basic facts. Yet another type of diversity is given in this case, which may be termed *coding space diversity.* It may be considered as being one level above the signal space diversity. Although symbols are faded out, the decoding might be able to recover the information symbol vectors $\underline{q}(k)$. In the noise-free case euclidean distances may remain non-zero, despite of the fades of the channel. It is to be expected that the higher the signal space dimension and the smaller the code rate, the more coding can help in this situation. For a simple (24,12) Golay block code it has been demonstrated already in [13] by an example, that with ML detection only a small performance loss remains in comparison with a single carrier transmission (which corresponds to full spreading). This means that the combination of signal and coding space diversity in case of SC is dominated by the coding space diversity, which is the only diversity given for COFDM.

Comparisons like these do not only depend on the code rate, the coding space dimension and the signal space dimension, but also on the channel and the complexity of the algorithms allowed at the receiving side. So, general statements with respect to practical applications cannot be made. But it can be stated, that both, codeword and symbol diversity together can only be achieved by a combination of coding and spreading (together with interleaving). This combination may be considered as a concatenation of conventional coding with rates less than 1 with a code in real or complex numbers with code rate 1. The latter code is identical with spreading.

3.3 TIME-VARYING FREQUENCY-SELECTIVE CHANNELS

The better all details are defined, the more R_{OFDM} will approach the ideal case $R_{OFDM} = I$ with I being the identity matrix. For time-varying frequency-selective channels $R_{OFDM,k}$ in eq. (2) can also approach $R_{OFDM,k} = I$, if spreading in time and frequency is made. There is no basic difference between spreading in frequency and spreading in time direction. For details concerning the combination of both see [11] or [2].

4. MULTIUSER CASE

The vector transmission model described in fig. 1 allows to relate the components of the vectors to subchannels of a transmission from one transmitter to one receiver with, e.g. OFDM (with $U_0 = I$ for OFDM). The subchannels are then defined by the OFDM scheme together with the properties of the impulse response or transfer function of the physical channel. This was considered in the last section. But the model allows also to relate the vector components to different users of a multiuser transmission scheme. In this case the number of physical channels will correspond to the number of users. Moreover, if multiple transmitting and/or receiving antennas are involved, the number of physical channels is increased further. Also, a combination is possible: One user can have more than one subchannel (or vector component) for his transmission.

Independent of all these possibilties there will be only *one* matrix $R = U_0^{T*} R_{OFDM} U_0$ or $R_k = U_0^{T*} R_{OFDM,k} U_0$ for the time-variant case. These matrices exhibit the performance potential of the transmission and multiple access method used. All details, including the physical channels and the transmitting and/or receiving antennas are contained in R. The better all details are defined, the more R or R_k tend to become an identity matrix.

Because coding and detection are not included in R or R_k, they form the final part which determines the performance of a whole system. But things which have been done wrong already – reflecting in a "bad" R – cannot be corrected any more by coding and/or good or optimum detection schemes.

For the difference between coding and spreading it is important to note, that in the multiuser case coding cannot be defined in user direction. Spreading can be defined over the whole physical bandwidth available for all users of a system. One extreme case is pure CDMA, where each user covers the whole bandwidth. In this case of pure CDMA U_0 contains the fourier transforms of the spreading sequences in its rows. But MC-CDMA users usually occupy only such a percentage of the whole channel bandwidth, which guarantees already the maximum frequency diversity.

All other conclusions made for the non-multiuser case hold here too. But while spreading can be done in user direction and coding cannot, there might be cases for a coded OFDM-OFDMA multiuser system, where a great variation in quality between users is given. For some users a large percentage of subchannels might be faded out and the coding cannot use possibly existing excellent OFDM subchannels of other users. With spreading, e.g. with MC-CDMA, this effect can be avoided. The example in the next section demonstrates that.

Another effect not present in the non-multiuser case is a non-perfect synchronization between users. For OFDM-OFDMA this is needed, but for MC-CDMA a certain amount of asynchronity may produce only little decrease in performance because the equalization in the MC-CDMA receiver can cope with it.

5. CODING VERSUS SPREADING, EXAMPLE

In [14] an example for an indoor scenario is described, which illustrates the explanations given so far. OFDM-OFDMA and MC-CDMA were compared for the case of a "fully loaded" system with 8 users. In total 32 subchannels were used for the OFDM-OFDMA scheme, i.e. 4 subchannels per user, with each subchannel being identical with 4 exclusive frequencies. For the MC-CDMA scheme, all 8 users had access to all 32 frequencies and spreading was done with 4 identical 8 by 8 walsh-hadamard (WH) submatrices. So, like in case of OFDM-OFDMA 4 symbols in parallel were transmitted by each user.

Depending on the the user considered for reception, best an worst cases were identified. Fig. 3 shows simulation results belonging to bad cases, caused by strong multipath propagation. The BER curves belong to a BPSK transmission and reception of user 2. ML equalization was used in case of MC-CDMA. The solid curves are without coding and the dashed ones for a coded transmission. A convolutional code of rate 2/3 and memory 6 was taken and soft decision ML decoding. BER and $\frac{E_b}{N_0}$ values were averaged over the 4 parallel subchannels of user 2. The AWGN curves might be considered as lower bounds.

The uncoded OFDM-OFDMA curve shows an irreducible error rate which is a result of the frequency-selective behavior of the transfer function of the channels. With coding there is an improvement, but the remaining distance to

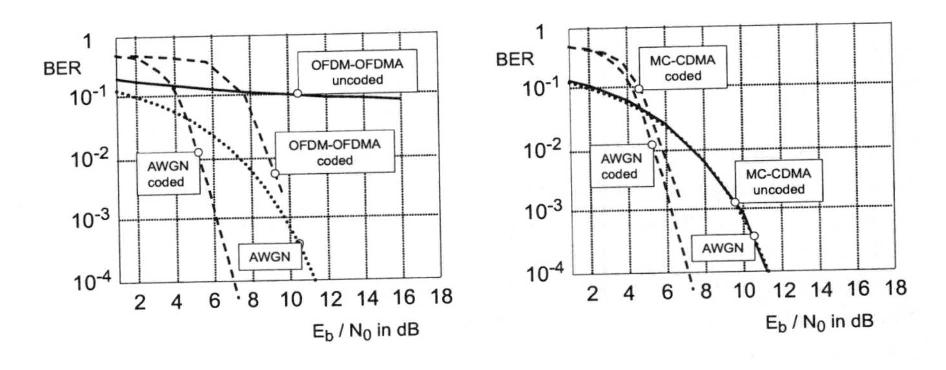

Figure 3 Bit error rates for OFDM-OFDMA and MC-CDMA, user 2; dashed curve: convolutional coding, memory 6, rate 2/3

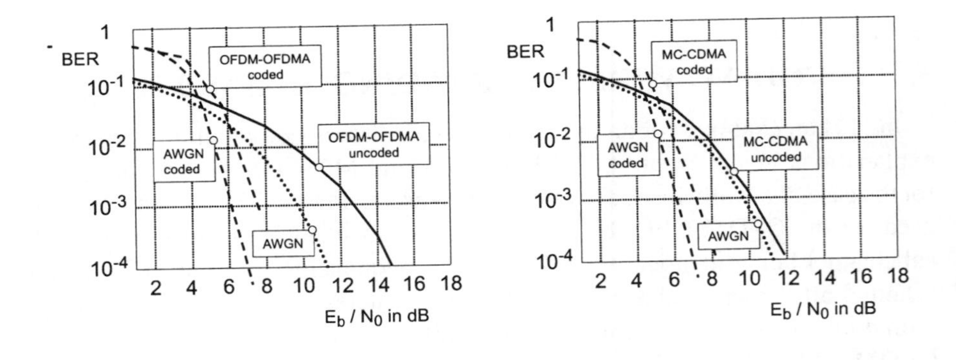

Figure 4 Bit error rates for OFDM-OFDMA and MC-CDMA, user 4; dashed curve: convolutional coding, memory 6, rate 2/3

the coded AWGN curve is of about 4 dB. For MC-CDMA the situation is much better. The coded and uncoded curves are close to the AWGN curves. This example demonstrates the advantage of spreading.

Another example, which is taken from the group of good cases (good for OFDM-OFDMA), is shown in Fig. 4. For the coded curves there are only small improvements due to spreading. The reason for this behavior is the different channel transfer function for user 4 which is considered here.

6. CONCLUSION

In a MC-CDMA system spreading gives *signal space diversity*, which is a diversity for individual symbols, while coding gives *coding space diversity*, which may be considered to be one level above signal space diversity. A coded OFDM-OFDM system – which can be defined as a coded MC-CDMA system with no spreading – uses coding space diversity only, with the advantage of lower complexity in the receiver. In any practical application one should consider a combination of both types of diversity, i.e. coded MC-CDMA. But general statements with respect to practical applications cannot be made, because the coding scheme, the physical channels, and the complexity of the algorithms allowed at the receiving side have an influence on the BER performance. But for code rates close to one, a better performance of coded MC-CDMA can be expected.

Because coding cannot be done in user direction while spreading can, there might be cases in practice, where MC-CDMA is a priori the better solution, because spreading can guarantee equal performance for different users better than OFDM-OFDMA. The indoor communication system example in the last section demonstrated that. Coding could be only over 4 subchannels while spreading could use 8. Additionally, the equalization part of a MC-CDMA receiver can cope to some extend with asynchronous users, which is not possible in case of no spreading.

References

[1] M. Reinhardt and J. Lindner. Transformation of a Rayleigh fading channel into a set of parallel AWGN channels and its advantage for coded transmission. *Electronics Letters*, Vol. 31, No. 25:2154–2155, 1995.

[2] J. Egle, M. Reinhardt, and J. Lindner. Equalization and Coding for Extended MC-CDMA over Time and Frequency Selective Channels. *Multi-Carrier Spread-Spectrum, K. Fazel and G.P. Fettweis (eds.), Kluver Academic Publishers, The Netherlands*, pages 127–134, 1997.

[3] J. Boutros, E. Viterbo, C. Rastello, and J. C. Belfiore. Good Lattice Constallations for Both Rayleigh Fading and Gaussian Channels. *IEEE Trans. Inf. Theory*, IT-42:502–518, 1996.

[4] J. Boutros and E. Viterbo. Signal Space Diversity: A Power- and Bandwidth-Efficient Diversity Technique for the Rayleigh Fading Channel. *IEEE Trans. Inf. Theory*, IT-44:1453–1467, 1998.

[5] J. Lindner, M. Reinhardt, and J. Hess. OCDM - Ein Übertragungsverfahren für lokale Funknetze. *ITG Fachtagung Codierung für Quelle, Kanal und Übertragung, München, 26.-28.9.1994*, ITG-Fachber. 130:401–409, 1994.

[6] J. Lindner. Channel Coding and Modulation for Transmission over Multipath Channels. *AEÜ, No. 3*, 49:110–119, 1995.

[7] S. Kaiser. Trade-off between channel coding and spreading in multi-carrier CDMA systems. *Proc. ISSSTA '96, Mainz, Germany, September 22-25*, pages 1366–1370, 1996.

[8] S. Kaiser. MC-FDMA and MC-TDMA versus MC-CDMA and SS-MC-MA: Performance Evaluation for Fading Channels. *ISSSTA '98, Sun City, South Africa, September 02-04*, pages 200–204, 1998.

[9] J. Lindner. MC-CDMA and its Relation to General Multiuser/Multisubchannel Transmission Systems. *Proc. ISSSTA '96, Mainz, Germany, September 22-25*, 1:115–120, 1996.

[10] J. Lindner. Multi-Carrier Spread Spectrum: An Attractive Special Case of General Multiuser/Multisubchannel Transmission Methods . *Multi-Carrier Spread-Sprectrum, K. Fazel and G.P. Fettweis (eds.), Kluver Academic Publishers, The Netherlands*, pages 3–12, 1997.

[11] M. Reinhardt. Kombinierte vektorielle Entzerrungs- und Decodierverfahren. *Fortschr. Ber. VDI Reihe 10 No. 519. VDI Verlag, Düsseldorf, 1997.*

[12] T. Huschka. Untersuchungen zum Funkkanal innerhalb von Gebäuden. *Dissertation, University of Ulm, Ulm, Germany, 1996.*

[13] J. Lindner, J. Hess, and M. Reinhardt. Übertragung und Teilungsverfahren bei Kanälen mit frequenzselektivem Fading. *ITG-Fachtagung Mobile Kommunikation, Neu-Ulm, 27.-29.9.1993*, ITG-Fachber. 124:175–186, 1993.

[14] J. Lindner. MC-CDMA in The Context of General Multiuser / Multisubchannel Transmission Methods. *European Transactions on Telecommunications (ETT)*, will appear in fall 1999.

MCM-DSSS with DPSK Modulation and Equal Gain Combining in Delay and Doppler-Spread Rician Fading

AUTHORS

Rodger E. Ziemer and Thaddeus B. Welch, III

Affiliation

University of Colorado at Colorado Springs and United States Naval Academy

Key words: Multicarrier modulation; direct-sequence spread spectrum; equal gain combining; Rician fading; delay and Doppler spread

Abstract: The bit error probability (BEP) of multi-carrier modulated (MCM) direct-sequence spread-spectrum (DSSS) using differentially coherent phase-shift keying (DPSK) with equal gain combining in Doppler spread Rician fading is given. Gauss-Chebyshev quadrature integration of the moment generating function of the decision statistic at the combiner output is used. The technique is extended to delay and Doppler spread fading, but with Rayleigh amplitude statistics. It is based on diagonalizaton of the subcarrier correlation matrix.

I. INTRODUCTION

Third generation (3G) personal communication systems (PCS) will invariably be wideband due to the demands of mixed traffic (voice, video, and data), the need for variable rate transmission, and the desire to provide finer channel multipath resolution in an attempt to achieve more diversity gain [1]. Indications are that the Rayleigh model now used in so many current analyses will no longer suffice for wideband PCS. Finer resolution will mean that each resolvable component will be composed of a smaller number of rays, which will result in less severe amplitude fading, but with a phase component that is somewhat random [2]. Furthermore, smaller cell sizes will mean that a line-of-site path will result in a specular component in the received signal [3].

It is important to consider the effects of both delay and Doppler-spread fading in such communications channels. For example, the transmission path between a low earth-orbit satellite and moving terrestrial vehicle in a clutter-producing environment, such a buildings or trees, may be better modeled as Rician fading with Doppler spread. Also, terrestrial communication links between moving vehicles over short distances may well be modeled by delay-Doppler spread Rician fading since, due to the short distance between transmitter and receiver, both a direct and diffuse component of the received signal may be present.

One way to combat the effects of such fading environments is to employ diversity transmission. In this paper we consider the use of multicarrier modulation (MCM) with direct-sequence spread spectrum (DSSS) on each carrier. Kondo and Milstein [4] showed that such a system produces the same diversity improvement as a Rake receiver provided that the number of resolved paths in the Rake system and the number of independently fading carriers in the DSSS/MCM system are the same. They gave conditions for choosing the bandwidth of each spread carrier relative to the coherence bandwidth assuming a rectangular power delay profile for the multipath to ensure the independence condition. This was later extended to the case of nonrectangular power delay profiles in [5, 6]. Kondo and Milstein also assumed coherent phase-shift keyed (PSK) modulation. In this paper, differential phase-shift keying (DPSK) is used, partly to facilitate analysis but also because DPSK is a more robust modulation scheme for fading channels. Block diagrams for the transmitter and receiver models are shown in Figures 1 and 2, respectively.

The Rician amplitude model for the multichannel fading components is used in this paper because selection of the so-called K factor allows modeling of channels with arbitrary apportioning of power between specular and diffuse multipath components. In addition, correlation between carriers due to a nonrectangular power delay profile is analyzed by diagonalization of the covariance matrix of the MCM carriers.

II. ANALYSIS

In [4], the result for the MGF, $\Phi(s|E_n,1)$, of the decision statistic at the matched filter output, ζ, of a DPSK receiver given the fading amplitude, E_n, on the reference bit and that a 1 was transmitted, is given as [7]

$$\Phi(s \mid E_n, 1) = \frac{\exp\left\{-0.5E_n^2\left[\dfrac{2\phi s - \left(\sigma_t^2 + \phi^2\sigma_n^2\right)s^2}{1-(\sigma_t\sigma_n s)^2}\right]\right\}}{1-(\sigma_t\sigma_n s)^2} \tag{1}$$

where ϕ is the correlation coefficient between bits due to fading, σ_n^2 is the mean-square noise at the detector output, and $\sigma_t^2 = \sigma_n^2 + \sigma_s^2\left(1-\phi^2\right)$ with σ_s^2 being the

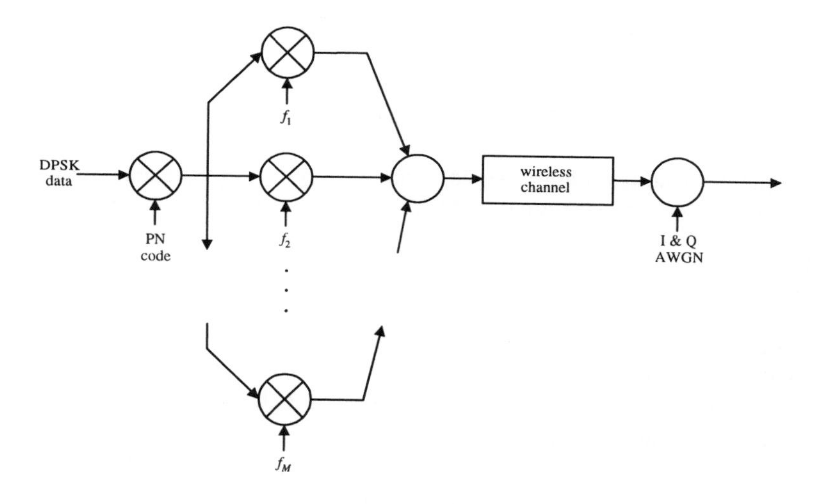

Figure 1. MCM-DSSS transmitter

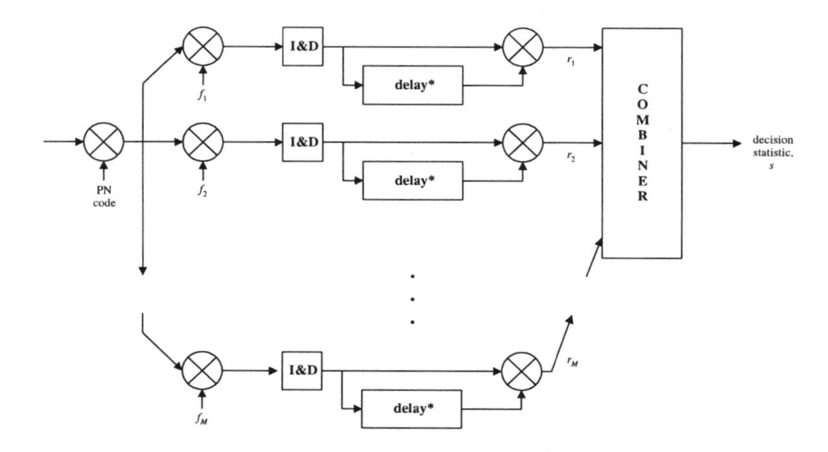

Figure 2. MCM-DSSS detector and combiner

mean square of a quadrature signal component at the detector output. For multichannel reception with equal gain combining, the decision statistic is

$$\varsigma = \sum_{n=1}^{L} \varsigma_n \tag{2}$$

where ς_n is the matched filter output of the nth receiver. Since $\Phi(s \mid E_n, 1)$ is a MGF, the MGF of the sum (2) is obtained simply as the product of MGFs like (1). Each can have different powers. To obtain the MGF for the nth carrier averaged with respect to the fading statistics, it is necessary to average the conditional MGF with respect to the appropriate probability density function (pdf) for E_n^2. For Rician amplitude statistics, it is the noncentral chi-square distribution of 2 degrees of freedom, which is [8]

$$p_{E_n^2}(y) = \frac{1}{2\sigma^2} \exp\left(-\frac{s_0^2 + y}{2\sigma^2}\right) I_0\left(\sqrt{y}\,\frac{s_0^2}{\sigma^2}\right) \tag{3}$$

where $s_0^2/2$ is the specular component (real signal) power and σ^2 is the diffuse component power of the fading envelope. The ratio $K = s_0^2/2\sigma^2$ (the specular-to-diffuse power ratio) is often referred to as the K-factor. Manipulation of the averaging integral gives

$$\Phi_n(s|1) = \frac{B(s)}{\left(1 + 2\sigma^2 A(s)\right)} \exp\left(-K \frac{2\sigma^2 A(s)}{1 + 2\sigma^2 A(s)}\right) \tag{4}$$

where

$$B(s) = \frac{1}{1 - (\sigma_I \sigma_n s)^2} \text{ and } A(s) = \frac{\phi\, s - 0.5\left(\sigma_I^2 + \phi^2 \sigma_n^2\right) s^2}{1 - (\sigma_I \sigma_n s)^2}$$

In [8], it is shown that the probability of decision error given a 1 was transmitted (which in this case is also the probability of bit error), is given by

$$P_b = \frac{1}{2\pi j}\int_{c_o - j\infty}^{c_o + j\infty} \Phi(s\,|\,1)\frac{ds}{s} \tag{5}$$

where $c_o > 0$ is chosen to avoid the pole at $s = 0$. In this paper, we use the Gauss-Chebyshev quadrature method of [9, 10] to numerically obtain the probability of error from the MGF, $\Phi(s\,|\,1) = \prod_{n=1}^{L} \Phi_n(s\,|\,1)$, of the decision statistic at the output of the combiner. The probability of bit error is approximated as [9, 10]

$$P_b = P(U < 0) = \frac{1}{n}\sum_{k=1}^{n/2} \mathrm{Re}\left[\Phi\left(s = c + jc\tau_k\,|\,1\right)\right] + \tau_k \,\mathrm{Im}\left[\Phi\left(s = c + jc\tau_k\,|\,1\right)\right] + R_n \tag{6}$$

where

$$\tau_k = \tan\left[(2k-1)\pi / 2n\right] \tag{7}$$

Methods for choosing c for fast convergence are discussed in [5], and the error $R_n \to 0$ as n gets large. Usually $n = 50 - 100$ suffices ($n = 150$ was used here); this can be determined by computing results for increasing n and stopping when the results change negligibly. A closed form was derived in [11] based on residue integration. This was possible because the amplitude statistics were assumed Rayleigh as opposed to Rician as assumed here.

To account for delay spread, it is tacitly assumed that the delay spread is an insignificant fraction of the symbol interval, but that its power delay profile imposes correlation between DSSS subcarriers. To account for nonrectangular power delay profiles, we diagonalize the subcarrier covariance matrix [11]

$$\mathbf{R} = \begin{bmatrix} \gamma_1 & \sqrt{\gamma_1\gamma_2}\,\rho_{12} & \cdots & \sqrt{\gamma_1\gamma_L}\,\rho_{1L} \\ \sqrt{\gamma_1\gamma_2}\,\rho_{21} & \gamma_2 & \cdots & \sqrt{\gamma_2\gamma_L}\,\rho_{2L} \\ \vdots & \vdots & \ddots & \vdots \\ \sqrt{\gamma_L\gamma_1}\,\rho_{L1} & \sqrt{\gamma_L\gamma_2}\,\rho_{L2} & \cdots & \gamma_L \end{bmatrix} \tag{8}$$

where [6] $\gamma_n = \frac{(PG)^2 E_c}{2\sigma_n^2}$, in which E_c is the chip energy, σ_n^2 is the noise variance at the nth carrier matched filter output, and ρ_{ij} is the covariance between carriers i and j. For example, for an exponential power delay profile, it is

$$\rho_{ij} = \frac{J_0^2\left(2\pi f_d T_b\right)}{1 + \left(2\pi\Delta f \tau_0\right)^2} \tag{9}$$

where f_d is the "bathtub" Doppler power spectrum width, T_b is the bit period, Δf Hz is the separation between carriers, and τ_0 is the delay spread standard deviation.

For Rayleigh fading, a closed form expression can be derived for the delay-Doppler spread case in terms of the eigenvalues of $\mathbf{R}$ [5, 6]. First, the power-envelope probability density function is obtained through the diagonalization process as [11]

$$f_\Gamma(\gamma) = \sum_{i=1}^{L} d_i \exp(-\gamma/\lambda_i),\ \gamma \geq 0 \tag{10}$$

where d_i is the residue associated with the eigenvalue λ_i of $\mathbf{R}$. Using this to average the conditional bit error probability for DPSK over the fading power gives

$$P_b = \int_0^\infty \frac{1}{2}\exp(-\gamma)\sum_{i=1}^{L} d_i \exp(-\gamma/\lambda_i)d\gamma = \frac{1}{2}\sum_{i=1}^{L} d_i \frac{\lambda_i}{\lambda_i - \lambda_p} \tag{11}$$

where the residues are given by

$$d_i = \lambda_i^{L-2} \prod_{\substack{p=1 \\ p\neq i}}^{L} \frac{1}{\lambda_i - \lambda_p} \tag{12}$$

To account for both delay and Doppler spread, the conditional MGF (1) must be averaged with respect to the power probability density function of the fading after the diagonalization process. Unfortunately, this is not possible analytically for Rician fading, so only the Rayleigh fading case is dealt with here for the doubly spread case. In this case, the probability of error takes the form of the double integral (inside one is the inverse transform of the MGF)

$$P_{b,L} = \sum_{i=1}^{L} \frac{d_i}{2\pi j}\int_0^\infty \int_{-\infty}^{\infty} \left[1-(\sigma_t\sigma_n s)^2\right]^{-L} \left[\left(\frac{\phi s - 0.5\left(\sigma_t^2 + \phi^2\sigma_n^2\right)s^2}{1-(\sigma_t\sigma_n s)^2}\right) + \frac{1}{\lambda_i}\right]^{-1} \exp(s\varsigma)ds d\varsigma \tag{13}$$

where all parameters have been previously defined. The inside integral can be evaluated using residue theory (done with a symbolic processor to handle large L here) or the Gauss-Chebyshev technique used previously can be applied. A similar procedure for the Rician case is difficult, if even possible, to carry out because of the singularity in the exponent of (1).

A more fruitful procedure for Rician fading may be to obtain the MGF of each fading subcarrier, take their product, and use the Gauss-Chebyshev procedure since each MGF is that of the envelope of a conditionally independent (complex) Gaussian process after the diagonalization process.

III. RESULTS

Results are compared in Fig. 3 for K = -10 dB with curves for purely Rayleigh fading computed from closed form expressions. The remarkably close agreement for this case, where only 10% of the total signal power is in the specular component and 90% is in the diffuse component, gives confidence in the numerical technique used for computing the curves for the Rician case. A more specular case is shown in Fig. 4 where K = 10 dB (90% of the total power in the specular component). Two characteristics of these curves are the error floor due to the Doppler spread and the crossovers which move right with increasing K. The crossover points of the curves of Fig. 4 are about 15 dB greater than those of Fig. 3.

With the crossover nature of the BEP curves, it is clear that for a range of $\overline{E}_b / N_0$ –values, an optimum order of diversity exists. This is illustrated by Fig. 5 where BEP is plotted versus L for K = 5 dB and $\overline{E}_b / N_0$ = 20 dB. The optimum order of diversity is somewhere in the range of L = 10 to 15. In Fig. 6, with $\overline{E}_b / N_0$ = 25 dB, the BEP is monotonically decreasing for $L < 30$.

Finally, a case where the correlation between MCM carriers due to nonrectangular power delay profile is shown in Fig. 7 where Doppler spread only curves are compared with doubly spread curves where $\tau_0 = 1\mu s$. Note the increased separation as the number of carriers increases.

IV. CONCLUSIONS

Bit error probabilities for multichannel DPSK signaling in delay and Doppler spread Rician fading have been computed for MCM/DSSS using Gauss-Chebyshev numerical integration of the MGF of the decision statistic conditioned on a 1 being sent. The method is general as shown in [9]; if one has the MGF, $\Phi(s \mid 1)$, and can find the residues of the singularities of $\Phi(s \mid 1)/s$ in the right-half plane, an analytical formula results. Alternatively, various bounds and exact expressions may be computed [12, 13] for the case where the MCM carriers are uncorrelated. The numerical approach used here is fast, accurate, and provides insight into the behavior of any multichannel system using DPSK modulation and equal-gain combining. Analysis of the effect of correlation between subcarriers due to nonrectangular power delay profiles uses diagonalization of the correlation matrix.

REFERENCES

[1] E. Dahlman, et al., "WCDMA – The radio interface for future mobile multimedia communication," *IEEE Trans. On Veh. Technol.*, Vol. 47, pp. 1105-1118, Nov. 1998.

[2] H. Iwai and Y. Karasawa, "The theoretical foundation and applications of equivalent transmission-path model for assessing wideband digital transmission characteristics in Nakagami-Rice fading environments," *IEICE Trans. Commun.*, Vol. E79-B, pp.1205-1214, Sept. 1996.

[3] Jenn-Hwan Tarng and Kung-Min Ju, "Statistical distributions of Rician factor for radio LOS propagation in urban microcells," *IEICE Trans. Commun.*, Vol. E81-B, pp.1283-1285, June 1998.

[4] S. Kondo and L. B. Milstein, "Performance of multicarrier DS CDMA systems," *IEEE Trans. on Commun.*, vol. 44, pp. 238-246, Feb. 1996.

[5] R. E. Ziemer and N. Nadgauda, "Effects of correlation between subcarriers of an MCM/DSSS communication system," *Proc. VTC'96*, pp. 146-150, Oct. 1996.

[6] W. Xu and L. B. Milstein, "Peformance of multicarrier DS CDMA systems in the presence of correlated fading," *Proc. VTC'967* pp. 2050-2054, Oct. 1997.

[7] H. B. Voelcker, "Phase-shift keying in fading channels," *IEE Proc.*, Vol. 107, Part B, pp.31-38, Jan. 1960.

[8] J. G. Proakis, *Digital Communications*, New York: McGraw Hill, 1995.

[9] E. Biglieri, G. Caire, G. Taricco, J. Ventura-Traveset , "Simple method for evaluating error probabilities," *Electr. Letters*, Vol. 32, pp. 191-192, Feb. 1, 1996.

[10] E. Biglieri, G. Caire, G. Taricco, J. Ventura-Traveset, "Computing error probabilities over fading channels: a unified approach," *European Trans. on Telecommun. and Related Technol.*, Vol. 9, pp. 15-25, Jan. – Feb. 1998.

[11] T. B. Welch, III, *Analysis of Reduced Complexity Direct-Sequence Code-Division Multiple-Access Systems in Doubly Spread Channels*," Ph. D. Dissertation, University of Colorado at Colorado Springs, Dec. 1997.

[12] P. Kam, "Tight bounds on Rician-type error probabilities and some applications," *IEEE Trans. on Commun.*, Vol. 42, pp.3119-3128, Dec. 1994.

[13] Y. Chow, J. McGeehan, and A. Nix, "Simplified error bound analysis for M-DPSK in fading channels with diversity reception," *IEE Proc.-Commun.*, Vol. 141, pp. 341-350, Oct. 1994.

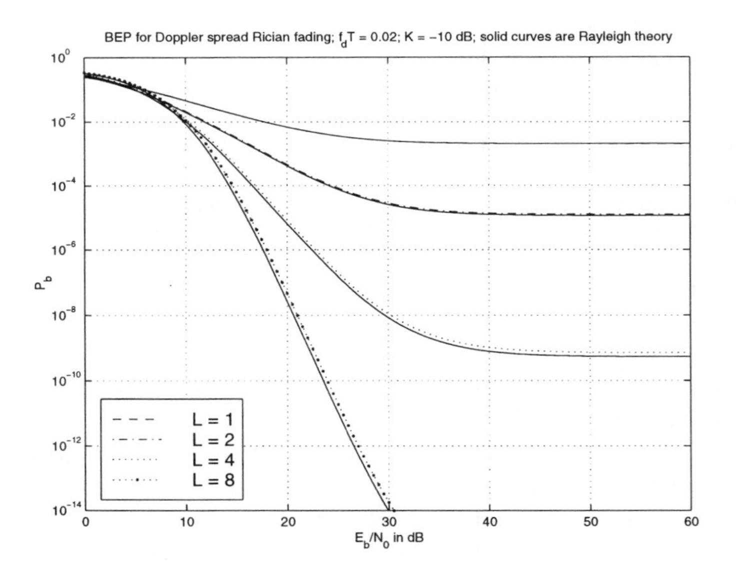

Figure 3. Bit error probability for K = -10 dB compared with the Rayleigh case.

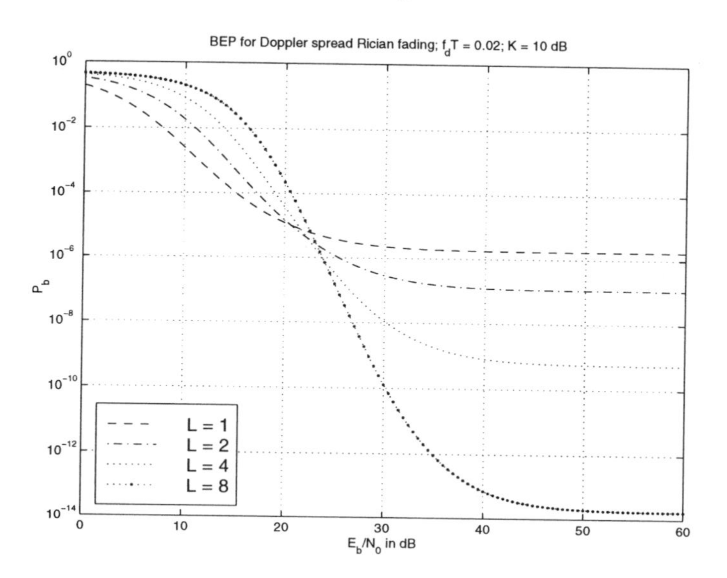

Figure 4. Bit error probability for $K = 10$ dB and $f_d T_b = 0.02$ and various numbers of carriers.

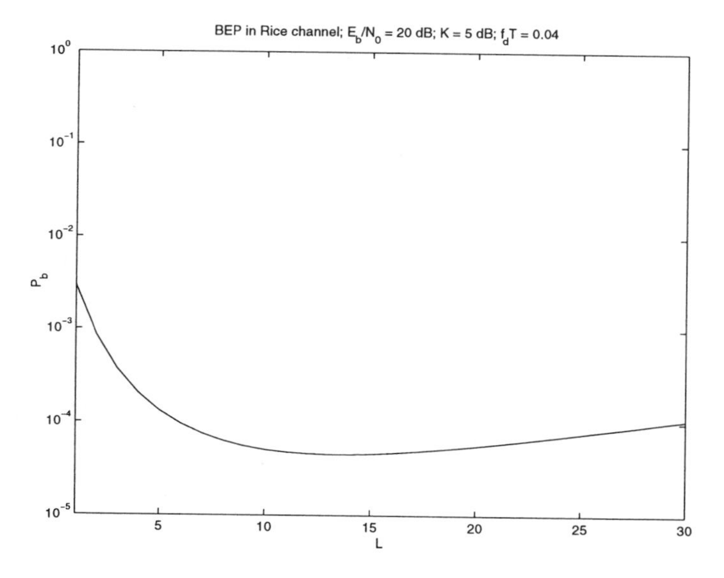

Figure 5. Bit error probability versus L for $\overline{E_b} / N_0 = 20$ dB and $f_d T_b = 0.04$

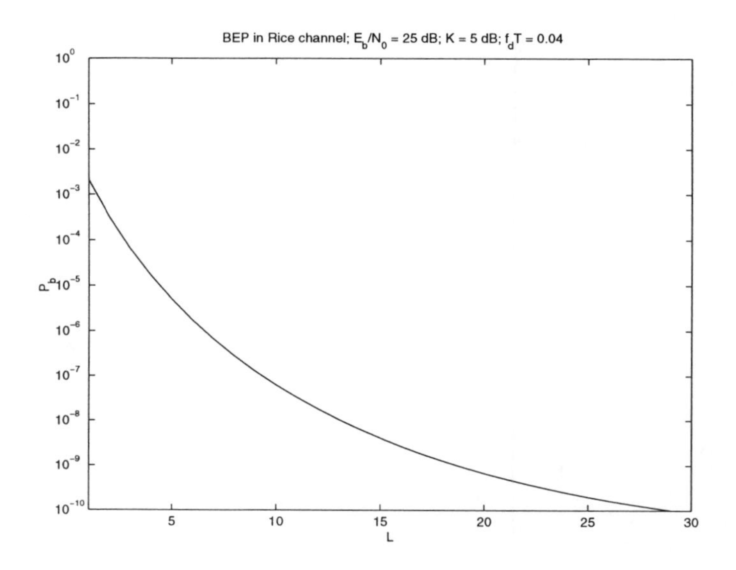

Figure 6. Bit error probability versus L for $\overline{E_b} / N_0 = 25$ dB and $f_dT_b = 0.04$

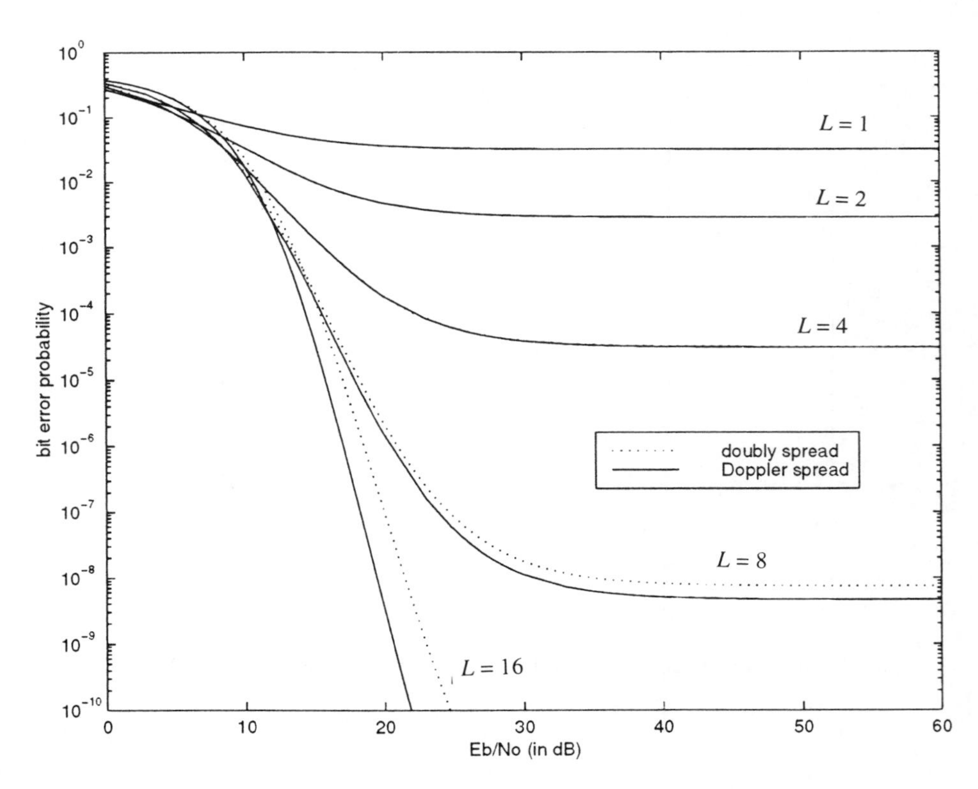

Figure 7. System performance in delay-Doppler spread channel; $\tau_0 = 1\,\mu$s and $f_dT_b = 0.08$.

INFLUENCE OF CODE SELECTION ON THE PERFORMANCE OF A MULTI-CARRIER SPREAD-SPECTRUM SYSTEM†

Jörg Kühne, Achim Nahler, and Gerhard P. Fettweis

Dresden University of Technology, Mobile Communications Systems D-01062 Dresden, Germany, Tel.: +49 351-463 5521, Fax.: +49 351-463 7255,
e-mail: kuehnej@ifn.et.tu-dresden.de

ABSTRACT

Multi-carrier spread-spectrum (MC-SS) modulation is a new kind of spread-spectrum modulation[1],[2]. The spreading code is designed in the frequency domain. Hence, the system is dual to a direct-sequence spread-system (DS-SS) where the spreading code is given in the time domain. There exist two different views: the OFDM view and the DS-SS view. Many publications deal with the frequency domain processing. Transmitter and receiver structure are similar to OFDM. Therefore all these implementations use a guard interval to avoid ISI. The disadvantage of this guard interval is the higher hardware complexity and the waste of symbol energy. In the following paper the effect of ISI on a MC-SS system is investigated. Both types of implementation (time or frequency domain) are compared and the dependencies of the performance from the spreading code are demonstrated. It is shown that with proper code selection and a time domain implementation without guard interval (RAKE receiver) it is possible to outperform equivalent frequency domain receivers with guard interval (MRC).

I. INTRODUCTION

During the last decade, much of *spread-spectrum* research has been focussed on *Multi-carrier Spread-Spectrum* (MC-SS). In the literature, two different views about MC-SS can be found. The first approach ([1],[3]) was to use a *Direct-Sequence* (DS) code in the frequency domain to create a spread-spectrum system dual to *Direct-Sequence Spread-Spectrum* (DS-SS). The second approach [2] is based on ideas of *Orthogonal Frequency Division Multiplex* (OFDM). In contrast to OFDM, all sub carriers transmit the same signal. This spectral diversity can be viewed as spreading in frequency whereby using a code to distinguish different users. To avoid *Inter-Symbol-Interference* (ISI) a guard interval is introduced similar to a OFDM system [4]. In [5], it the feasibility of a MC-SS system with a RAKE receiver in the time domain instead a *Maximum-Ratio-Combiner* (MRC) in the frequency domain was shown. But there was nothing said about guard interval. As it is known from

†This work is partially supported by the German National Science Foundation (Deutsche Forschungsgemeinschaft contract Fe 423/2)

conventional DS-SS systems a RAKE receiver does not need a transmit signal with guard interval. In this paper we will point out the differences and similarities of MRC and RAKE. We will show that it is not always necessary to make use of a guard interval. If the code fulfill properties, which will be explained below, one can omit it. The paper is organized as follows. In section II, the MC-SS transmit signal is shown and in section III the receiver structures in time and frequency domain are analyzed. Section IV deals with the influence of the guard interval on the performance and in V numerical results are presented.

II. SYSTEM MODEL

We can describe a single user MC-SS transmit signal $s(t)$ without guard interval by:

$$s(t) = d(t) \sum_{k=0}^{K-1} C(k) e^{2\pi ikt/T_S} \quad \text{with: } c(t) = \sum_{k=0}^{K-1} C(k) \cdot e^{2\pi ikt/T_S} \tag{1}$$

$d(t)$ is the (pulse shaped) *Binary Phase Shift Keying* (BPSK) data signal, $C(k)$ is the spreading code vector in the frequency domain, K is the number of carriers, $c(t)$ is the analog code signal in the time domain and T_S is the symbol duration. In a multi path environment the designed orthogonality between the sub carriers is lost. Therefore, every symbol is expanded by a so called cyclic extension (Fig. 1), which is removed in the receiver. Usually, the guard time T_G is about 20% of the symbol time T_S [6],[4]. This leads to a longer symbol duration ($T_S' = T_S + T_G$), which is equivalent to a higher transmit power or a lower *Signal-to-Noise-Ratio* (SNR) of approximately 0.8 dB. In an OFDM system, the guard interval is essential. Every carrier transmits

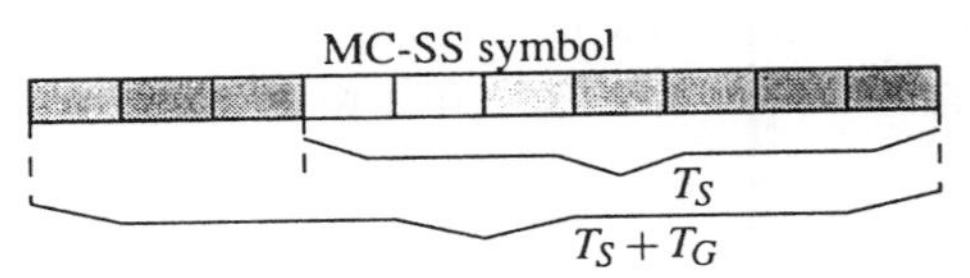

Fig. 1: Principle of the guard interval

a different symbol. In a MC-SS system, every carrier transmit the same symbol at the same time and all carriers are combined in the receiver in a convent manner before making a symbol decision. Therefore, it is not clear which effect is caused by the *Inter-Carrier-Interference* (ICI). We will compare three different systems with different codes (code properties will be explained in section IV).

- System Ⓐ: with cyclic extension
- System Ⓑ: with empty guard interval (same duration as Ⓐ)
- System Ⓒ: without guard interval.

System Ⓑ is interesting, because it has no ISI and no additional transmit power is needed. As we will show later, the performance of all systems depends on code properties. We chose two codes of length $K = 32$.

- Code ①: Row 13 of a Walsh-Hadamard matrix of the dimension 32
- Code ②: A crest factor optimized Newman phase code, see [7]

Walsh-Hadamard codes are normally used in a multi user system to orthogonalized the different users. The disadvantages of these codes are the large crest factor and bad correlation properties of the time signal. Code ② was originally developed to reduce the crest factor and hence to increase the efficiency of the power amplifier. The multi-path fading channel can be described as follows:

$$h(t) = \sum_{f=0}^{F-1} h_f \cdot \delta(t - \tau_f) \tag{2}$$

F is number of relevant paths, τ_f is the delay of path f and h_f is the appropriate path weight, which is Rayleigh distributed but assumed to be constant over the frame length T_{frame}

III. TIME-FREQUENCY VIEW OF MC-SS

Fig. 2 and 3 show the two receiver structures for a slow multi path fading channel for time and frequency domain processing, respectively. The algorithm for the MRC

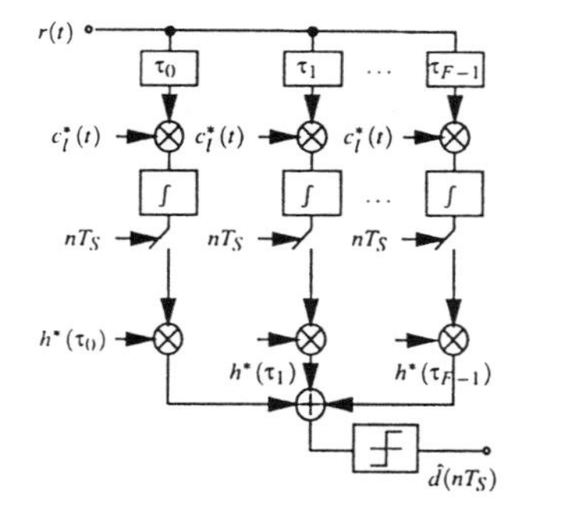

Fig. 2: RAKE receiver

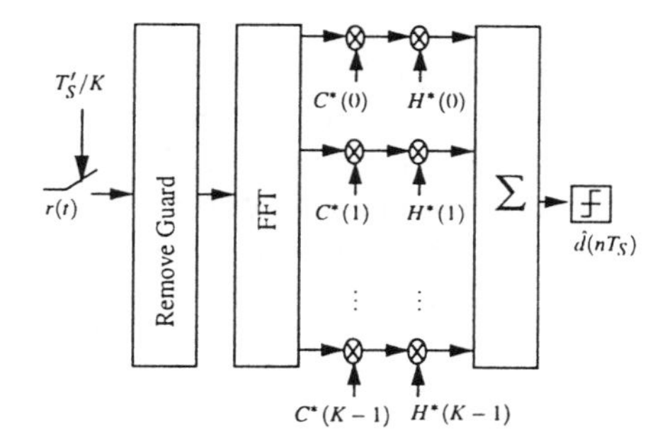

Fig. 3: MRC receiver

receiver can described by:

$$\hat{d}(n) = \operatorname{sgn}\left(\sum_{k=0}^{K-1} R(k) \cdot C^*(k) \cdot H^*(k)\right) \tag{3}$$

$\hat{d}(n)$ is the BPSK signal after the slicer, $R(k)$ is the received signal vector after a *Discrete Fourier Transform* (DFT) and $H(k)$ is the discrete Channel coefficient vector in the frequency domain. Eq. 3 can be transformed, see [8]

$$\hat{d}(n) = \operatorname{sgn}\left(\sum_{n=0}^{K-1} r(n) \cdot (c(n) \circledast h(n))^*\right) \text{ with: } h(i) = \sum_{f=0}^{F-1} h_f \cdot \delta(i - \tau_f) \tag{4}$$

$$\hat{d}(n) = \text{sgn}(\sum_{f=0}^{F-1} h_f^* \cdot \sum_{n=0}^{K-1} r(n) \cdot c^*((n-\tau_f)_{modK})) \tag{5}$$

By $\circledast$, we denote the operator for a cyclic convolution. Eq. 5 is the algorithm for a RAKE receiver with a cyclic correlation of each correlator. In practical systems, this is not usual but shows that a MRC performs the same algorithm as a RAKE with cyclic correlation. This fact is very important for further performance considerations of MC-SS systems. It is therefore only necessary to investigate the RAKE receiver with different types of correlations. The advantage of the RAKE receiver is the lower complexity. A RAKE needs $(K+1) \cdot F$ complex elementary operations and a MRC receiver $\frac{K}{2} \cdot \log_2(K) + 2 \cdot K$ operations. Note that the number of relevant paths F is usually much smaller than the number of carriers K.

IV. PERFORMANCE EVALUATION

To evaluate the performance of a RAKE the universal principle of diversity reception of statistically independent signals can be used. In [9] the *Bit-Error-Rate* is given as:

$$P_b = \frac{1}{2} \sum_{f=1}^{F} \pi_f \left(1 - \sqrt{\frac{\bar{\gamma}_f}{1+\bar{\gamma}_f}}\right) \qquad \text{with} \qquad \pi_f = \prod_{m=1;m\neq f}^{F} \frac{\bar{\gamma}_f}{\bar{\gamma}_f - \bar{\gamma}_m} \tag{6}$$

$\bar{\gamma}_f$ is the mean SNR of path f, see Eq. 2. This is a lower bound for P_b for the RAKE receiver. There are two reasons why this bound cannot be achieved:

1. The fingers of the RAKE are not statistically independent due to non ideal autocorrelation properties of the spreading code.
2. Not all fingers can perform a complete correlation caused by limiting the symbols in a cyclic RAKE/MRC implementation.

Point 1 was examined in [10]. The non-ideal autocorrelation function leads to interference of one RAKE finger to another RAKE finger, which causes an error floor dependent on the spreading code and the channel model. One can distinguish between *Self-Interference* (SI) and ISI, where SI specifies the interference part of the current and ISI the interference part of the previous or following symbol, respectively. Both terms can be combined to special correlation functions listed in tables 1 and 2.
$I_{1/2}(\tau_1, \tau_2)$ is the interference of the RAKE finger with delay τ_1 on the finger with delay τ_2, where $\tau_1 < \tau_2 < T_G$. The performance degradation can be evaluated by modeling this interference as *additional white Gaussian noise* (AWGN),see [10].
Even more important is point 2. This effect occurs only with a cyclic RAKE/MRC receiver due to limiting the symbol and performing a cyclic correlation. In System Ⓑ, the finger with delay τ_1 can only perform a partial correlation, see Eq. 7. This causes a mismatch of this finger which leads to a loss of *symbol energy* E_b.

$$\hat{d}_f = \frac{1}{T_S} \int_{0}^{T_S-\tau_f} c(t)c^*(t)\,dt \tag{7}$$

Table 1: Types of Interference for the MRC receiver

	Interference
Ⓐ	$I(\tau_1) = \frac{1}{T_S}\int_0^{T_S} c(t \pm \tau_1)c^*(t)\,dt$ = even autocorrelation
Ⓑ	$I_{2/1}(\tau_1,\tau_2) = \frac{1}{T_S}\int_0^{T_S-\tau_2} c(t)c^*(t-(\tau_1-\tau_2))\,dt$ $I_{1/2}(\tau_1,\tau_2) = \frac{1}{T_S}\int_0^{T_S-\tau_2} c(t-(\tau_1-\tau_2))c^*(t)\,dt + \frac{1}{T_S}\int_0^{\tau_2-\tau_1} c(t)c^*(t-(\tau_1-\tau_2))\,dt$
Ⓒ	$I_{2/1}(\tau_1,\tau_2) = \frac{1}{T_S}(\underbrace{\int_0^{T_S-\tau_2} c(t)c^*(t-(\tau_1-\tau_2))\,dt}_{SI}$ $\pm \underbrace{\int_0^{\tau_2-\tau_1} c(t+(\tau_1-\tau_2))c^*(t)\,dt \pm \int_0^{\tau_1} c(t-\tau_2)c^*(t-\tau_1)\,dt)}_{ISI}$ $I_{1/2}(\tau_1,\tau_2) = \frac{1}{T_S}(\underbrace{\int_0^{T_S-\tau_2} c(t-(\tau_1-\tau_2))c^*(t)\,dt + \int_0^{\tau_2-\tau_1} c(t)c^*(t-(\tau_1-\tau_2))\,dt}_{SI}$ $\pm \underbrace{\int_0^{\tau_1} c(t-\tau_1)c^*(t-\tau_2)\,dt)}_{ISI}$

Table 2: Types of Interference for the RAKE receiver

	Interference
Ⓑ	$I(\tau_1,\tau_2) = \frac{1}{T_S} \underbrace{\int_0^{T_S+(\tau_1-\tau_2)} c(t)c^*(t \pm (\tau_1-\tau_2))\,dt}_{SI}$
Ⓒ	$I(\tau_1,\tau_2) = \frac{1}{T_S}(\underbrace{\int_0^{T_S+(\tau_1-\tau_2)} c(t)c^*(t \pm (\tau_1-\tau_2))\,dt}_{SI} \pm \underbrace{\int_{T_S+(\tau_1-\tau_2)}^{T_S} c(t)c^*(t \pm (\tau_1-\tau_2))\,dt)}_{ISI}$

In system Ⓒ the a second part is added because the correlation is done with a part of the previous symbol (Eq. 8). It depends on the previous symbol phase whether it occurs a loss of E_b or not.

$$\hat{d}_f = \frac{1}{T_S}\int_0^{T_S-\tau_f} c(t)c^*(t)\,dt \pm \frac{1}{T_S}\int_{T_S-\tau_f}^{T_S} c(t)c^*(t)\,dt \tag{8}$$

V. NUMERICAL RESULTS

Figs. 4 and 5 shows the dependence of E_b loss of a single RAKE finger with delay τ_1 with a cyclic RAKE implementation. Code ① has a dramatic loss if $\tau_1 > 0.08T_S$. Code ② has only a linear drop-off. This comes from the almost constant envelope of this code. The best code would be a code with a zero tail. But this is equivalent to the insertion of a new empty guard interval. The bad behavior of the code ① is the reason for the very large performance degradation (see Fig. 6). The optimized Newman code has only a very slight degradation if $\tau_1 < 0.2T_S$.

Figs. 6 and 7 show comparisons of the BER performance of both codes for System Ⓐ, Ⓑ and Ⓒ. The channel is a two path Rayleigh fading model with the same mean path power. The delay between the paths is equally distributed in the interval $0 \leq \tau_1 \leq 0.2T_S$. The dashed line is the diversity combiner bound and the solid line is the performance of the MRC receiver with guard interval. This performance is independent of the implemented spreading code since there is no ISI and the SI is the same for each code with constant envelope coefficients, see [10]. It is interesting to note, that one can get a slight gain in BER by using a RAKE receiver with both codes and all systems. A RAKE has no loss as described in point 2. The total interference

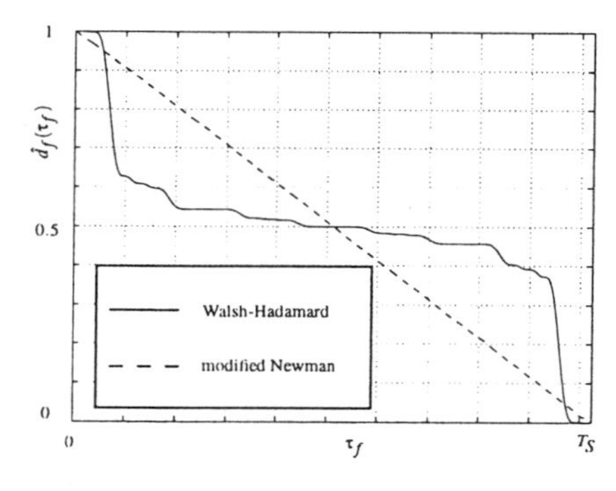

Fig. 4: $\hat{d}_f$ with System Ⓑ

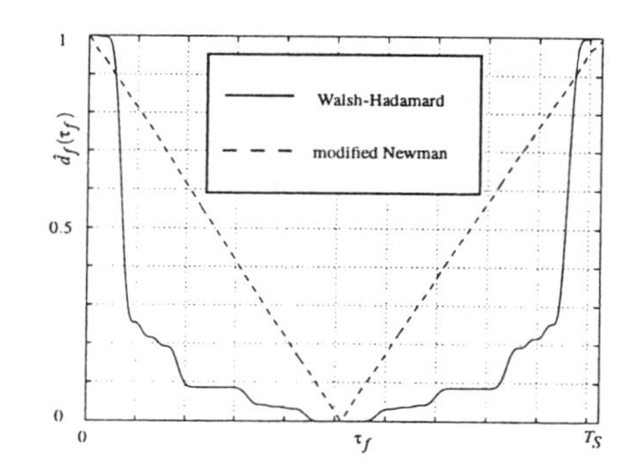

Fig. 5: $\hat{d}_f$ with System Ⓒ

between the diversity paths (see table 2) which is the sum of the SI and ISI is not larger or even smaller than the SI of a MRC. Moreover, an additional gain can be obtained by leaving out the guard interval.

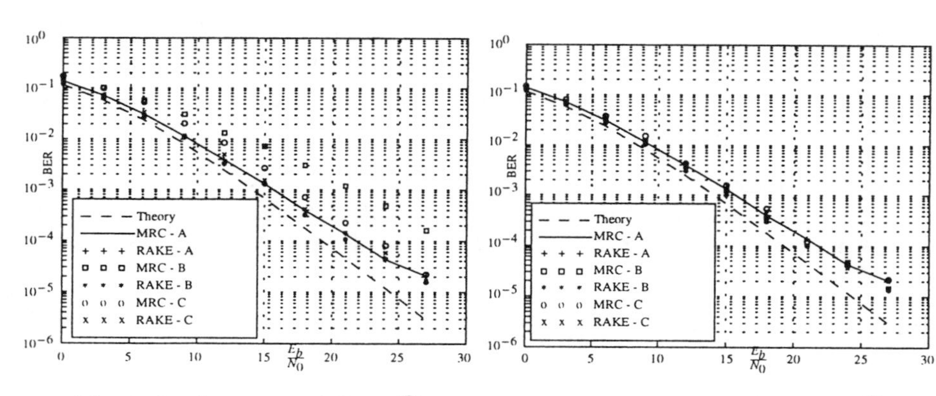

Fig. 6: Performance of Code ①

Fig. 7: Performance of Code ②

VI. CONCLUSIONS

In this paper, the necessity of the guard interval for a MC-SS system was investigated. It was shown, that the frequency domain MRC receiver is equivalent to a special kind of RAKE receiver. A lower bound for the performance of these receivers in a multi path fading channel was given and the reasons for the performance degradation of real implementations were given. It can be summarized that there is no need for a guard interval to avoid ISI with a RAKE receiver. With a MRC receiver the performance depends strongly on the code. Using an optimized Newman code with almost constant envelope time signal, there is no loss compared to a system with guard interval. The complexity of a RAKE receiver is much lower. The number of operation increases with the number of significant paths but not with the number of sub carriers like a MRC receiver. Furthermore, the effort for channel estimation for a few paths in time is lower than the effort for estimating all channel coefficients in the frequency domain. Concluding, we may stay that the RAKE receiver is more suitable than a MRC receiver for the channels considered here.

REFERENCES

[1] N. Yee, J.-P. Linnartz, and G. P. Fettweis, "Multi-Carrier CDMA in Indoor Wireless Radio Networks," in *Proc. of the 4th IEEE International Symposium on Personal, Indoor and Mobile Radio Communications*, (Yokohama), pp. 109–113, 1993.

[2] K. Fazel, "Performance of CDMA/OFDM for Mobile Communication Systems," in *Proc. of the 2nd IEEE International Conference on Universal Personal Communications*, (Ottawa), pp. 975–979, 1993.

[3] G. P. Fettweis, A. S. Bahai, and K. Anvari, "On Multi-Carrier Code Division Multiple Access (MC-CDMA) Modem Design," in *Proc. of the 44th IEEE Vehicular Technology Conference*, vol. 3, (Stockholm), pp. 1670–1674, 1994.

[4] K. Fazel and G. P. Fettweis, eds., *Multi-Carrier Spread-Spectrum*. Kluwer Academic Publishers, 1997.

[5] A. Nahler and G. P. Fettweis, "An Approach for a Multi-Carrier Spread Spectrum System with RAKE-Receiver," in *Proc. of the First International Workshop on Multi-Carrier Spread-Spectrum*, pp. 97–104, Kluwer Academic Press, April 1997.

[6] W. Y. Zou and Y. Wu, "COFDM: AN OVERVIEW," *IEEE Transactions on Broadcasting*, vol. 41, pp. 1–8, March 1995.

[7] V. Aue and G. P. Fettweis, "Multi-Carrier Spread Spectrum Modulation with Reduced Dynamic Range," in *Proc. of the 46th IEEE Vehicular Technology Conference*, pp. 914–917, April 1996.

[8] H. D. Lüke, *Signalübertragung*. Berlin: Springer-Verlag, 6 ed., 1995.

[9] J. G. Proakis, *Digital Communications*. New York: McGraw-Hill, 3rd ed., 1995.

[10] A. Nahler, J. Kühne, and G. P. Fettweis, "Investigations of Bit Error Floors for Multi-Carrier Spread-Spectrum CDMA Systems with RAKE Receiver in Frequency-Selective Channels," in *Proc. of the 5th IEEE International Symposium on Spread Spectrum Techniques and Applications*, (Sun City, South Africa), pp. 169–173, Sept. 1998.

COMPLEX VALUED BLOCK CODES FOR OFDM-CDMA APPLICATION

Armin Dekorsy and Karl-Dirk Kammeyer

University of Bremen, FB-1, Department of Telecommunications
P.O. Box 33 04 40, D-28334 Bremen, Germany,
*Fax: +(49)-421/218-3341**

dekorsy@comm.uni-bremen.de

Abstract This paper introduces the extension of Walsh-Hadamard block codes in order to obtain complex valued block codes. These complex valued block codes lead to an improved spectral efficiency with additional lower error rate. Applied for low rate concatenated coding schemes required for CDMA uplink transmission, the gain in spectral efficiency due to the new codes can be used for strengthening other codes participating. We observe a gain of about 1.7 dB for BER=10^{-3} for OFDM-CDMA transmission.

1. INTRODUCTION

In the asynchronous uplink of CDMA based systems inherent *multiple access interference* (MAI) requires efficient low rate channel coding under the constraint of low decoding complexity. In order to meet these requirements serial concatenated coding schemes using a $(M, \log_2(M), M/2)$ *Walsh-Hadamard* (WH) code as inner block code followed by a repetition code can be applied [1; 2; 3]. With WH codes, the decoding can be carried out by the *Fast*

*Compressed postscript files of our publications are also available from our WWW server http://www.comm.uni-bremen.de.

Hadamard Transform (FHT).
Nevertheless, conventional coding schemes, e.g. with WH codes as inner block codes, represent real valued signal processing systems without exploitation of the complex signal space.

In this paper, we present an extension of block codes by an additional M-PSK modulation in order to obtain *complex valued block codes* (CBC), where the extension is based on multiplication of code words with M-PSK symbols. In case of a WH block code as constituent block code, the entire code represents a *complex valued WH block code* (CWC). This scheme was presented for the first time as *hybrid modulation* in [4], where non-coherent detection and no *maximum-likelihood detection* (MLD) in a single-carrier CDMA system was performed. The new aspects of this paper are *(i)* a complete new analysis based on the interpretation as CWC, *(ii)* MLD of the CWC, and *(iii)* the application in an OFDM-CDMA system [5; 6].
The paper is structured as follows: In section 2, we describe the construction and MLD of the CWCs. Section 3 presents analytical derivations for fully interleaved Rayleigh fading channels. The application in OFDM-CDMA will be discussed in section 4, and section 5 concludes the paper.

2. CONSTRUCTION AND MLD FOR CWC

Fig. 1 illustrates the construction of CWCs. The data bits, $b \in \{0,1\}$ each of duration T_b are serial-to-parallel converted to groups of $(\log_2(M_1)+\log_2(M_2))$ data bits each. Each group is divided into parts of $\log_2(M_1)$ and $\log_2(M_2)$ data bits, respectively. A $(M_1, \log_2(M_1), M_1/2)$ *Walsh-Hadamard* (WH) block code with additional antipodal modulation maps the $\log_2(M_1)$ data bits to vector $\boldsymbol{w}^{m_1} = [w_0^{m_1}, w_1^{m_1}, \ldots, w_{M_1-1}^{m_1}]^T$. Vector $\boldsymbol{w}^{m_1}$ can be interpreted as a code word in the Euclidean signal space including M_1 code symbols $w_\mu^{m_1} \in \{\pm 1\}$, $\mu = 0, \ldots, M_1-1$. This mapping is also well-known as M_1-ary Walsh modulation [3].
Afterwards, each complete codeword $\boldsymbol{w}^{m_1}$ is multiplied by an M_2-PSK symbol $e^{j\varphi_{m_2}}$, $m_2 \in \{0, \ldots, M_2-1\}$, determined by the group of $\log_2(M_2)$ bits. This multiplication results in code words $\boldsymbol{u}^{m_1,m_2} = \boldsymbol{w}^{m_1} e^{j\varphi_{m_2}}$ with code symbols $u_\mu^{m_1,m_2} = w_\mu^{m_1} e^{j\varphi_{m_2}}$. In order to simplify notation, we can denote the code words by $\boldsymbol{u}^i = \boldsymbol{w}^i e^{j\varphi_i}$, where subscript i indicates the i-th word with $i \in \{0, \ldots, M_1 M_2 - 1\}$.

Thus, the proposed scheme is to multiply the complete codeword of a WH block code with an M_2-PSK symbol in order to construct a new entire block code with rate

$$R_c^{cwc} = \frac{\log_2(M_1 \cdot M_2)}{M_1}. \tag{1}$$

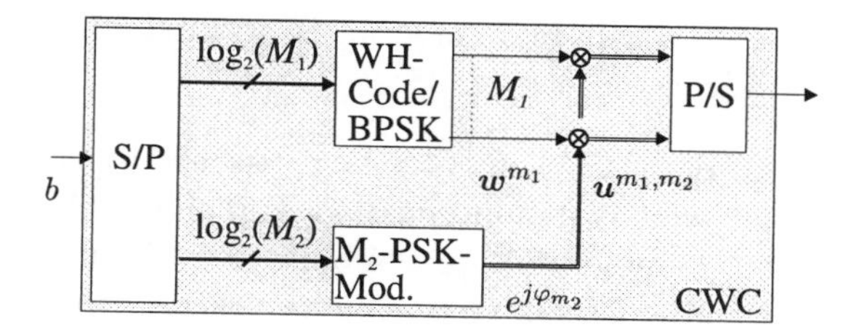

Figure 1 Construction of a complex valued block code based on a WH code

In case of M_2-PSK modulation with $M_2 \geq 4$, this new block code represents a *complex* valued WH block code. Therefore, the analysis in terms of error correction capability should be performed in the Euclidean space instead of the Galois-Field GF(2).

Generally, the quadratic Euclidean distances of an arbitrary CBC can be expressed by

$$e_w^2(i,j) = \sum_{\mu=0}^{M_1-1} e_\mu^2(i,j) = 2M_1 - 2\cos(\varphi_i - \varphi_j) \sum_{\mu=0}^{M_1-1} w_\mu^i w_\mu^j, \qquad (2)$$

where $e_w^2(i,j)$, $e_\mu^2(i,j) = |u_\mu^i - u_\mu^j|^2$ are the distances between the two code words $\boldsymbol{u}^i$, $\boldsymbol{u}^j$ and the two symbols u_μ^i and u_μ^j of these code words, respectively. Note that the distances (2) are increased in comparison with the distances $e_w^2(i,j) = 2M_1 - 2\sum_{\mu=0}^{M_1-1} w_\mu^i w_\mu^j$ of an arbitrary real valued block code, and this increase depends on the used M_2-PSK modulation. As an example, we will confine our discussion on QPSK ($M_2 = 4, \Delta\varphi = (\varphi_i - \varphi_j) \in \{0, \pm\pi/2, \pi\}$) combined with WH codes ($\sum_{\mu=0}^{M_1-1} w_\mu^i w_\mu^j = 0,\ \forall i \neq j$) throughout the paper.

In this case, the quadratic Euclidean distance for different code words ($i \neq j$) lead to

$$e_w^2(i,j) = \begin{cases} 4M_1, \ \boldsymbol{w}^i = \boldsymbol{w}^j \ \wedge \Delta\varphi = \pi \\ 2M_1, \ \text{else}, \end{cases} \qquad (3)$$

where the quadratic Euclidean distance between two code symbols determined by the QPSK results in

$$e_\mu^2(i,j) = \begin{cases} 0, \ \Delta\varphi = 0 \\ 2, \ \Delta\varphi = \pm\pi/2 \\ 4, \ \Delta\varphi = \pi. \end{cases} \qquad (4)$$

In summary, the proposed combination of WH codes with QPSK modulation benefits from a higher code rate (1), and, hence, from an improved spectral efficiency. This is done without loss of error correction capability. We can also observe by (1) that CWCs gain more in terms of spectral efficiency than lower the code rate of the WH code.

Fig. 2 shows the MLD of the CWC that can be performed by the *Fast Hadamard Transform* (FHT) followed by a multiplication with all possible alternatives of transmitted conjugate complex QPSK symbols. Finally, the ML decoder is to take the real part with consecutive maximum decision and decoding of the decided codeword to $\log_2(M_1 M_2)$ data bits. Since the used WH block code and QPSK are combined in a way that a *complete* WH code word is multiplied with a QPSK symbol, it is possible to interchange the FHT and the QPSK detection in the receiver. This leads to complex valued input signals for the FHT. Nevertheless, the conventional FHT can further be applied since the complex symbol acts in a constant factor manner.

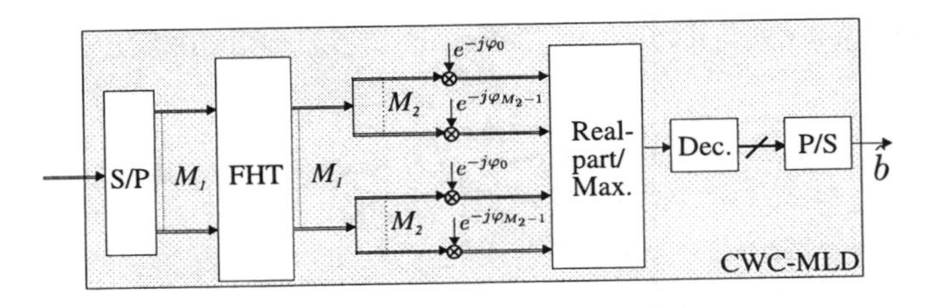

Figure 2 Maximum-Likelihood-Detection of CWC

3. ANALYTICAL PERFORMANCE

With regard to the application of CWC in OFDM-CDMA systems, we will briefly derive the bit error probability P_b by using the well known union bound for 1-Path fully interleaved Rayleigh fading. We assume perfect channel state information, equalization with the conjugate complex channel coefficients and a non-dissipative channel. Since the distances given by (3) are independent of the chosen reference word, we can use $\boldsymbol{u}^0 = [1, \ldots, 1]$ as reference.
Let us start with the code word specific pairwise error probability. First, we define $L(0,j)$ as the effective length of error events in code symbols which means that code words differ in $L(0,j)$ positions. At these positions, we obtain identical distances $e^2(0,j) = e^2_\mu(0,j)$ given by (4) (except zero) of the corresponding codeword $\boldsymbol{u}^j$. With (2), the lengths $L(0,j)$ are directly related to the distances,

$$e^2_w(0,j) = \sum_{\mu=0}^{M_1-1} e^2_\mu(0,j) = L(0,j)e^2(0,j) = \begin{cases} 4M_1, \; \boldsymbol{w}^i = \boldsymbol{w}^j \wedge \Delta\varphi = \pi \\ 2M_1, \; \text{else.} \end{cases} \tag{5}$$

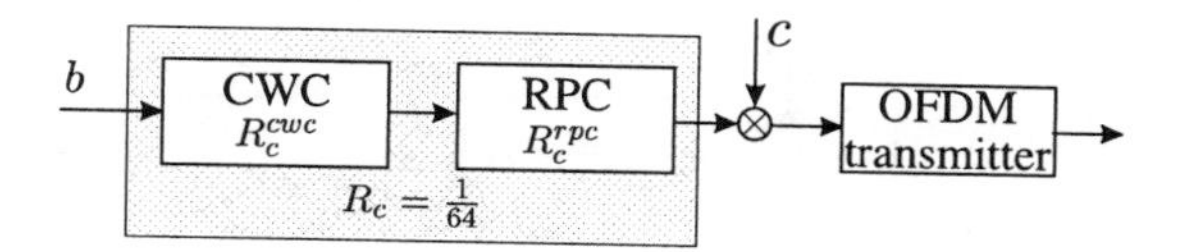

Figure 3 OFDM transmitter with serial concatenated coding scheme

Thus, the codeword specific pairwise error probability is given by

$$P(e^2(0,j)) = \left(\frac{1-\nu}{2}\right)^{L(0,j)} \sum_{l=0}^{L(0,j)-1} \binom{L(0,j)-1+l}{l} \left(\frac{1+\nu}{2}\right)^l \qquad (6)$$

with

$$\nu = \sqrt{\frac{\bar{\gamma}}{1+\bar{\gamma}}}, \quad \bar{\gamma} = \frac{1}{4} R_c^{cwc} \frac{E_b}{N_0} e^2(0,j). \qquad (7)$$

In order to accomplish the derivation, we sum all code words with identical input weight w as well as distance $e^2 \equiv e^2(0,j)$, and, hence, with identical $P(e^2(0,j))$ up in sets. Denoting the number of code words in the sets by the *input-output weight enumerating function* (IOWEF) coefficient A_{w,e^2}, we finally obtain with (6) the upper bounded bit error probability

$$P_b \leq \sum_w \frac{w}{k} \sum_{e^2} A_{w,e^2} P(e^2), \quad k = \log_2(M_1 M_2). \qquad (8)$$

From the analysis we note for fully-interleaved Rayleigh fading the absence of the codeword distance e_w^2 in (6) which is used to play the main role for the correction capability of codes in case of AWGN [7]. Instead, the distance between code symbols e^2 has to be taken into account due to (6) and (7). Moreover, it turns out by (5) that the higher the code symbol distance e^2 the lower the effective length L associated with code diversity. This attribute is similar to trellis-coded modulation [8], and it is contrary to the well-known knowledge that coding schemes with subsequent antipodal modulation maximizing the Euclidean distance ($\equiv$ Hamming distance) also have the largest built-in degree of diversity [7; 9]. Finally, it is remarkable that QPSK reveals to exploit the highest degree of diversity (M_1) in case of Rayleigh fading (4), (5).

4. APPLICATION TO OFDM-CDMA

In this section the introduced CWC will be analyzed in combination with OFDM-CDMA. Fig. 3 shows the transmitter with the CWC of rate R_c^{cwc} (1) followed by a *repetition code* (RPC). The repetition encoder has the code rate $R_c^{rpc} = \frac{1}{N_p}$ ensuring a constant entire code rate

$$R_c = R_c^{cwc} \cdot R_c^{rpc} \stackrel{!}{=} 1/64 \qquad (9)$$

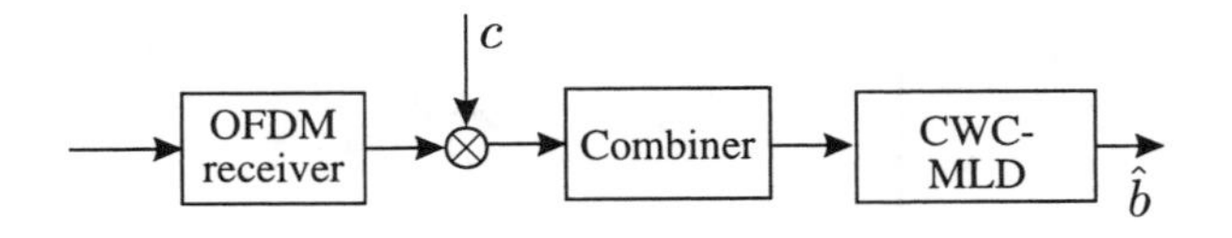

Figure 4 OFDM receiver with serial concatenated decoding scheme

of the serial concatenated coding scheme. Note that the whole spreading is incorporated in the serial coding scheme. After encoding, the sequence is scrambled by a user specific code c. Due to asynchronous transmission in the uplink, we use *pseudo-noise* (PN) codes for scrambling. The resulting signal is fed to the OFDM transmitter consisting of a serial-parallel converter, a *perfect frequency interleaver*, an IFFT, and a parallel-serial converter. Furthermore, it also includes the insertion of the guard interval.
The number of carriers for OFDM transmission is chosen to be $N_M = M_1\, N_p$. This choice leads to a scenario where in the average $\log_2(M_1 M_2)$ information bits are mapped to one OFDM symbol, and it will result in the advantage of less mismatching if the product $M_1\, N_p$, i.e. the number of subcarriers, is raised (the guard time is assumed to remain unchanged).

For one user, the coherent OFDM receiver with subsequent decoding is shown in Fig. 4. We assume conventional OFDM reception [3], *maximum-ratio-combining* (MRC) [6], and perfectly known channel state information. After OFDM reception and descrambling by the user specific code c, the signal enters the combiner that correlates N_p successive samples (decoding the RPC). Finally, the above described MLD of the applied CWC is performed. In case of fully interleaved Rayleigh fading, the coding gain of the RPC in terms of diversity, the number of active users as well as the mismatching caused by the guard time have to be taken into account in the following way:
With (6), (7), and (9) the new average signal-to-noise ratio and diversity are

$$\bar{\gamma} = \frac{1}{4} \cdot \frac{R_c S/N}{1 + (J-1) R_c S/N} \cdot e^2(0,j), \qquad L'(0,j) = L(0,j) \cdot N_p, \quad (10)$$

where J is the number of simultaneously active users[1], and S/N the signal-to-noise ratio with mismatching [3; 6].

4.1 RESULTS

Monte-Carlo simulation as well as union bounded analytical results for P_b with $J = 8$ active users and $M_1 = 16$ are shown in Fig. 5. Besides the *bit error rate*

[1]We assumed delay times of the other active users that equals an integer multiple of the OFDM sampling time (worst case). This is in contrast to the well-known formula [5] with user interference $\frac{1}{3}(J-1)$.

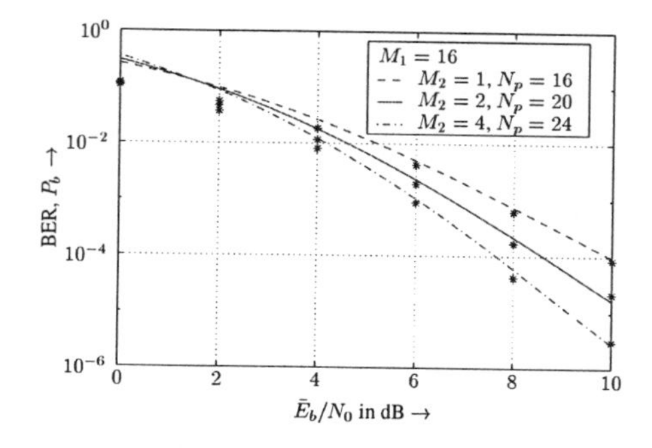

Figure 5 Comparison of WH code with CWC for $M_2 = 2, 4$ and $M_1 = 16$ in OFDM-CDMA with $J = 8$ active users

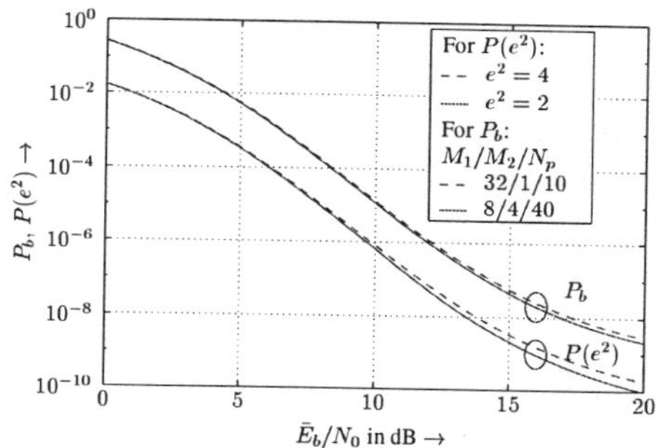

Figure 6 Pairwise error and bit error probability for $M_1/M_2/N_p = 32/1/10$ and $8/4/40$ with $J = 8$ active users

(BER) of the WH block code ($M_2 = 1$) and the CWC with QPSK ($M_2 = 4$), we also plotted results for a "CWC" with BPSK ($M_2 = 2$) instead of QPSK[2]. First, Fig. 5 indicates conformity and the expected divergence between simulation (marked by $*$) and analytical results for high and low $\bar{E}_b/N_0$, respectively. We also observe the best performance for the CWC with QPSK. Note a gain of about 1.7 dB at BER=10^{-3}. Thus, raising M_2 in conjunction with an unchanged M_1 results in a higher code rate R_c^{cwc} (improved spectral efficiency) for the CWC, and, hence, allows lower rate repetition coding (lower R_c^{rpc}). Due to the latter fact, the entire systems degree of diversity $L'(0, j)$ being decisive in case of perfectly interleaved Rayleigh fading is also raised.

In order to analyse diversity aspects, Fig. 6 shows P_b as well as $P(e^2)$ for $M_1/M_2/N_p$ parameter constellations $32/1/10$ and $8/4/40$ with a mismatching of 0.17 dB in both cases. For the WH block code with RPC we have $e^2 = 4$ and $L' = 160$, where for CWC with RPC we obtain $e^2 = 4$, $L' = 160$ as well as $e^2 = 2$, $L' = 320$, and one code word with $e^2 = 4$, $L' = 320$ which can be neglected. Thus, there exists an additional slightly lower $P(e^2)$ for the CWC which results with (8) in a slightly improved P_b.

5. CONCLUSION

The main objective of this paper has been the extension of block codes by M-PSK modulation in order to perform *complex valued block coding*. In this context, we have focused on *Walsh-Hadamard* block codes with QPSK modulation. In general, the extension of WH codes to CWCs results in an improved

[2]The mismatching is about 0.26 dB, 0.21 dB and 0.18 dB for $M_2 = 1, 2$ and 4, respectively. Thus, the influence of different mismatching can be neglected.

spectral efficiency with additional lower bit error rate. Thus, the spectral efficiency gained can be used for strengthening codes participating in a concatenated coding scheme if a constant entire code rate is assumed. Here, we have considered RPC, where future analysis has to be carried out concerning more powerful codes. Moreover, it would be an interesting task to investigate the extension of arbitrary block codes to CBCs.

References

[1] R. Herzog, A. Schmidbauer, and J. Hagenauer. Iterative Decoding and Despreading improves CDMA-Systems using M-ary Orthogonal Modulation and FEC. In *Proc. IEEE International Conference on Communications (ICC)*, volume 2, pages 909–913, Montreal, June 8–12 1997.

[2] A. Dekorsy and K.D. Kammeyer. M-ary Orthogonal Modulation for Multi-Carrier Spread-Spectrum Uplink Transmission. In *Proc. IEEE International Conference on Communications (ICC)*, volume 2, pages 1004–1008, Atlanta, June 7–11, 1998.

[3] A. Dekorsy and K.D. Kammeyer. A new OFDM-CDMA Uplink Concept with M-ary Orthogonal Modulation. *European Trans. on Telecommunications*, 1999. Accepted.

[4] D. Nikolai and K.D. Kammeyer. BER Analysis of a Novel Hybrid Modulation Scheme for Noncoherent DS-CDMA Systems. In *Proc. IEEE Int. Symp. on Personal, Indoor and Mobile Radio Communications (PIMRC)*, volume 2, pages 256–260, Helsinki, Finland, September 1997.

[5] K. Fazel, S. Kaiser, and M. Schnell. A Flexible and High Performance Cellular Mobile Communications System Based on Orthogonal Multi-Carrier SSMA. *Wireless Personal Communications*, 2:121–144, 1995.

[6] Stefan Kaiser. *Multi-Carrier CDMA Mobile Radio Systems – Analysis and Optimization of Detection, Decoding and Channel Estimation*. PhD thesis, German Aerospace Center, VDI, January 1998.

[7] J.G. Proakis. *Digital Communications*. McGraw–Hill, 3-rd edition, 1995.

[8] E. Biglieri, D. Divsalar, P. McLane, and M. Simon. *Introduction to Trellis-Coded Modulation with Applications*. Maxwell MacMillan International Editions, Republic of Singapore, 1991.

[9] V. Kühn, A. Dekorsy, and K.D. Kammeyer. Channel Coding Aspects in an OFDM-CDMA System. In *3rd ITG Conference Source and Channel Coding*, Munich, Germany, January 2000. Accepted.

TWO-DIMENSIONAL DIFFERENTIAL DEMODULATION OF OFDM FRAMES

Erik Haas and Stefan Kaiser
German Aerospace Center (DLR) - Institute for Communications Technology
82234 Oberpfaffenhofen, Germany
Erik.Haas@dlr.de, Stefan.Kaiser@dlr.de

Abstract In this contribution, the possibility to combine differential demodulators in time direction and frequency direction for Orthogonal Frequency Division Multiplexing (OFDM) frames is investigated. The proposed algorithm uses not only the direction in which the information is modulated, but also joins it with the additional second direction that is differentially demodulated as well. In the special case that is investigated here, the second direction does not carry any modulated information, but can be used for the two-dimensional demodulation process without changing the transmitter. Therefore, the proposed algorithm is applicable for existing digital transmission systems, like for example Digital Audio Broadcasting (DAB), as well as new systems.

1. INTRODUCTION

As commonly used for example in the European Digital Audio Broadcasting (DAB) standard [1], differential demodulation is only performed in one direction (here by OFDM symbols in time t, see Fig. 1(b)) with the help of one reference symbol at the beginning of the frame. This means that only the closest neighbour in negative direction of time is used for demodulation. It is obvious that there are additionally seven other closest neighbours (see Fig. 2). It is shown that these neighbours can also be used for differential demodulation without changing the transmitter.

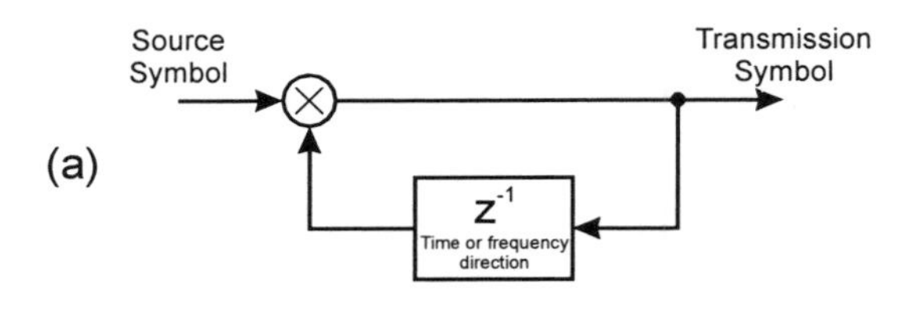

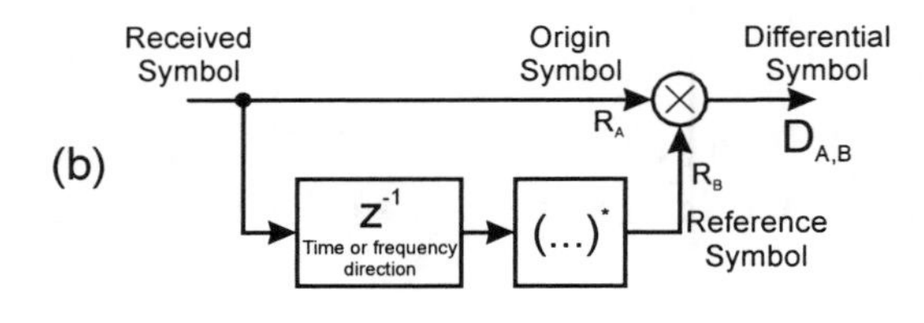

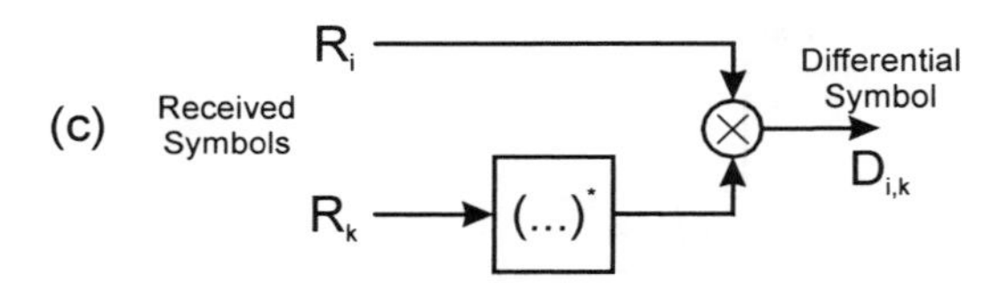

Figure 1 (a) One-dimensional differential modulation, (b) Conventional one-dimensional differential demodulation, (c) Differential demodulation between two arbitrary received symbols

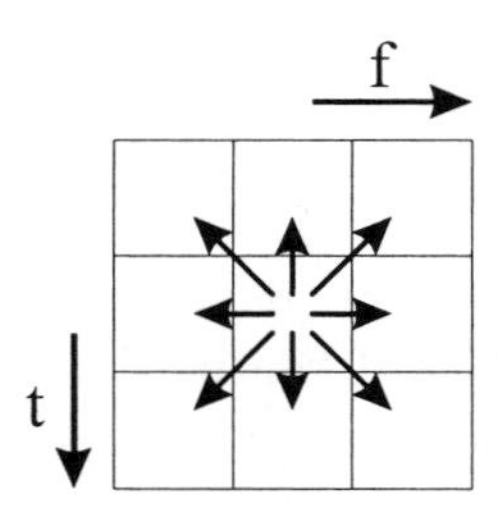

Figure 2 Differential neighbourhood demodulation of one OFDM symbol

2. BASIC IDEA FOR TWO-DIMENSIONAL DIFFERENTIAL DEMODULATION

Two-Dimensional Differential Demodulation (2D-DD) is done by not necessarily choosing the closest path with distance one (here in negative direction of time t), but by using a detour path with distance larger than one that results from the combination of several differential demodulation steps (which means hopping to one or several intermediate neighbours). These additional steps are chosen according to an algorithm that assures that the possibility of a bit error is decreased by choosing the detour path rather than the direct path. Differential demodulation between two arbitrary symbols within the frame is possible since all transmitted differentially modulated symbols in general come from the same transmission symbol-alphabet (see Fig. 1(a)). Thus, the demodulation between two arbitrary symbols again delivers a valid source symbol (see Fig. 1(c)).

3. FRAME-WORM ALGORITHM FOR DETOUR PATH DETERMINATION

In Fig. 3, the outline of the upper frequency half of an OFDM frame (128 subcarriers, 24 OFDM symbols) with a grid is shown. The frequency axis is the horizontal direction and the time axis is the vertical direction with origin in the top left corner. The used modulation scheme can be chosen to be Differential Quaternary Phase Shift Keying (DQPSK) [2] in time direction, for example. Like discussed above, all transmitted symbols themselves come from the same alphabet, here Quaternary Phase Shift Keying (QPSK) symbols, and differential demodulation can theoretically be performed on every selected pair of received symbols. Due to stronger statistical correlations, especially in fading channel environments, only the eight surrounding symbols are regarded as relevant here. For each of the eight neighbours of one symbol in the center, now a differential demodulation step as shown in Fig. 2 and Fig. 1(c) is performed according to:

$$D_{i,k} = R_i \cdot R_k^* = a_i \cdot a_k \cdot e^{j(\varphi_i - \varphi_k)}, \tag{1}$$

where R_i is the current symbol under investigation, R_k is one of the eight neighbour symbols, $D_{i,k}$ is the symbol after the differential demodulation, $(\ldots)^*$ is the conjugate complex of the expression, a_i, a_k are the amplitudes and φ_i, φ_k the arguments of the symbols.

If DPSK is considered, the information is included in the differential phase

$$\Delta\varphi_{i,k} = \varphi_i - \varphi_k. \tag{2}$$

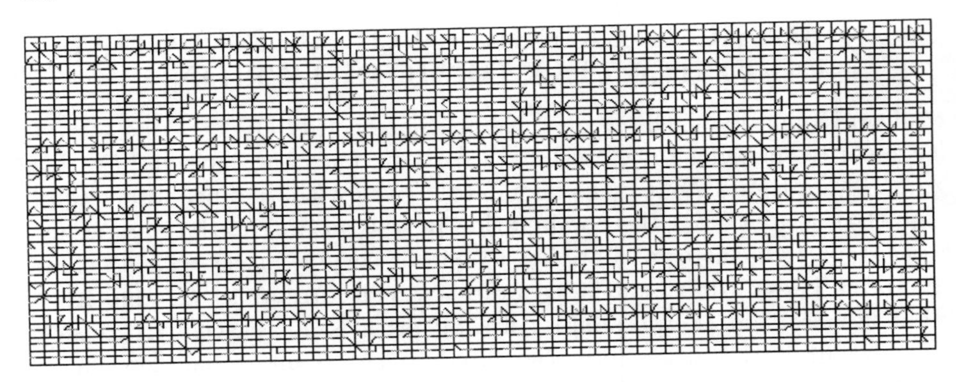

Figure 3 Two best neighbours (1st:grey, 2nd:black) of each QPSK symbol

This means that the information-carrying phase of the original one-dimensional differential demodulation is

$$\Delta\varphi_{A,B} = \varphi_A - \varphi_B. \tag{3}$$

If now instead of the direct path, a path with several iteration steps is chosen, it results in differential phases $\Delta\varphi_{A,1}, \Delta\varphi_{1,2}, \ldots \Delta\varphi_{(N-1),B}$ for each step (N is the iteration path depth, $N \geq 2$). If those differential intermediate phases would be summed up, the result would be

$$\begin{aligned} \Delta\varphi_{A,1} + \left[\sum_{i=1}^{N-2} \Delta\varphi_{i,i+1}\right]_{N>2} + \Delta\varphi_{(N-1),B} = \\ = (\varphi_A - \varphi_1) + (\varphi_1 - \varphi_2) + \ldots + (\varphi_{N-1} - \varphi_B) = \\ = \varphi_A - \varphi_B = \Delta\varphi_{A,B}, \end{aligned} \tag{4}$$

which means that the phase difference $\Delta\varphi_{A,B}$ remains unchanged and a simple adding of the intermediate phases wouldn't bring any advantage, but on the other hand also wouldn't influence the phase of the original path $A \rightarrow B$. This offers the possibility to take a closer look on the intermediate steps individually. Each received differential phase can be divided into a valid phase $\varphi_{source\ i,k}$ and a phase error $\varphi_{error\ i,k}$, that corresponds to the phase offset of the closest matching source-symbol phase:

$$\Delta\varphi_{i,k} = \varphi_{source\ i,k} + \varphi_{error\ i,k}. \tag{5}$$

The common one-dimensional differential demodulation directly concludes from $\Delta\varphi_{A,B}$ to the phase of the source-symbol $\varphi_{source\ A,B}$.

In contrast to this, the new two-dimensional differential demodulation algorithm proposed here decides for every individual intermediate phase $\Delta\varphi_{A,1}, \Delta\varphi_{1,2}, \ldots \Delta\varphi_{(N-1),B}$ on a possible phase of the source-alphabet $\varphi_{source\ A,1}, \varphi_{source\ 1,2}, \ldots \varphi_{source\ (N-1),B}$. Additionally, the phase errors $\varphi_{error\ A,1}, \varphi_{error\ 1,2}, \ldots \varphi_{error\ (N-1),B}$ and the amplitudes $a_A a_1, a_1 a_2, \ldots a_{(N-1)} a_B$ can be used for the path criteria and the resulting estimation.

Whenever a decision is made about a possible symbol of the source alphabet, it can't be guaranteed that this is the true source symbol that has been modulated and transmitted. For this reason, the path criteria has to make sure that choosing a detour path rather than the direct path is more reliable even for the fact that the detour path with its intermediate steps is "longer". Therefore, at first the direct path is considered the most reliable and chosen as the reference path (iteration depth $N = 1$). For a certain criteria now all paths or a subgroup of all possible paths up to a definable iteration path length N that reach the destination symbol B are investigated. If, considering the criteria, one of the alternative paths is more reliable than the reference path, this alternative path becomes the new reference path and so on.

Considering the criteria, the most reliable path therefore has a length L of $1 \leq L \leq N$. For the final decision about the possible phase of the source symbol, now the differential phase sum of the chosen path is calculated as follows:

$$\Delta\varphi'_{A,B} = \begin{cases} \varphi_{source\ A,B} & for\ L = 1 \\ \varphi_{source\ A,1} + \varphi_{source\ 1,B} & for\ L = 2 \\ \varphi_{source\ A,1} + \sum_{i=1}^{L-2} \varphi_{source\ i,(i+1)} + \varphi_{source\ (L-1),B} & else \end{cases} \tag{6}$$

4. MINIMUM PHASE ERROR CRITERIA

4.1 PATH DETERMINATION

The phase error $\varphi_{error\ i,k}$ of the intermediate steps is now used as a decision variable for the optimum path criteria. At first, the absolute value of the phase error $\varphi_{error\ A,B}$ of the direct path is calculated. This value is used as the reference value. Now all possible paths up to the chosen maximum iteration length N that lead to the destination symbol B are investigated. For each intermediate step of the current alternative path under investigation, the phase error $\varphi_{error\ i,k}$ is determined. If the absolute value of one of these phase errors is larger than the reference value, this alternative path is rejected. If on the other hand all individual absolute phase errors in the alternative path are smaller than the reference value, the maximum absolute value of the phase errors in this alternative path becomes the new reference value and this path becomes the new optimum path.

To easier explain the algorithm, for each symbol of the OFDM frame in Fig. 3, the two neighbours with the smallest phase errors to the closest source symbol for differential demodulation are shown. The channel that has been used here for the transmission was Rician. Lines within one symbol point into the direction of the two best neighbours (1st:grey, 2nd:black). Horizontal lines through several symbols indicate a non-frequency selective fade during the transmission of the corresponding OFDM symbol. Vertical lines through several symbols indicate that the phase offset on the corresponding subcarrier hasn't changed very much throughout these symbols. An enlarged version of a part of the three bottom lines is shown in Fig. 4.

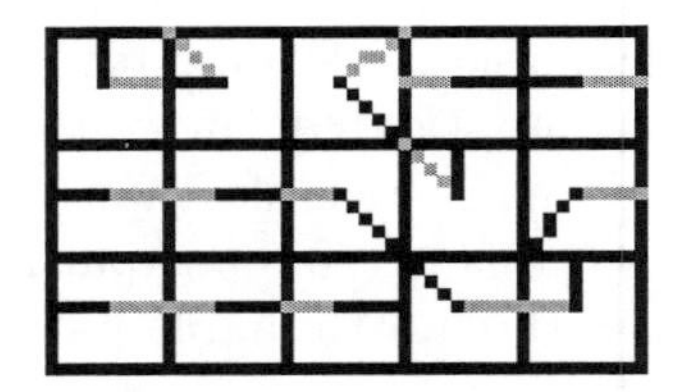

Figure 4 Detail of bottom lines from Fig. 3

First a look at the symbol in the right bottom corner is taken. The black line to the symbol above indicates that the second best neighbour is the desired reference symbol for the DQPSK demodulation so that it is likely that the differential demodulation can be performed in the common way.

Now the symbol in the center of the bottom line of Fig. 4 is investigated. Here the best neighbour is to the left and the second best neighbour is to the right of this symbol. This means that a decision in one of these directions is less likely to be wrong than making a decision to the top. If now the path to the right hand side is followed, it can be seen that again the second best neighbour is the symbol to the top left corner. This symbol on the other hand is the needed reference symbol for the regular differential demodulation path. The resulting detour path therefore can be written as *right-topleft.* This detour path is only one and not necessarily the best of all possible detour paths of a maximum length N.

4.2 PATH EVALUATION

To get the final corresponding bit sequence of the demodulation step, the information from the obtained path has to be extracted. Therefore the phase sum $\Delta\varphi'_{A,B}$ for the optimum path is calculated and DQPSK-demodulation is performed on this phase.

5. SIMULATION RESULTS

The transmission of the system under investigation uses OFDM-frames with 24 OFDM-symbols consisting of 128 subcarriers each. The information is differentially QPSK modulated in time-direction on subcarrier level. The first OFDM-symbol of the OFDM-frame is a reference symbol that is used for the first differential step. In Fig. 5, the performance of the OFDM system with two-dimensional differential demodulation using the Minimum Phase Error Criteria (MPEC) in the Additive White Gaussian Noise (AWGN) channel is shown with different iteration path depths. It can be seen that for iteration path depth $N = 1$, the system performs like common differential demodulation. For iteration path depths $N \geq 2$, the required Signal-to-Noise-Ratio (SNR) can be decreased significantly. For example to achieve a Bit Error Rate (BER) of 10^{-4}, the SNR can be reduced by more than 1 dB already for an iteration depth of $N = 2$.

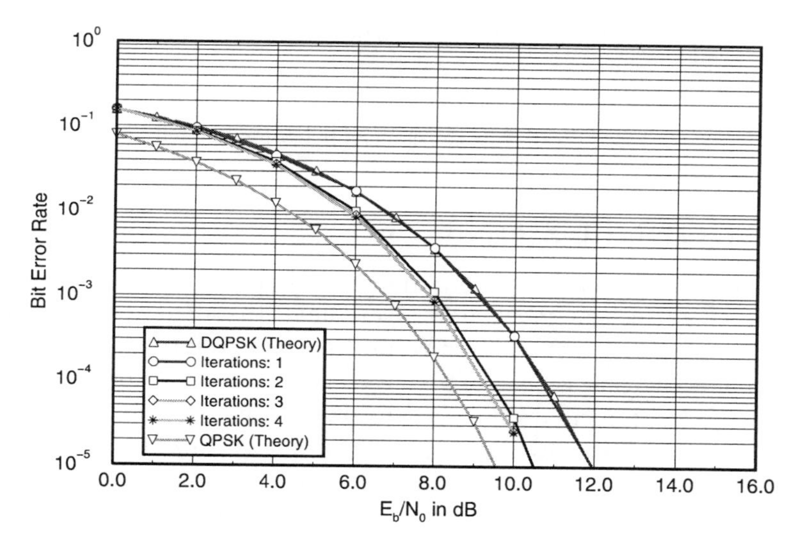

Figure 5 BER performance for AWGN channel

Fig. 6 shows the performance of the system in a Rician fading channel environment. The system parameters have been chosen as follows: subcarrier spacing $F_S = 3.85\text{kHz}$, maximum Doppler frequency $f_{D_{max}} = 2500\text{Hz}$, line-of-sight path (LOS) at $f_{D_{max}}$, scattered components with classical (Jakes) distribution (positive frequencies only), Rice-factor $K = 15\text{dB}$, exponential decreasing delay spread with $\tau_{max} = 7\mu\text{s}$, guard interval $T_g = 10\mu\text{s}$, carrier lock on LOS. Again it can be seen, that already for low SNRs and a short iteration path depth, the performance of the system improves significantly. For high SNRs, the BER can be decreased by approximately one decade.

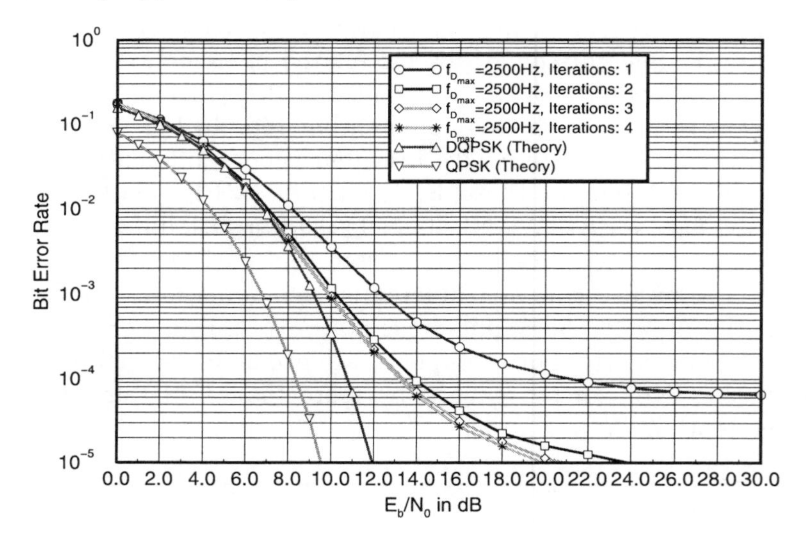

Figure 6 BER performance for Rician channel

6. CONCLUSIONS

A novel two-dimensional differential demodulation algorithm has been proposed that for the most part overcomes the degrading effects of DQPSK demodulation compared to QPSK demodulation. For the proposed scheme, neither a change in the transmitter nor knowledge about the transmitted information is necessary. Moreover, no information about the channel (in other words no channel estimation) is required. The efficiency of the algorithm can be adjusted by the iteration depth and the chosen criteria. With the new algorithm, significant performance improvements for AWGN and fading channels can be obtained even for a short iteration path depth compared to conventional one-dimensional DQPSK demodulation.

References

[1] ETSI ETS 300 401, "Radio broadcasting systems; digital audio broadcasting (DAB) to mobile, portable and fixed receivers," Feb. 1995.

[2] John G. Proakis, *Digital Communications*, McGraw-Hill, 3rd edition, 1995.

Block Turbo Coding in OFDM-based HIPERLAN/2 Systems

R. Gaspa Maynou[1], A. Hinrichs[2] and R. Mann Pelz[2]
[1]Universitat Politécnica de Catalunya, Barcelona, Spain: [2]Corporate Research and Development, Robert Bosch GmbH, D-31139 Hildesheim, Germany

Key words: Block turbo codes, OFDM, HIPERLAN/2

Abstract: Future local area networks operating in the 5 GHz frequency band will allow wireless, broadband access to a wide range of services supported by a variety of core networks world-wide. The harmonised radio transmission technique under consideration is based on a multicarrier scheme in conjunction with forward error correction based on convolutional codes, whereas the application of advanced coding schemes such as turbo codes are not precluded in the second phase of the standardisation process. This paper presents corresponding results regarding the applicability of two-dimensional block turbo codes in the case of a multicarrier-based wireless access system.

1. INTRODUCTION

High Performance Radio Local Area Networks Type 2 (HIPERLAN/2) is intended to provide high speed wireless communication between portable computing devices and broadband IP-, ATM- and UMTS-based core networks in the 5 GHz frequency band. HIPERLAN/2 is being standardised in Europe within ETSI BRAN [1], while equivalent standardisation activities can be found in the USA (IEEE802.11a) and Japan (MMAC). The future standard considers private or public systems with a centralised pico cellular network structure characterised by a coverage range of 50 m, user data rates in the range 6-54 Mbit/s and support of restricted user mobility (3 m/s) within the local service area. The typical operating environment is indoor.

The relevant layers of the air interface protocol being specified within ETSI BRAN are the Physical Layer (PHY), the Data Link Control Layer

(DLC) and parts of the Network Layer (Convergence Sublayer). As a medium access scheme Time Division Multiple Access (TDMA) in conjunction with Time Division Duplex (TDD) has been chosen. Radio transmission is based on an Orthogonal Frequency Division Multiplexing (OFDM) scheme in conjunction with various modulation techniques and forward error correction (FEC) based on punctured convolutional codes. In a later phase of the standardisation process advanced coding schemes such as turbo codes could be taken into account.

In [2] a convolutional turbo code (CTC) based on two systematic, recursive, parallel concatenated convolutional component codes in conjunction with a suboptimal decoding scheme characterised by an independent, soft-decision decoding of each component code within an iterative process were proposed. It was shown, that CTCs are capable of achieving a BER=10^{-5} at an E_b/N_0=0.7 in the case of an AWGN channel. This requires individual Soft-Input/Soft Output (SISO) decoders, which share reliability information, also called extrinsic information, in form of soft values. Examples of SISO algorithms for soft-decision decoding of convolutional codes are the maximum a-posteriori (MAP) algorithm [3], the SOVA [3] and the MAP suboptimal versions Log-MAP and Max-Log-MAP [4].

Analogous to CTCs systematic block codes (e.g. Hamming codes, BCH codes, RS codes) can be applied as component codes in a serial or parallel concatenation scheme. Of course the MAP algorithm can be applied for decoding if a trellis representation of the block code is considered [5]. Recently, a new SISO decoding algorithm based on the Chase algorithm has been proposed [6]. Although it performs suboptimal ML decoding, it is characterised by a relative low decoding complexity in comparsion to the MAP technique.

This paper deals with the evaluation of the performance of two-dimensional BTCs based on binary BCH codes in the case of a packet oriented, OFDM-based transmission system according to the future HIPERLAN/2 standard. Section 2 describes the considered HIPERLAN/2 air interface and the main characteristics of the applied BTCs. Corresponding performance results in the case of the considered FEC schemes and various system constellations are presented in Section 3, where the current HIPERLAN/2 FEC scheme consisting of punctured convolutional codes is taken as a reference.

2. TRANSMISSION SYSTEM

2.1 HIPERLAN/2 air interface

A TDMA/TDD medium access control scheme with a frame period of 2 ms has been specified within ETSI BRAN. Each frame includes a Downlink (DL) interval and an Uplink (UL) interval with a variable switching point for the support of asymmetric data traffic. The DL interval consists of a broadcast channel (BCH) and point-to-point user channels (DLCHs) with resource reservation, while the UL interval is subdivided in point-to-point user channels (ULCHs) with resource reservation and a Random Access Channel (RACH) based on S-Aloha. *Figure 1* depicts the underlying TDMA frame structure.

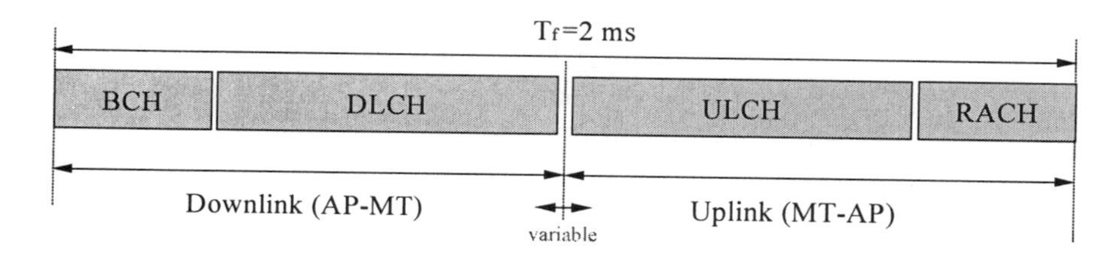

Figure 1. TDMA frame structure

Data transmission within a TDMA frame is accomplished in time bursts consisting of a preamble and payload. The preamble (payload) consists of a fixed (variable) number of OFDM symbols for the transmission of a training sequence (DLC data). Four different burst are specified, namely a broadcast burst, a DL burst and a UL burst with a short/long preamble. In the case of the DL/UL PHY burst 9 or 54 bytes of binary DLC data are mapped to N OFDM symbols of duration T_s. Radio transmission is based on an OFDM-based multicarrier scheme, where N_u=48 out of 64 subcarriers (FFT length) are used for data transmission. Channel estimation is assisted by means of 4 pilot subcarriers, while the remaining 12 subcarriers are set to zero. The subcarrier spacing equals 1/T=312.5 kHz. A guard interval of duration T_G=0.8 μs per OFDM symbol consisting of a cyclic prefix is inserted after the IFFT for compensation of intersymbol interference. The OFDM symbol duration equals $T_s=T+T_G$=4 μs. *Figure 2* depicts the key functional elements of the HIPERLAN/2 physical layer (signal generation).

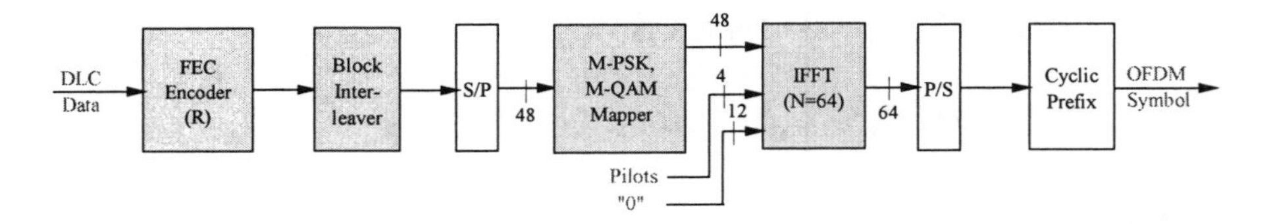

Figure 2. Functional elements of the HIPERLAN/2 physical layer

Link adaptation is accomplished through M-ary modulation in conjunction with punctured convolutional code of codes with code rate R and a block channel interleaver with a depth equal to one OFDM symbol. *Table 1* summarises the foreseen system modes, where the user data rate is given by $R_u=N_u(\log_2 M)R/T_s$.

Table 1. Link adaptation parameters (*54 byte DLC packet)

M-ary modulation	FEC code rate R	User data rate [Mbit/s]	OFDM symbols / payload*
2-PSK	1/2	6	18
	3/4	9	12
4-PSK	1/2	12	9
	3/4	18	6
16-QAM	9/16	27	4
	3/4	36	3
64-QAM	3/4	54	2

A channel raster of 20 MHz has been chosen in the 5 GHz frequency band. Additional techniques for enhancement of the system performance (e.g. reduction of the peak-to-average signal power ratio) have been proposed by some of the authors elsewhere [7].

2.2 Block turbo codes

In the considered case of a block interleaver and two component block codes the resulting BTC represents a product code, where the two linear, systematic block codes, (N_1,K_1,d_1) and (N_2,K_2,d_2), are applied to the rows and columns, respectively, of a two-dimensional array of binary information symbols. This representation is valid for the serial and parallel concatenation scheme, whereas in the latter case the checks on checks are not present. The structure also considers the case of two extended component block codes.

As a SISO decoding algorithm we consider in the sequel the suboptimal modified Chase algorithm [6], which minimises the probability of word or sequence error instead of the bit error probability as in the case of the MAP algorithm. The well known Chase algorithm allows soft-decision decoding of block codes in conjunction with an algebraic decoder (e.g. Berlekamp-Massey algorithm). An additional soft output computation is mandatory in order to apply this technique in an iterative (turbo) decoding process. It can be shown [6], that the soft-output equals the soft-input plus an additive term, which plays the role of the extrinsic information in CTCs. Each iteration corresponds to the decoding of rows (columns) followed by the decoding of columns (rows) of the product code. Instead of selecting the code word having the minimal Euclidian distance to the received code word as proposed in [6], we chose the code word minimising the corresponding correlation product. This metric yields a performance enhancement for the considered BTC schemes.

3. PERFORMANCE RESULTS

3.1 Simulation model

The performance of the BTC-based HIPERLAN/2 physical layer has been assessed by computer simulation of the point-to-point transmission link characterised by *Figure 2*, whereas ideal time and frequency synchronisation as well as ideal channel equalisation was assumed. Typical profiles for a time- and frequency-selective Rayleigh fading channel specified within ETSI BRAN with delay spreads in the range 50-250 ns and a maximum Doppler frequency f_{Dmax}=50 Hz (v=3 m/s) were considered. Since $f_{Dmax}T<<1$ we assume a time-invariant channel within a DLC data packet of length 54 bytes. An additive white Gaussian noise (AWGN) channel is additionally considered for reference purposes. For comparison of the different FEC schemes the $SNR=10\log_{10}(E_b/N_0)$ required to achieve a packet error rate $PER=10^{-2}$ is applied as a criteria. Iterative decoding is carried out without performance loss with 4 iterations.

3.2 Shortened and punctured block turbo codes

According to *Table 1* only specific user data rates in conjunction with dedicated code rates are allowed in the HIPERLAN/2 specification. This requirement can only be fulfilled through shortening and puncturing of BTCs obtained by concatenation of existing (extended) BCH codes. Herefore, a (N',K') BTC based on a (N,K) BTC is applied with K'<K and N'<N, respectively. *Table 2* lists the relevant BTCs considered and the underlying concatenation scheme.

Table 2. Considered block turbo codes

User data rate [Mbit/s]	(N_1,K_1,d_1) BCH (N_2,K_2,d_2) BCH	(N,K) BTC, R	(N',K') BTC, R'	Encoder structure
6	(31,21,6)	(961,441), R = 0.45	(864,432), R' = 0.5	Serial
27	(32,26,4)	(1024,676), R = 0.66	(768,432), R' = 9/16	Serial
36	(31,26,3)	(961,676), R = 0.70	(576,432), R' = 0.75	Serial

The necessary code matching affects the performance of the applied FEC schemes, since code shortening (puncturing) yields a reduction of the code rate (design distance) and therefore an enhancement (degradation) of the code performance. *Figure 3* shows the resulting degradation on an AWGN channel in the case of different punctured codes obtained from the serial concatenation of two identical extended (32,21,6) BCH codes. For a SNR=3 dB one can observe that the (824,512) BTC without puncturing achieves a BER=2.5×10^{-5}. As the code rate increases, and thus the puncturing increases, a significant degradation of the BER performance results. For the same SNR a BER=10^{-3} is obtained in the case of the (731,512) BTC characterised by 93 punctured symbols, while a (682,512) BTC with 192 punctured symbols yields a BER=1.5×10^{-2}.

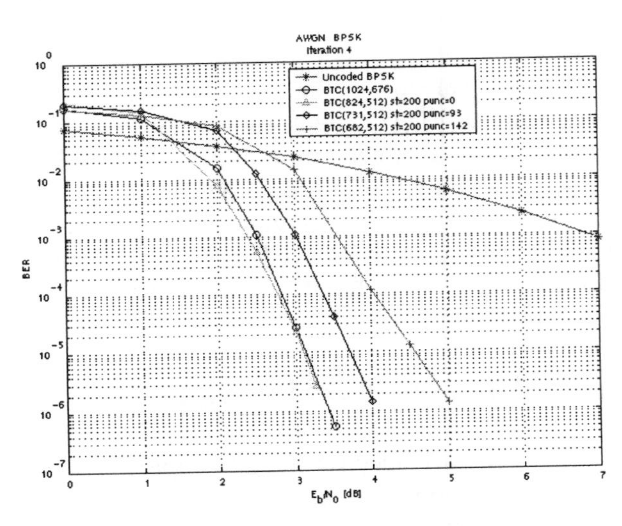

Figure 3. Performance degradation due to puncturing of BTCs

3.3 AWGN channel

As a reference the PER performance of BTCs and convolutional codes (CCs) on an AWGN channel with user data rates of 6, 27 and 36 Mbit/s is summarised in *Table 3*.

Table 3. Required SNR for achieving a PER = 10^{-2} on an AWGN channel

Net user data rate [Mbit/s]	Code rate R	SNR [dB], CC	SNR [dB], BTC
6	1/2	4.4	3.4
27	9/16	7.6	7.0
36	3/4	10.0	9.7

A coding gain of approximately 1, 0.4 and 0.3 dB can be achieved in the case of a user data rate of 6, 27 and 36 Mbit/s, respectively, through application of BTCs instead of a 64-state CC with the same code rate.

3.4 Rayleigh channel

The PER performance of BTCs and convolutional codes on a Rayleigh fading channel characterised by a delay spread of 150 ns for user data rates of 6, 27 and 36 Mbit/s is shown in *Figure 4*. In the case of a data rate of 36 Mbit/s, which implies a code rate equal to 3/4, the BTC outperforms the convolutional code. On the other hand the convolutional code outperforms the BTC in the case of low code rates.

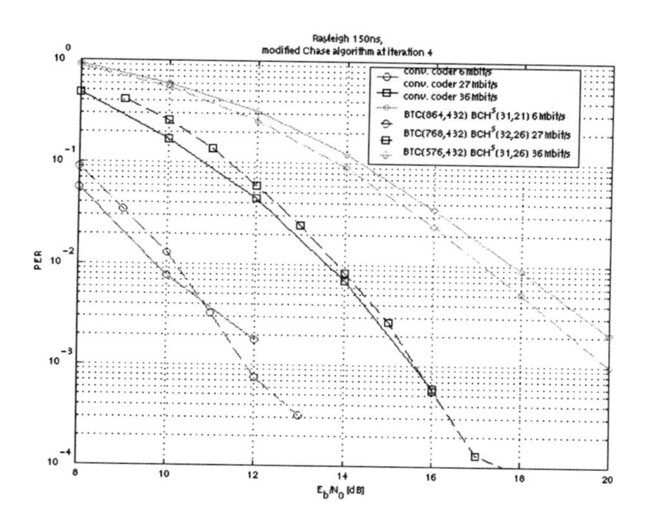

Figure 4. Performance on Rayleigh fading channel, delay spread = 150ns

4. CONCLUSIONS

The applicability of Block Turbo Codes based on extended BCH codes for future HIPERLAN/2-based wireless access systems has been investigated by means of computer simulations for various user data rates and propagation conditions. In the case of an AWGN channel BTCs exhibit a better performance than the standard convolutional codes with 64-states specified within ETSI BRAN at a low BER. The shortening (puncturing) required to meet the HIPERLAN/2 specifications yields an enhancement (degradation) in the BTC performance. Simulations results for different user data rates on Rayleigh fading channels showed, that BTCs outperform the specified convolutional codes only in the case of a high code rate (e.g. R=3/4). This characteristics of BTCs have been reported elsewhere. A strong constrain is the information packet length of 54 byte specified within ETSI BRAN. A fair comparison of BTCs and convolutional codes with the same code rate is not feasible, since changing the number of information symbols implies a different BTC code rate. It can be concluded, that the bad performance of BTCs on Rayleigh fading channels is due to the specified short packet length which precludes the application of turbo codes for HIPERLAN/2-based access systems.

REFERENCES

[1] DTS/BRAN030003, HIPERLAN Type 2 Functional Specification, ETSI BRAN, 1999

[2] C. Berrou, A. Gavieux, P. Thitimajshima, "Near Shannon limit error–correcting coding and decoding: Turbo–codes (1)", Proc. IEEE Int. Conf. on Communications, ICC'93, Geneva, pp. 1064-1070, May 1993

[3] J. Hagenauer, P. Hoeher, J. Huber, "Soft-Output Viterbi and Symbol-by-Symbol MAP Decoding: Algorithms and Applications", IEEE Trans. on Comm. Theory, pp. , June 1991

[4] P. Hoeher, "New Iterative ("Turbo") Decoding Algorithms", IEEE Int. Symp. Turbo Codes & Related Topics, Brest, pp. 63-70, September 1997

[5] J. Hagenauer, E. Offer, L. Papke, "Iterative Decoding of Binary Block and Convolutional Codes", IEEE Trans. Inf. Theory, vol. 42, no. 2, pp. 429-445, March 1996

[6] R. Pyndiah, "Iterative Decoding of Product Codes: Block Turbo Codes", IEEE Int. Symp. Turbo Codes & Related Topics, Brest, pp. 71-79, September 1997

[7] R. Mann Pelz, H. Schmidt, A., Hinrichs, "Proposal for HIPERLAN Typ 2 Physical Layer", ETSI BRAN, Sophia Antipolis, July 1998

The Performance Analysis of Multi-Carrier CDMA Systems Using Different Spreading Codes in Frequency Selective Fading Environment

Hongnian Xing[1], Jukka Rinne[2], and Markku Renfors[2]
1. Nokia Mobile Phones, Ltd. PL 407, 00045 Nokia Group, Finalnd Email: Hongnian. Xing@nokia.com, hongnian@cs.tut.fi : 2. Tampere University of Technology, Digital Media Institute, Telecommunications, P. O. Box 553, 33101, Tampere, Finland

Key words: multi-carrier CDMA, guard interval, crest factor.

Abstract: In this paper, we investigate the system performance of multi-carrier CDMA (MC-CDMA) systems in both downlink and uplink cases. Some simulation results of using different spreading codes for an MC-CDMA system with $\pi/4$ QPSK modulation are given in urban mobile environment. Two kinds of spreading codes, Walsh sequence and Zadoff-Chu sequence are investigated. From the simulation results, we find that in the downlink and synchronized uplink cases; the Walsh sequences have a good performance since they are completely orthogonal in the code set. In asynchronized uplink case, the Zadoff-Chu codes may be selected.

1. INTRODUCTION

The idea of multi-carrier CDMA (MC-CDMA) appeared around 1993 [1]. In this technique, multi-carrier modulation (also called orthogonal frequency division multiplexing- OFDM) is combined with spread spectrum idea (direct sequence spread spectrum) to realize the multiple access function. In mobile environments, each subcarrier is affected by the inter-symbol interference (ISI) and inter-channel interference (ICI). In downlink cases (data symbols from different users are synchronized and combined together), the output sample from a subcarrier is only attenuated by a complex factor, if the guard interval is used and longer than the maximum delay spread of the multipath slow fading channel. That is, there is no ISI in the subcarrier and no ICI between subcarriers. So the equalization for the complex attenuation factor can be avoid by using differential PSK (DPSK) signals, since the phase difference between two consecutive symbols, which brings the data information, is not severely distorted. From this point of view, DPSK signal is suitable for synchronized multi-carrier CDMA systems in slow fading environment.

Like in the case of traditional direct sequence CDMA (DS-CDMA) systems, spreading code design is an important aspect, which affects the system performance significantly. The use of guard intervals can relax the restriction of auto-correlation property of the codes for MC-CDMA systems, since the ISI due to the multipath propagation can be "absorbed". On the other hand, the codes for MC-CDMA systems should be able to reduce the crest factor; a unique drawback of OFDM based signal [2].

In Section 2, we describe briefly the multi-carrier CDMA transmitter, receiver, and the channel structures. A brief analysis of the system performance is also included in this section. The spreading code design issue is discussed in Section 3. The descriptions of simulations and simulation results are given in Section 4. Section 5 is the conclusions.

2. THE SYSTEM DESCRIPTIONS

2.1 Transmitter and Receiver Models

The baseband MC-CDMA transmitter model is shown in Fig. 1. The binary data is first mapped to DQPSK symbols, $X_{i,j}$. Here i is the user index and j is the index of data symbol. Then $X_{i,j}$ is replicated into N parallel copies. Each branch of the parallel stream is multiplied by a chip of a user-specific sequence P_i of length N, and then modulated to a subcarrier spaced $1/T_s$ apart from its neighboring subcarriers. T_s is defined as the symbol duration in time domain. IDFT is the multicarrier modulation. The duration of T_s is usually called OFDM symbol duration. N is the total number of subcarriers (length of IDFT).

After multi-carrier modulation and P/S conversion, OFDM symbols are generated. Before sending them to the channel, a cyclic extension is added at the head of the symbol as a guard interval. The aim of guard intervals is to remove ISI and preserve the orthogonality of the subcarriers.

The sampled version of transmitted OFDM symbol (when jth data symbol $X_{i,j}$ is transmitted) is given as

$$x_{i,j} = IDFT(X_{i,j}P_i(k)) = \sum_{k=0}^{N-1} X_{i,j}P_i(k)e^{\frac{j2\pi kn}{N}} \quad (1)$$

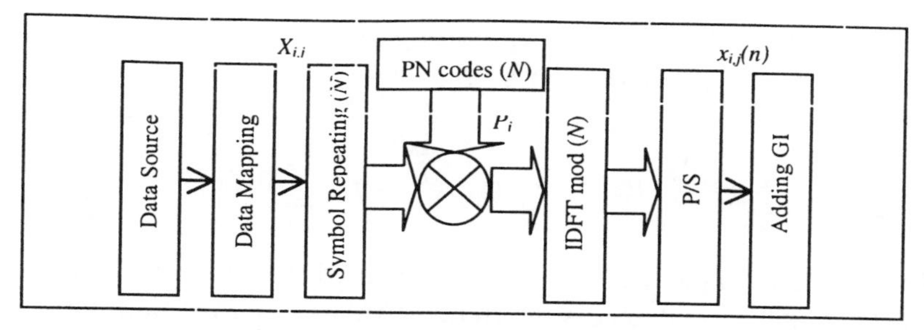

Figure 1. The simplified baseband MC-CDMA transmitter model.

At the receiver side, DFT is the multi-carrier demodulator. After despreading operation (with local sequence P_l^*), the data symbol from all subcarriers are summed together and then fed to the DQPSK slicer. This acts just as a correlator, not only for separating different users, but also for collecting the same symbol information from all subcarriers.

2.2 Channel Model

The channel is described as a multipath fading channel with the discrete impulse response as

$$h(n) = \sum_{m=0}^{M-1} a_{m,n} \delta(m-n) \tag{2}$$

In this case, the channel has a finite impulse response of length M, and $a_{m,n}$ is the attenuation factor for path m at time moment n. For slow fading channels, $a_{m,n}$ can be viewed as a constant during several (or at least one) OFDM symbols, given as,

$$a_{m,n} = a_{m,0} = a_{m,1} = \ldots = a_m \tag{3}$$

2.3 Performance Analysis of a MC-CDMA System

The received signal of the 0th mobile user ($I=0$) can be given as

$$r_{0,j}(n) = (x_{0,j}(n) + \sum_{i=1}^{N-1} x_{i,j}(n)) * h(n) + w(n) \tag{4}$$

where $w(n)$ is the sampled version of AWGN noise and $*$ is the convolution operator. After sum operation, the recovered signal is then

$$X'_{0,j} = \sum_{k=0}^{N-1} (P_0^*(k) DFT(r_{0,j}(n))) \tag{5}$$

If (3) is hold, the (5) can be rewritten as [3]

$$X'_{0,j} = X_{0,j} \sum_{k=0}^{N-1} (P_0(k) P_0^*(k) \sum_{m=0}^{M-1} a_m e^{\frac{-j2\pi mk}{N}})$$
$$+ \sum_{i=1}^{I-1} (X_{i,j} \sum_{k=0}^{N-1} (P_i(k) P_0^*(k) \sum_{m=0}^{M-1} a_m e^{\frac{-j2\pi mk}{N}})) + \sum_{k=0}^{N-1} W(k) \tag{6}$$

The first term in the pervious equation is the distorted desired signal. The second term is the multi access interference (MAI), and the third term is the AWGN noise, which is given as

$$W(k) = \frac{1}{N} \sum_{n=0}^{N-1} w(n) e^{\frac{-j2\pi nk}{N}} \tag{7}$$

Without MAI, the symbol error rate is very low, due to RAKE receiver structure. MAI is the main term to affect the system performance, and it depends on the cross-correlation property of the code sets, which could be represented as

$$R = \sqrt{\frac{1}{N-1} \sum_{i=1}^{N-1} \left| \frac{\sum_{k=0}^{N-1} (P_0(k) P_0^*(k) \sum_{m=0}^{M-1} a_m e^{\frac{-j2\pi mk}{N}})}{\sum_{k=0}^{N-1} (P_i(k) P_0^*(k) \sum_{m=0}^{M-1} a_m e^{\frac{-j2\pi mk}{N}})} \right|} \tag{8}$$

Without distortion, $R=\infty$ if the completely orthogonal code sets are used. As we have seen from the equation, the frequency selectivity of the multipath channel destroys the code orthogonality. This kind of distortion is difficult to be estimated and compensated by a single filter. So the equalization may be necessary for each subcarriers. Also the spreading codes should be chosen to minimize the R, when there exists distortions.

The case is different in uplink. The received signal in uplink is given by

$$r_{0,j}(n) = x_{0,j}(n) * h_0(n) + \sum_{i=1}^{I-1} x_{i,j}(n - \tau_i) * h_i(n) + w(n) \tag{9}$$

where each user has its own channel impulse response $h_l(n)$ and propagation delay τ_l (in samples). In slow fading channels, the MAI can be given as

$$MAI = \sum_{i=1}^{I-1} X_{i,j} \sum_{k=0}^{N-1} (P_i(k) P_0^*(k) e^{\frac{-j2\pi k\tau_i}{N}} \sum_{m=0}^{M_i-1} a_{m,i} e^{\frac{-j2\pi mk}{N}} \tag{10}$$

It is shown that both the user dependent channel impairments and the delays between the desired user and other users destroy the code orthogonality. For synchronized uplink case, there is no delay effect, so the code orthogonality is only affected by the user dependent channel impairments.

3. THE SPREADING CODE SELECTION FOR MC-CDMA SYSTEMS

The code selection for MC-CDMA systems is based on several aspects. First of all, the codes should have good cross-correlation property. If we do not consider the crest factor effect, this is then the main requirement for the downlink codes (also for the synchronized uplink codes).

For the asynchronized uplink case, the periodic cross-correlation properties of the codes should be considered. That is, the cross-correlation between two codes should have low peak value, not only for the synchronized case, but also for the case when there is a random phase shift between two codes.

Unlike DS-CDMA systems, in either downlink or uplink case, the auto-correlation property is not critical since the ISI due to multipath propagation could be "absorbed" by guard intervals. This is an obvious advantage of MC-CDMA systems; since even the set of codes is optimal for both auto-correlation and cross-correlation properties (reaching the Welch band), it can not be completely orthogonal.

Minimizing the crest factor is another aspect of the code design for MC-CDMA systems. Due to the high peak-to-average power ration (PAPR), the MC-CDMA (or other OFDM based) signal is very sensitive to nonlinear effect at the power amplifier. In MC-CDMA systems, the signal is repeated to all subcarriers, a possibility is then given for the code design to reduce the PAPR [4].

Two kinds of codes are investigated by simulations. A brief description for them is given in the following subsections.

3.1 Walsh Code

Walsh code is a well-know orthogonal code set, generated by the basic Walsh matrix. Although its auto-correlation and periodic cross-correlation properties are not optimal, it is chosen due to its completely orthogonal property. The code is also easy to be implemented.

3.2 Zadoff-Chu Code

The Zadoff-Chu codes are the special case of the Generalized Chirp-Like sequences having the ideal periodic auto-correlation function and the optimum periodic cross-correlation function, given as

$$P_i(n) = \begin{cases} e^{\frac{j2\pi i}{N}(\frac{n^2}{2}+qn)} & N \quad even \\ e^{\frac{j2\pi i}{N}(\frac{n(n-1)}{2}+qn)} & N \quad odd \end{cases} \tag{11}$$

where N is the length of the code, and it should be a prime number, q is any integer, and $r=1,2,\ldots,N-1$ is the code index. Any pair of codes in such a set has the optimum periodic cross-correlation function, with the constant magnitude equal to the square root of N. An another main advantage is that the crest actor can be reduced significantly by using this kind of codes [4].

4. SIMULATION RESULTS

4.1 Simulation Conditions

The performance of an MC-CDMA system based on Fig. 1 and Fig. 2 is investigated by simulations in multipath fading environment. The main simulation parameters are given in the following table.

Table 1. The main simulation parameters.

Items	Parameters and values
System sampling frequency f_s	9.8304 MHz
Chip rate	9.8304 Mcps
Symbol rate	76.8 Ksps (T_s=13.02̇μs)
Carrier number	128 (same as the spreading factor)
Data mapping	π/4-QPSK
Spreading codes	Walsh code (length 128) and Zadoff-Chu code (length 127: first subcarrier is empty)
Guard interval	T_g=2.034μs
Multicarrier modem	IDFT-DFT pair (length 128)
Simulation length (number of symbols)	10000

The channel in the simulation is built according to the UMTS channel models, main parameters are given below.

Table 2. The main channel parameters.

Tap	Delay (μs)	Average Power (dB)	Doppler Spectrum
1	0.00	0.0	CLASSIC
2	0.31	-1.0	CLASSIC
3	0.71	-9.0	CLASSIC
4	1.09	-10.0	CLASSIC
5	1.73	-15.0	CLASSIC
6	2.50	-20.0	CLASSIC

The "CLASSIC" means that the channel is a Reylaigh fading channel. The simulation tool is Signal Processing Workstation - SPW™.

4.2 Simulation Results

Some simulation results are given in Figure 2. Two uplink cases, synchronized uplink and asynchronized uplink, are considered in the simulation. For uplink multi-user simulations, each user has an independent channel (the noise seed to generated Doppler spectrum is independent of users and paths), although the channel structures are the same. Users are chosen randomly, and their delays are assumed to have uniform distributions over an extended OFDM symbol duration.

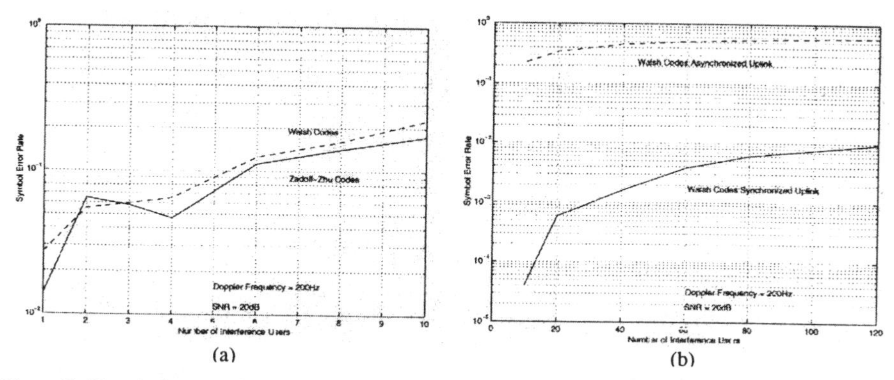

Figure2. Simulation results: (a) the comparison of two code sets in small MAI, asynchronized uplink. (b) the comparison of Walsh codes in synchronized and asynchronized uplink.

Figure 2 (a) shows the asynchronized uplink performance of two different code sets. They have quite similar performance, although the reasons of causing this kind of performance are quite different. For Walsh code set, the performance degradation is caused by the bad periodic cross-correlation property (due to the delays). The performance degradation in Zadoff-Chu codes is mainly due to the non-complete orthogonality between different codes. In fact, Zadoff-Chu codes have kind of the same performance in either asynchronized uplink or synchronized uplink cases. Considering the ability of reducing the crest factor, Zadoff-Chu codes may be selected for the asynchronized uplink transmission, when there is not a large number of active users.

When IDFT-DFT pair is used as a modem, there comes a problem of using Zadoff-Chu codes, since the length of Zadoff-Chu codes must be a prime, which is in general not equal to the DFT length. In this case, some of the subcarriers are not used. But due to the random channel delay, the values in those subcarriers are possibly joined to the code correlation computation at the receiver, which will destroy the original code cross-correlation property.

Since the periodic auto-correlation property is not critical in MC-CDMA systems, the uplink codes are selected mainly based on the periodic cross-correlation property, and the ability of reducing the crest factor.

From Figure 2(b), we find that the Walsh codes have a rather bad performance in asynchronized case with strong MAI interference. The Zadoff-Chu codes, which are not plotted in the figure, have a very similar performance (in either asynchronized or synchronized uplink cases). So in general, it can not be used for the synchronized uplink applications if MAI exists. The significant improvement can be found by using the Walsh code, when all users are synchronized in uplink case. Combined with results given in [3], it can be conclude that Walsh code is a suitable code candidate for MC-CDMA downlink and synchronized uplink applications, regardless of the consideration of the crest factor reduction.

5. CONCLUSIONS

An MC-CDMA-DQPSK system with guard intervals is investigated for two different code sets. The Walsh codes, which are completely orthogonal with each other, are a suitable code candidate for the synchronized link applications. The Zadoff-Chu codes, which are able to reduce the crest factor of the MC-CDMA signal, can be used for asynchronized uplink, with a small MAI. Due to the use of guard intervals, the optimum codes for MC-CDMA asynchronized uplink applications may be the codes which have the optimum periodic cross-correlation property. When considering the crest factor reduction, the best codes are the trade-off of periodic cross-correlation property and crest factor reduction ability.

6. REFERENCE

[1]. N.Yee, J.P.Linnartz and G.Fettweis, "Multi-Carrier CDMA in Indoor Wireless Radio Networks," Proceedings PIMRC'93,pp.109-113, Yokohama, Japan.

[2]. S. Boyd, "Multitone Signals with Low Crest factor," IEEE Trans. on Circuits and Systems, Vol. CAS-33, No.10, pp.1018-1022, Oct., 1986.

[3]. Hongnian Xing, Petri Jarske,and Markku Renfors, "The Performance Analaysis of Multi-Carrier CDMA systems in Multipath Fading Environments," Proceedings of SPAWC, pp.245-248, France, 1997.

[4]. Branislav M. Popovic, "Spreading Sequences for Multi-Carrier CDMA Systems," Printed and published by the IEE, Savoy Place, London WC2R 0BL, UK.

Section IV

DETECTION AND MULTIPLEXING

EACH CARRIER TRANSMISSION POWER CONTROL WITH ANTENNA-CARRIER DIVERSITY FOR OFDM DS-CDMA

Sigit P.W. Jarot and Masao Nakagawa
Department of Electrical Engineering
Graduate School of Science and Technology, Keio University
3-14-1 Hiyoshi, Kohoku-ku, Yokohama 223-8522 Japan
sigit@nkgw.ics.keio.ac.jp ; nakagawa@nkgw.ics.keio.ac.jp

Abstract An Each Carrier Transmission Power Control with Antenna-Carrier Diversity is investigated for the reverse link of OFDM DS-CDMA system. Power control is used to compensate for non uniform attenuation level at each subcarrier due to frequency selective fading by assigning different transmission level in each subcarrier. And antenna-carrier diversity is used to minimize reception of strong attenuated signals which unable to be compensated by power control because of power limitation in mobile station. It was confirmed through computer simulation that proposed system achieved better performance in terms of bit error rate probability.

Keywords: power control, frequency selective fading, diversity, OFDM, multicarrier CDMA

1. INTRODUCTION

Combinations of CDMA(Code Division Multiple Access) and multicarrier modulation have achieved increasing attention for mobile communication application, in particular for next generation system which enable to scope not only voice communication but also image and data communications. This is partly due to its possibility to respond the need of high speed, high reliability and high flexibility mobile communication system. High speed transmission may be realized by transmitting information symbol over several low speed (narrow band) subcarriers which independently fading. While, high flexibility may be realized by changing the number of subcarriers, depending on its contents. These systems allow to combine the advantages of multi-carrier modulation, as mentioned above, with the advantages of CDMA system, such as high multiple access capability and robustnest against fading.

Basically Multi-Carrier(MC) combined CDMA can be classified into two type. One is "Copy type" which is widely known as MC-CDMA. And the other one is "Serial to Parallel(S/P) type" which is widely known as MC-DS-CDMA. We deal in this paper with the later one, MC-DS-CDMA. In this system, serial to parallel (S/P) converted data are DS-SS modulated in time domain using a user specific spreading code, and these signals are transmitted over several low rate (narrow-band) subcarriers in parallel. This system is effective for establishment of a quasi-synchronous channel, which is suitable for reverse link. Many methods have been proposed in order to improve system performance of this system, with many different approaches, such as novel schemes [1; 2], coding[3], multicode[4].

In our previous work, we have investigated transmission power control applications in MC-DS-CDMA system[5; 6]. This work was motivated by the fact that power control is a crucial method to realize the high multiple access capability of conventional CDMA system[7] and this method is also known as a powerful anti-fading strategy[8]. On the other hand, we considered that power control is very potential to upgrade system performance of MS-DS-CDMA, through compensating for non uniform attenuation level at each subcarrier due to frequency selective fading, which is one of the most important characteristics of MS-DS-CDMA system. It was confirmed through computer simulation that Each Carrier Transmission Power Control (TPC) can improve the system performance in terms of bit error rate(BER) performance. However, the major problem in applying this method is that, there is a limitation in power amplification factor due to power limitation in mobile station. Thus, in the case of subcarriers subject to strong fading, power control is unable to counter measure by transmitting signals with large power. This condition is potential to degrade system performance.

In this paper, we consider to solve such problems by introducing Antenna-Carrier Diversity reception[9]. This system allows base station (receiver side) to select the best combination between the carriers and antennas after receiving OFDM signals from mobile station (transmitter). With multi-antenna, and also multi-carrier reception, the probability of strong fading occurance in a subcarrier will be smaller. It will bring to improvement of measurement accuracy and effectiveness of Each Carrier TPC. We investigated system performance in term of Bit Error Rate (BER) Performance as a function of Doppler frequency, number of users, number of branches, and updating period.

This paper is organized as follows. Section 2 describes system model and the mechanism of proposed system. In Section 3., the results of performance evaluation through computer simulation are presented. And finally, a conclusion of this paper is given in Section 4.

2. SYSTEM DESCRIPTION

2.1 OFDM DS-CDMA SYSTEM

In this paper, we consider the reverse link of OFDM-DS-CDMA system. The transmitter scheme is depicted in Fig.1(a). In transmitter side, user's binary data stream of rate R and bit duration T_b is serial to parallel (S/P) converted into M substreams. The rate and bit duration in each stream are $\frac{R}{M}$ and $T(=M \cdot T_b)$, respectively. Differential Binary Phase Shift Keying (DPSK) is utilized in each substream for data modulation. The modulated stream is then DS-SS modulated using a user specific spreading code in time domain, with N chips length, such that $T = NT_c$ where T_c is chip duration. And then, DS-SS signals are transmitted over one of the narrowband subcarrier. We assume that each subcarrier is orthogonally overlapped, which is known as OFDM(Orthogonal Frequency Division Multiplexing). And we use IDFT (Inverse Discrete Fourier Transform) for OFDM signal generator.

It is important to be note, that transmission power of DS-SS signals in each subcarrier are adjusted periodically, based on transmission power control(TPC) commands received from Base Station. And the updating period T_{tpc} is decided by base station. The process to produce this TPC commands will be described in following subsection.

2.2 EACH CARRIER TRANSMISSION POWER CONTROL

In OFDM-DS-CDMA system, transmitted signals are often subjected to frequency selective fading due to multipath transmission. However, since OFDM system uses several narrowband sub-carriers, instead of single wideband carrier, each sub-carrier is subjected to frequency non-selective fading. And fluctuation in each subcarrier is monitorable for

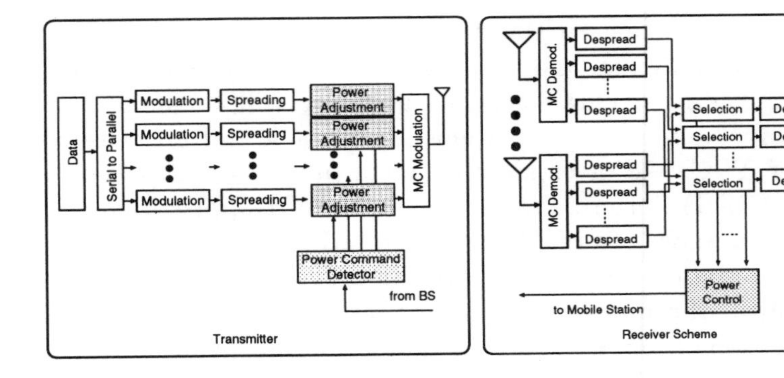

(a) Transmitter Scheme

(b) Receiver Scheme

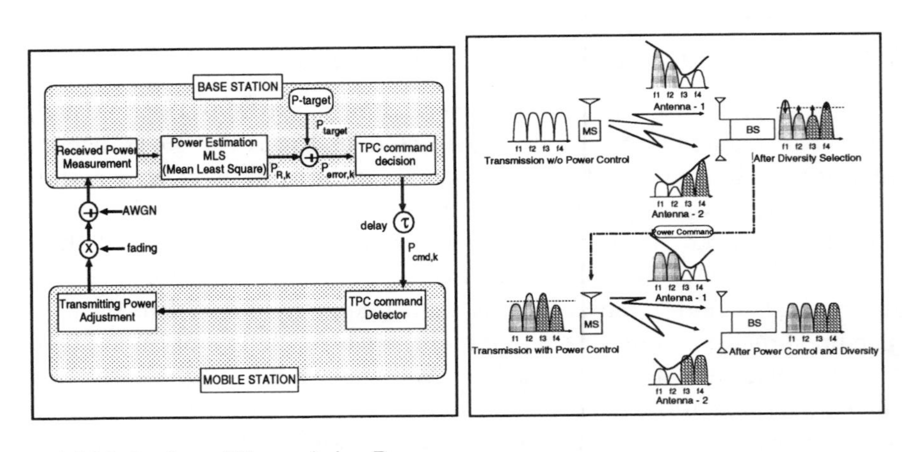

(c) Mechanism of Transmission Power Control

(d) Illustration of Proposed System

Figure 1 System Model and Proposed System

receiver side. We consider that power control which compensates for attenuation in each subcarrier is very potential to improve system performance. The block diagram is depicted in Fig.1(c) and its position in transmitter and receiver are depicted in Fig.1(a) and (b) respectively.

In Each Carrier Transmission Power Control(TPC) system, receiver always monitors and measures the power level of received signals. Based on this measurement result the power level of the reception is estimated using Mean Least Square(MLS) algorithm. This algorithm is also to minimize the effects of background noise in power measurement process. All these process should be performed after despreading, in order to determine target user's signals from interference signals.

Finally, TPC commands are produced based on comparison results between estimated power and target level. Target level is a power level of received signal when there is no distortion due to fading, and this value is maintained at base station. If estimated power is smaller than target level, then the TPC command is to update the next transmission power by increasing it 1 dB, and vice versa. Because this system controls transmission power at each sub-carrier, TPC commands should contain power increment or decrement for every subcarriers. In transmitter side, these power commands will be received by Power Command Detector.

However, because of power limitation in mobile station the dynamic range of transmission power is limited, there will be a condition in which mobile station is unable to increase transmission power anymore, though a subcarrier is subjected to relatively strong attenuation. This condition is potential degrade system performance if it occurs frequently. In this papar, we introduce antenna-carrier diversity to minimize such condition.

2.3 ANTENNA-CARRIER DIVERSITY

Antenna-carrier diversity is realized through multi-antenna and multi-carrier reception, as depicted in Fig.1(b). Base station receives OFDM signal from multiple antenna, which space far enough to obtain independent fading. Thus, different spectrum pattern will be observed at each antenna. Base station selects the best combination between antennas and subcarriers using a simple Selection Combining algorithm. For instance, in condition as showed in Fig.1(d), the best combination is $\{(f_1, a_1), (f_2, a_1), (f_3, a_2), (f_4, a_s)\}$, where (f_k, a_n) is the signals of subcarrier k received from antenna n.

With antenna-carrier diversity, the effectiveness of Each Carrier TPC will be improved, because TPC only compensates for attenuation at subcarriers which subject to relatively smaller.

3. PERFORMANCE EVALUATION

Montecarlo simulations have been performed to evaluate the performance of proposed system. And conventional system is defined as an Each Carrier TPC system without diversity reception. And Error Correction Coding is not used in this evaluation. We consider a reverse link with frequency selective fading channel due to multipath propagation. It is modelled as two-ray Rayleigh fading channel with uniform power delay profile. Over an observation period, each transmitter is assumed to move continuously within a small geographic area, so the effects of long term fading due to path loss and shadowing are negligible. And for multiuser environment, we assume that each user is subjected to independent fading, with asynchronous transmission. Simulation model is described in Table 1.

Table 1 Simulation Model

Data modulation	DBPSK
Spreading sequence	15 chip Gold Sequence
No of sub-carriers	8 carriers
Bit rate per sub-carrier	64 Kbps
Channel model	2-ray Rayleigh fading
Number of branches	1,2,4
Combining method	Selection Combining

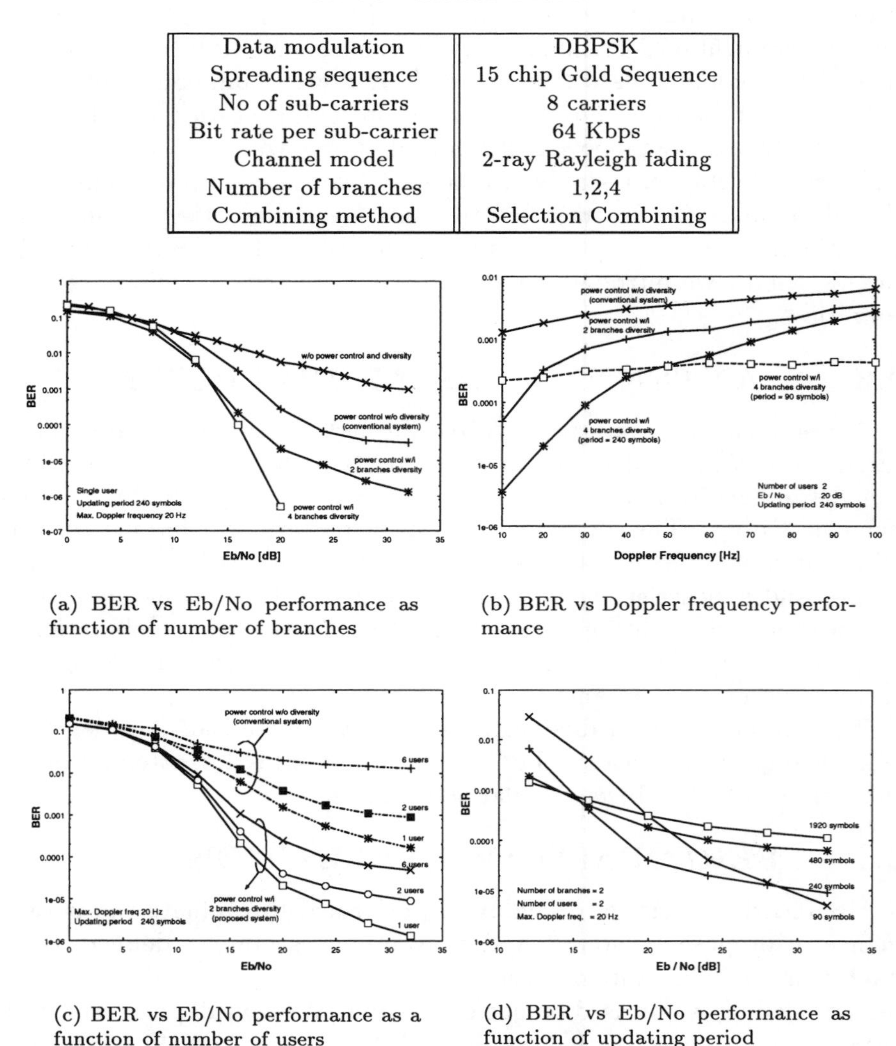

(a) BER vs Eb/No performance as function of number of branches

(b) BER vs Doppler frequency performance

(c) BER vs Eb/No performance as a function of number of users

(d) BER vs Eb/No performance as function of updating period

Figure 2 Simulation Results

In Fig.2(a), BER vs Eb/No performance of the proposed system with 2 and 4 branches diversity are compared with conventional system and also with a system without power control and diversity strategy. In this simulation, maximum Doppler frequency is assumed to be 20 Hz, and transmission power updating period is updated every 240 symbols transmission. It is shown that proposed system performed better in particular for larger Eb/No. This result can be expained by the effect of background noise at the measurement process of received signals. Poor accuracy in measurement may affect to accuracy of power command decision. The effects of background noise can be minimized by lengthening updating period, as presented in Fig.2(d). The longer updating period will lead to minimization of noise effects by averaging the power of received signals. However, the robustness against fading will degrade. Because, system with long updating period may unable to adapt channel condition which changes fastly.

In Fig.2(b), BER vs Doppler frequency performance is presented. In this simulation, updating period of conventional system is 90 symbols. For updating period 240 symbols, performance of the proposed system degraded as Doppler frequency increased. While for updating period 90 symbols, degradation became not significant. It can be seen that proposed system outperformed conventional system for both slow dan fast fading. Moreover, the shorter updating period are more suitable for fast fading. In Fig.2(c), BER vs Eb/No performance as a function of number of users is presented. This simulation is to evaluate the effects of multiple access interference(MAI) in performance of proposed system. In this simulation, updating period is assumed as 240 symbols and maximum Doppler frequency is 20 Hz. It can be seen that the existence of MAI affects performance of both proposed and conventional system, however proposed system always outperformed conventional system.

4. CONCLUSION

In this paper, we have proposed Each Carrier Transmission Power Control(TPC) with Antenna-Carrier Diversity for the reverse link of OFDM DS-CDMA system, in order to compensate for non uniform attenuation of each subcarrier power level due to frequency selective fading. And Antenna-Carrier diversity is introduced to minimize reception of strong attenuated signals which unable to be counter-measured by Each Carrier TPC. System performance has been evaluated for various number of users, various Doppler frequency and various updating period. The result of computer simulation clearly shows that proposed system improved system performance significantly in terms of BER performance.

Finally, we conclude that Antenna-Carrier diveristy is required to enhance Each Carrier Transmission Power Control, and proposed system is effective to improve the performance of OFDM DS-CDMA system.

For further works, it is important to investigate the application of proposed system in other Multi-Carrier combined CDMA schemes.

References

[1] Essam A. Sourour and Masao Nakagawa. "Performance of Orthogonal Multicarrier CDMA in a Multipath Fading Channel". *IEEE Trans. on Commun*, 44(3):356–367, March 1993.

[2] H. Matsutani and M. Nakagawa. "Multicarrier DS-CDMA with Frequency Spread Coding". *Proceeding of IEEE ICPWC*, pages 244 – 248, February 1999.

[3] H. Atarashi and M. Nakagawa. "Packet Combining ARQ Scheme with Adaptive Data Order Rearragement for Multi-Carrier CDMA Systems". *Proc. of 3rd Asia Pacific Communications Conference*, pages 41 – 45, December 1997.

[4] D.Takeda, H. Atarashi and M. Nakagawa. "Orthogonal Multicode OFDM DS-CDMA System Using Partial Bandwidth Transmission". *IEICE Trans. on Commun*, E.81-B(11):2183 – 2190, November 1998.

[5] Sigit P.W Jarot and Masao Nakagawa. "Transmission Power Control for OFDM-DS-CDMA System". *Technical Report of IEICE*, SST-98(53):29–35, Desember 1998.

[6] Sigit P.W Jarot and Masao Nakagawa. "Transmission Power Control Techniques for the Reverse Link of OFDM-DS-CDMA System". *Proceeding of IEEE Symposium on Computer and Communiations*, pages 331 – 337, July 1999.

[7] Savo Glisic, Blanka Vucetic. *"Spread Spectrum Code Division Multiple Access Systems for Wireless Communication"*. Artech House Publisher, 1997.

[8] Sirikiat Ariyavisitakul, Li Fung Chang. " Signal and Interference Statistics of a CDMA System with Feedback Power Control". *IEEE Trans. on Commun.*, 41(11):1626–1634, November 1993.

[9] H Takahashi and Masao Nakagawa. "Antenna and Multi-Carrier Combined Diversity System". *IEICE Trans on Communication*, E79-B(9):1221–6, September 1996.

[10] V. DaSilva and Elvino Soussa. "Performance of Orthogonal CDMA codes for quasi-synchronous communication systems ". *Proc. ICUPC'93*, 2:995–999, October 1993.

A WAVELET-BASED MULTICARRIER SPREAD SPECTRUM SYSTEM WITH CONSTANT POWER

Samir Attallah

Centre for Wireless Communications, National University of Singapore, Kent Ridge Crescent, Singapore 119260, Singapore.

cwcsa@leonis.nus.edu.sg

Teng Joon Lim

As above

cwclimtj@leonis.nus.edu.sg

Abstract A new method for designing sequences which lead to a constant envelope in Wavelet-based Multicarrier spread spectrum systems is presented. It is shown that this method is valid for any unitary or orthogonal matrix and can generate sequences with quantized coefficients, which are suitable for fixed-point DSP implementations.

1. INTRODUCTION

Orthogonal frequency division multiplexing (OFDM) is well known [1] for its robustness to frequency-selective fading and its ease of implementation, but is not suitable for providing multiple access on the uplink of a cellular system, where users transmit in an unco-ordinated manner. Direct-sequence spread spectrum (DS-SS) [2, 3] modulation on the other hand naturally allows for multiple access through the use of different spreading codes for each user, giving rise to the code division multiple access (CDMA) concept. However the very high sampling rates required in DS-CDMA devices operating at high bit rates create potential implementation problems, which may be solved by parallel transmission of the chips over several orthogonal channels. Hence there has been a fair bit of interest [4] in hybrid multicarrier CDMA (MC-CDMA) systems.

Of the several possible MC-CDMA configurations, a well-known one [5] is to send the same data symbol over all subchannels. Each user in the system is assigned a unique identification code sequence, equal in length to the number of subchannels. The signal at the output of the MC modulator is then "scrambled" by the code before transmission. This is the system that is studied in this paper.

A major problem that remains to be effectively tackled in all multicarrier systems is that of high peak to average power ratio (PAR). Briefly, the transmitted signal has an envelope with a very large dynamic range because it is a sum of many sinusoids with random phases. High PAR makes such systems very sensitive to the nonlinearity of the power amplifier [6] and may saturate the digital to analog (D/A) converter in the system and cause clipping [7], leading to a degradation in performance.

In the literature, there have been few articles addressing this issue in a satisfactory manner. Chow *et al.* in [8] proposes two methods – block scaling and spectral shaping – which are ad-hoc and, in the case of block scaling, reduces the probability of clipping without reducing the PAR. Furthermore, their implementation requires the transmission of scaling information, which thus reduces bandwidth efficiency or approximate spectral shaping using a non-linear clipping function, which introduces information loss. The method to be proposed in this paper suffers from neither of these problems.

The proposed method is based on assigning to each symbol a vector which is designed to ensure that the transmitted signal has a constant envelope (CE). To transmit M bits, we would need 2^M sequences and symbol modulation is accomplished through a look-up table. The methods used to design such sequences are in general intimately related to the type of orthogonal transform[1] used in the system [9, 10], and are either difficult to derive for certain transforms or the number of useful CE sequences obtained is very limited by practical considerations. In this paper we give a general method for designing any number of CE sequences which can be used with any unitary transform. Moreover, the sequences may be constrained to satisfy additional constraints, such as maximizing the minimum distance between two signal constellation points, while retaining their CE property.

Finally, the use of the discrete cosine transform (DCT) and other subband filter banks in OFDM systems has recently been shown to offer some advantages over FFT in terms of intersymbol and interchannel interferences [11, 12]. In this paper we will use the discrete orthogonal wavelet transform to illustrate the proposed algorithm, but extension to any other transform is trivial.

2. ALGORITHM DESCRIPTION

2.1 A METHOD TO GENERATE CE VECTORS

We start by considering a DCT-based OFDM system and then generalize this result to any other orthogonal transform. Let $\mathbf{T}$ be the orthogonal DCT matrix of order N, $\mathbf{x}(i)$ be the length-N transmitted vector sequence and $\mathbf{y}(i)$ the input to the DCT that produces $\mathbf{x}(i)$. i is the symbol index. Then we can write

$$\mathbf{y}(i) = \mathbf{T}\mathbf{x}(i). \tag{1}$$

We are interested in $\mathbf{y}(i)$ vectors which generate constant-envelope (CE) $\mathbf{x}(i)$ vectors, that is

$$R_x = \frac{\max\{|x_0(i)|, |x_1(i)|, \ldots, |x_{N-1}(i)|\}}{\min\{|x_0(i)|, |x_1(i)|, \ldots, |x_{N-1}(i)|\}} \tag{2}$$

where $x_n(i)$ denotes the nth element of $\mathbf{x}(i)$, should be equal to one. Since i does not affect the algorithm in any way, we will drop it from here onwards for convenience without loss of generality.

When $R_x = 1$, $\mathbf{y}$ is called a CE vector. It is clear that if $R_x = 1$, for example $\mathbf{x}$ is generated [10] from a vector $\mathbf{y}$ whose nth element is

$$y_n = e^{\frac{2\pi}{4N}(n^2+n)}, \qquad n = 0, 1, \cdots, N-1, \tag{3}$$

then $\mathbf{S}\mathbf{x}$, where

$$\mathbf{S} = \mathrm{diag}[e^{j\phi(0)}, e^{j\phi(1)}, \cdots, e^{j\phi(N-1)}] \tag{4}$$

also leads to $R_x = 1$ for any arbitrarily chosen set of phases $\phi(j), j = 0, 1, \cdots, N-1$. In other words, $\mathbf{y}' = \mathbf{T}\mathbf{S}\mathbf{x} = \mathbf{T}\mathbf{S}\mathbf{T}^T\mathbf{y}$ is also a CE vector. It is therefore a simple matter to generate any number of CE vectors from one "basis" CE vector $\mathbf{y}$ (or equivalently $\mathbf{x}$) – we only need to find matrices $\mathbf{S}_m$ of the form given in (4).

Note that this method is general and can be applied to any orthogonal or unitary matrix $\mathbf{T}$, such as the Discrete Fourier Transform (DFT), the Walsh Hadamard Transform (WHT), the Discrete Orthogonal Wavelet Transform (DWT) and so on. We just need one CE vector to generate a very big number of vectors. For example we can choose $\mathbf{x} = [+1 - 1 + 1 - 1 \cdots]^T$ which has the CE property. Then construct $\mathbf{y} = \mathbf{T}\mathbf{x}$, where $\mathbf{T}$ can be any orthogonal matrix, and $\mathbf{T}\mathbf{S}\mathbf{T}^T\mathbf{y}$ is also a CE vector for any set of phases constituting $\mathbf{S}$. In the following we show how wavelets and subband filterbanks can be used.

2.2 OPTIMAL DESIGN OF CE VECTORS

Of course, not all the CE vectors generated using the technique proposed here will be interesting in practice. We would usually want the CE vectors to satisfy some additional constraints in order to maximize performance in an additive noise channel, for instance; or perhaps the $\mathbf{y}$ vectors should be orthogonal so that more than one symbol can be transmitted simultaneously without performance loss. In this section, we outline a method to maximize the minimum distance between pairs of CE vectors, so that the set of vectors designed in this way will lead to minimum probability of error in an additive noise channel. It should be noted that the concept of numerical optimization used here can be adapted to suit other constraints as well.

We define the objective function

$$f_{m,n}(\mathbf{S}_m, \mathbf{S}_n) = ||\mathbf{y}_m - \mathbf{y}_n||^2 \tag{5}$$

where $m = 0, \cdots, 2^M - 1$, $n = m + 1, \cdots, 2^M - 1$,

$$\mathbf{S}_m = \text{diag}[e^{j\phi_m(0)}, \ldots, e^{j\phi_m(N-1)}], \qquad m = 0, \ldots, 2^M - 1$$

$\mathbf{y}_m$ is a CE vector, and M is the number of bits per symbol (or CE vector).

Then the optimal set of CE vectors is found by solving the following maximin nonlinear optimization problem (see e.g. [15] for details):

$$\max_{\mathbf{S}_m, \mathbf{S}_n} \left\{ \min_{m,n} f_{m,n}(\mathbf{S}_m, \mathbf{S}_n) \right\}. \tag{6}$$

To ensure that each sub-channel carries the same energy, so that the transmitted energy is evenly "spread" over the entire available bandwidth, we introduce the constraints:

$$R_y = \frac{\max_k |y_{m,k}|}{\min_k |y_{m,k}|} = 1. \tag{7}$$

where $y_{m,k}$ is the kth component of $\mathbf{y}_m = \mathbf{T}\mathbf{S}_m\mathbf{x}$, and $\mathbf{x}$ is an arbitrary CE transmitted vector. For example we can choose $\mathbf{x} = [+1 \ -1 \ +1 \ -1 \ \cdots]^T$.

We can also design CE vectors which have integer real and imaginary coefficients. This constraint ensures that the CE property would not be spoilt when the coefficients are quantized.

3. THE DISCRETE WAVELET TRANSFORM

If $x(t)$ is any square integrable function, then it can be decomposed onto a set of square integrable basis functions, constructed by dilating and translating a single wavelet function, say $\varphi(t)$, as follows [13]:

$$x(t) = \sum_{j,k} \sqrt{2^j} x_{j,k} \varphi(2^j t - k) \tag{8}$$

where

$$\varphi(\omega) = H_1(e^{j\frac{\omega}{2}})\phi(\frac{\omega}{2}) \quad \text{and} \quad \phi(\omega) = \prod_{k=1}^{\infty} H_0(e^{j\frac{\omega}{2^k}}) \tag{9}$$

$\phi(\omega)$ is the Fourier transform of the scaling function. $H_0(z)$ and $H_1(z)$ are respectively the low-pass and high-pass filters whose discrete impulse responses, which are given by $h_0(n)$ and $h_1(n)$, should satisfy the 2-band FIR PR-QMF bank conditions [14], that is

$$\sum_k h_i(k)h_j(k+2n) = \begin{cases} 0, & i \neq j \\ \delta(n), & i = j \end{cases} \tag{10}$$

Any 2-band filter bank having several subband levels, be it dyadic or uniform, can be transformed into an equivalent M-band filter bank [14]. In this case, the wavelet coefficients due to each subband filter are computed as follows:

$$Z_i(l) = \sum_k h_i(k)x(Ml-k) \tag{11}$$

where the discrete signal $x(l)$ is the input to the M-band filter bank and $h_i(k)$, where $i = 0, 1, \cdots, M-1$, is the impulse response of the ith filter.

4. CONSTRUCTION OF ORTHOGONAL MATRICES USING THE ORTHOGONAL WAVELET TRANSFORM

The wavelet transform matrix can be easily constructed from the wavelet low-pass and high-pass filters. For example, consider the 8×8 matrix T_{W_8} associated with a 4-coefficient wavelet low-pass filter $\mathbf{h}_0$ whose coefficients are given by $(h_0(0), h_0(1), h_0(2), h_0(3))$. We denote by $\mathbf{h}_1$ the QMF filter of $\mathbf{h}_0$ whose coefficients are given by $(h_1(0), h_1(1), h_1(2), h_1(3))$. Then, by using (11), where $M = 2$, we can construct T_{W_8} as follows:

$$T_{W_8} = \begin{bmatrix} h_0(0) & h_0(1) & h_0(2) & h_0(3) & 0 & 0 & 0 & 0 \\ 0 & 0 & h_0(0) & h_0(1) & h_0(2) & h_0(3) & 0 & 0 \\ 0 & 0 & 0 & 0 & h_0(0) & h_0(1) & h_0(2) & h_0(3) \\ h_0(2) & h_0(3) & 0 & 0 & 0 & 0 & h_0(0) & h_0(1) \\ h_1(0) & h_1(1) & h_1(2) & h_1(3) & 0 & 0 & 0 & 0 \\ 0 & 0 & h_1(0) & h_1(1) & h_1(2) & h_1(3) & 0 & 0 \\ 0 & 0 & 0 & 0 & h_1(0) & h_1(1) & h_1(2) & h_1(3) \\ h_1(2) & h_1(3) & 0 & 0 & 0 & 0 & h_1(0) & h_1(1) \end{bmatrix} \tag{12}$$

The decimation of the signal (see (11)) is taken into account in this matrix by shifting twice to the right the coefficients of the filters $\mathbf{h}_0$ and $\mathbf{h}_1$, and the boundary conditions, that is the fourth and the last rows, are handled by wrapping around these two filters. This is equivalent to periodizing the data vector to be transformed and leads to an orthogonal matrix. The orthogonality property of T_{W_8} can easily be proved by using (10). When we apply this matrix to a signal, it splits its frequency spectrum into two equal frequency bands, called L for low frequency and H for high frequency, respectively. By applying it again to the L band and H band, we can get as many frequency bands as we want, as long as the length of the signal after each decimation is greater than

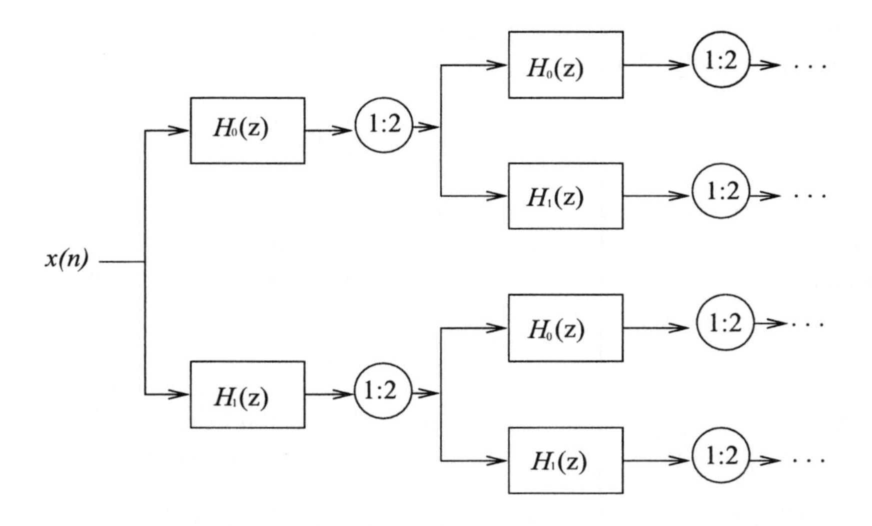

Figure 1 A 2-Level uniform subband tree structure.

or equal to the length of the filter. This is done by multiplying the length-N data vector $\mathbf{X}(n)$ to be transformed with a sequence of orthogonal matrices of order $N \times N$ each, that is

$$\begin{aligned} \mathbf{Z}(n) &= T_k \cdots T_2 T_1 T_0 \mathbf{X}(n) \\ &= T\mathbf{X}(n). \end{aligned} \tag{13}$$

where the matrices T_k, $k \geq 0$, depend on the type of subband tree structure and the number of subband levels. In the following, we restrict our selves to the uniform tree structure case.

4.1 UNIFORM SUBBAND TREE

If a uniform subband tree structure (Fig.1) is used, then the matrices $T_0, T_1, \cdots, T_k$ are given by $T_0 = T_{W_N}$, and T_k, where $k \geq 1$, is given by the block diagonal matrices $T_2 = diag[T_{W_{\frac{N}{2}}} T_{W_{\frac{N}{2}}}]$, $T_3 = diag[T_{W_{\frac{N}{4}}} T_{W_{\frac{N}{4}}} T_{W_{\frac{N}{4}}} T_{W_{\frac{N}{4}}}]$, etc.

5. RESULTS AND COMMENTS

In this paper, we have presented a new method for designing CE vectors for a multicarrier spread spectrum system which is valid for any orthogonal or unitary transform.

In the following, we give as an example the results obtained for 4 vectors of length 4 each which are designed for a Wavelet-based OFDM system (Table 1). We have actually used the simplest wavelet filter which is the 2-coefficient Haar filter. In figure 2 we give the theoretical upper and lower bounds on the probability of a symbol error (SER). We can notice that these two bounds are rather tight. CE vectors with quantized coefficients can also be easily obtained by adding more constraints to the optimization problem given above. In fact, the method can be easily used to design bigger sets of CE vectors which can be used in a real system.

The novel approach outlined here in which a numerical constrained optimization algorithm is used for signal design holds much promise, and the authors are currently investigating its use to improve the performance of MC-CDMA systems in frequency selective channels. The exploitation of the multi-dimensionality of the MC-CDMA signal space to design improved modulation techniques is also an area of interest.

Notes

1. OFDM systems traditionally use the discrete Fourier transform (DFT) but all other orthonormal transforms are admissible.

References

[1] J. A. C. Bingham. Multicarrier modulation for data transmission: An idea whose time has come. *IEEE Commun. Mag.*, 28(5):5–14, May 1990.

[2] R. E. Ziemer R. L. Peterson and D. E. Borth. *Introduction to Spread Spectrum Communications.* Simon & Schuster, Singapore, 1995.

[3] A. J. Viterbi. *CDMA – Principles of Spread Spectrum Communications.* Addison-Wesley, 1995.

[4] K. Fazel and G. P. Fettweis. *Multi-Carrier Spread-Spectrum.* Kluwer Academic, 1997.

[5] N. Yee, J.-P. M. G. Linnartz, and G. Fettweis. Multi-carrier CDMA in indoor wireless radio networks. *IEICE Trans. Communications*, E77-B(7):900–904, July 1994.

[6] J. Tellado and J. Cioffi. Efficient algorithms for reducing PAR in multi-carrier systems. In *Proceedings International Symposium on Information Theory*, Boston, MA, August 1998.

[7] R. Gross and D. Veenman. Clipping distortion in DMT ADSL systems. *Electronics Letters*, 29(24):2080–2081, November 1993.

[8] J. S. Chow, J. A. C. Bingham, and M. S. Flowers. Mitigating clipping noise in multi-carrier systems. In *the Proc. IEEE Int'l Conf. Communications (ICC)*, pages 715–719, 1997.

[9] J. E. M. Nilsson. Spectrum and waveform relations of multicarrier communications. In *Proc. IEEE Military Comms. Conf. (MILCOM)*, pages 255–259, McLean, VA, October 1996.

[10] S. Attallah and J. E. M. Nilsson. Sequences leading to minimum peak-to-average power ratios for dct-based multicarrier modulation. *Electronics Letters*, 34(15):1469–1470, July 1998.

[11] A. N. Akansu *et al.*. Orthogonal transmultiplexers in communications: a review. *IEEE Trans. Signal Proc.*, 46(4):979–995, April 1998.

Table 1 Optimal CE vectors designed using Haar filter.

$\mathbf{y}_1$	$\mathbf{y}_2$	$\mathbf{y}_3$	$\mathbf{y}_4$
0.0816 - 0.9967i	0.9806 + 0.1963i	0.9970 - 0.0780i	0.5141 + 0.8578i
0.0816 - 0.9967i	0.9997 - 0.0228i	0.9970 + 0.0780i	0.5141 + 0.8578i
-0.0816 + 0.9967i	-0.9806 - 0.1963i	0.9970 - 0.0780i	0.5141 + 0.8578i
0.0816 - 0.9967i	0.9997 - 0.0228i	0.9970 - 0.0780i	-0.5141 - 0.8578i
R_x=1.0000	R_x=1.0000	R_x=1.0000	R_x=1.0000
R_y=1.0000	R_y=1.0000	R_y=1.0000	R_y=1.0000

Figure 2 Upper and Lower bounds on SER for CE vectors with 2 bits/symbol.

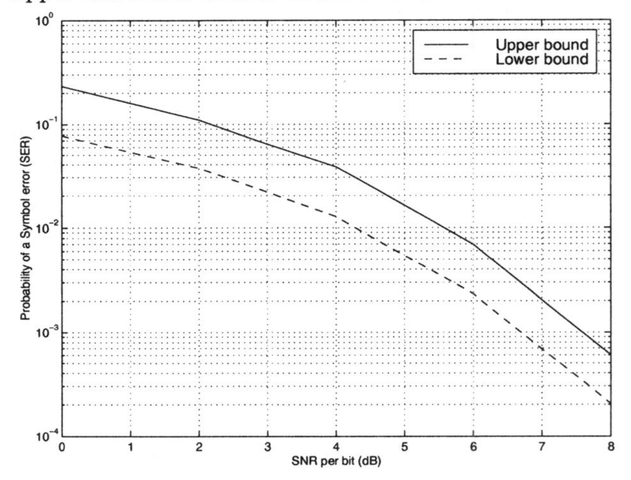

[12] G. W. Wornell. Emerging applications of multirate signal processing and wavelets in digital communications. *Proc. of the IEEE*, 84(4):586–603, April 1996.

[13] I. Daubechies. Orthonormal bases of compactly supported wavelets. *Comm. in Pure and Applied Math.*, 41:909–996, 1988.

[14] A. N. Akansu and R. A. Haddad. *Multiresolution Signal Decomposition.* Academic Press, New York, NY, 1992.

[15] R. Fletcher. *Practical Methods of Optimization.* John Wiley & Sons, 2nd edition, 1987.

A SUBSPACE METHOD FOR BLIND CHANNEL IDENTIFICATION IN MULTI-CARRIER CDMA SYSTEMS

Daniel I. Iglesia, Carlos J. Escudero, Luis Castedo
Department of Electronics and Systems. University of A Coruña
Campus de Elviña s/n, 15.071 La Coruña, SPAIN
Tel: ++ 34-981-167150, e-mail: escudero@des.fi.udc.es

Abstract

Blind channel estimation in Multi-Carrier CDMA systems using subspace decomposition techniques is considered. We demonstrate that the performance of these techniques is severely limited by the number of users and show that this limitation is considerabily reduced if we use signature codes that span several transmitted symbols (log codes). Finally, we propose a channel identification algorithm specifically developed for MC-CDMA systems using long codes.

1. INTRODUCTION

Multi-Carrier (MC) transmission methods for Code Division Multiple Access (CDMA) communication systems [1][2] split the transmitted digital information in multiple carriers and achieve high transmission speeds using large symbol periods. As a consequence, MC-CDMA systems can simply remove the Inter-Symbol Interference (ISI) introducing a short guard time between symbols and avoid the utilization of a channel equalizer. Although ISI is considerably reduced, MC-CDMA systems still suffer from Multiple Access Interference (MAI) due to the loss of orthogonality between codes caused by channel distortion.

A large number of existing techniques for MAI suppression require the estimation of the channel parameters. In this paper we introduce a new blind channel estimation technique that is based on a subspace decomposition [3]. We will derive a particular algorithm to identify the channel parameters and we will show how the system capacity can be improved by using signature codes with lengths larger than the spreading gain.

The paper is organized as follows. Section 2 presents the signal model of a synchronous MC-CDMA system. Section 3 presents the subspace decomposition and shows the need of using longer codes for increasing the system capacity. Section 4 introduces the modifications in the signal model to consider the large codes and Section 5 presents the algorithm to implement this technique. Section 6 shows the results of several computer simulations and, finally, Section 7 is devoted to the conclusions.

2. SIGNAL MODEL

Let us consider a discrete-time baseband equivalent model of a synchronous MC-CDMA system with N users using L-chip signature codes. The k-th chip corresponding to the n-th symbol transmitted by the i-th user is given by

$$v_n^i(k) = s_n^i c_i(k) \qquad k = 0, \cdots, L-1 \qquad n = 0, 1, 2, \cdots \tag{1}$$

where $c_i(k)$ is the k-th chip of the i-th user code. In a MC-CDMA system the modulator computes the L-IDFT (Inverse Discrete Fourier Transform) of (1) to obtain the following multicarrier signal

$$V_n^i(m) = IDFT[v_n^i(k)] = \frac{1}{L} \sum_{k=0}^{L-1} v_n^i(k) e^{j\frac{2\pi}{L}km} \tag{2}$$

This signal is transmitted through a dispersive channel with an impulse response $h_i(m)$; $m = 0, ...M-1$. At the receiver the observed signal is a superposition of the signals corresponding to N users plus an additive white Gaussian noise (AWGN). Therefore, the received signal for the n-th symbol is the following

$$X_n(m) = \sum_{i=1}^{N} V_n^i(m) * h_i(m) + r_n(m) \tag{3}$$

where $*$ denotes discrete convolution and $r_n(m)$ represents a white noise sequence.

To recover the transmitted symbols, the receiver applies a L-DFT (Discrete Fourier Transform) to the received signal (3). Assuming perfect synchronization and a sufficiently large guard time between symbols, the resultant signal is

$$x_n(k) = DFT[X_n(m)] = \sum_{i=1}^{N} v_n^i(k)H_i(k) + \Gamma_n(k)$$
$$= \sum_{i=1}^{N} s_n^i c_i(k)H_i(k) + \Gamma_n(k) \quad k = 0, \cdots, L-1 \quad (4)$$

where $H_i(k)$ and $\Gamma_n(k)$ are the DFT's of $h_i(m)$ and $r_n(m)$, respectively. Rewriting (4) in vector notation we obtain

$$\mathbf{x}_n = [x_n(0), \cdots, x_n(L-1)]^T = \sum_{i=1}^{N} s_n^i \mathbf{C}_i \mathbf{H}_i + \mathbf{\Gamma}_n \quad (5)$$
$$= \sum_{i=1}^{N} s_n^i \mathbf{C}_i \mathbf{F} \mathbf{h}_i + \mathbf{\Gamma}_n = \sum_{i=1}^{N} s_n^i \tilde{\mathbf{c}}_i + \mathbf{\Gamma}_n$$

where $\mathbf{C}_i$ is a diagonal matrix whose elements are the L chips of the code corresponding to the i-th user, $\mathbf{H}_i = [H_i(0), \cdots, H_i(L-1)]^T$, $\mathbf{\Gamma}_n = [\Gamma_n(0), \cdots, \Gamma_n(L-1)]^T$ and we have used the relationship $\mathbf{H}_i = \mathbf{F}\mathbf{h}_i$ where $\mathbf{F}$ is the DFT matrix and $\mathbf{h}_i = [h_i(0), \cdots, h_i(M-1)]^T$. Note that (5) can be interpreted as a CDMA signal where the user codes are perturbed versions of the transmitted one, i.e., $\tilde{\mathbf{c}}_i = \mathbf{C}_i\mathbf{F}\mathbf{h}_i$.

3. SUBSPACE DECOMPOSITION

Assuming statistical independence between users and noise, the autocorrelation matrix of the observations vector (5) can be decomposed as

$$\mathbf{R} = E[\mathbf{x}_n \mathbf{x}_n^H] = \sum_{i=1}^{N} \tilde{\mathbf{c}}_i E[s_n^i s_n^{i^H}] \tilde{\mathbf{c}}_i^H + E[\mathbf{\Gamma}_n \mathbf{\Gamma}_n^H]$$
$$= \sum_{i=1}^{N} \mathbf{C}_i \mathbf{F} \mathbf{h}_i E[s_n^i s_n^{i^H}] \mathbf{h}_i^H \mathbf{F}^H \mathbf{C}_i^H + E[\mathbf{\Gamma}_n \mathbf{\Gamma}_n^H]$$
$$= \sum_{i=1}^{N} \sigma_i^2 \mathbf{C}_i \mathbf{F} \mathbf{h}_i \mathbf{h}_i^H \mathbf{F}^H \mathbf{C}_i^H + L\sigma_r^2 \mathbf{I} \quad (6)$$

where $E[\cdot]$ is the expectation operator, ^{H} denotes conjugate transpose and σ_i^2 and σ_r^2 are the i-th user signal and noise power, respectively.

Let us consider the eigendecomposition of (6). There are L eigenvalues that we sort as $\lambda_0 \geq \lambda_1 \geq \cdots \geq \lambda_{L-1}$. It is wee-known that the eigenvectors associated to the N most significants eigenvalues ($\mathbf{u}_l$, $l = 0, \cdots, N-1$) span the signal subspace where the perturbed user codes, $\tilde{\mathbf{c}}_i$, lie. The remaining $L-N$ eigenvectors ($\mathbf{u}_l$, $l = N, \cdots, L-1$) span the noise (orthogonal) subspace and their associated eigenvalues are equal to the noise power, i.e., $\lambda_N = \cdots = \lambda_{L-1} = L\sigma_r^2$ [3].

As we have seen, the perturbed user codes lie in the signal subspace and, therefore, are orthogonal to the noise subspace. This property can be used to state the following system of equations for the i-th user

$$\tilde{\mathbf{c}}_i^H \mathbf{u}_l = (\mathbf{C}_i \mathbf{F} \mathbf{h}_i)^H \mathbf{u}_l = \mathbf{0} \quad l = N, \cdots, L-1 \tag{7}$$

Recall that this system of equations has M unknowns and $L-N$ equations. It will be solvable if and only if the number of equations is greater or equal than the number of unknowns, that is,

$$M \leq L - N \quad \Rightarrow \quad N \leq L - M \tag{8}$$

Note that we have obtained a limit for the system capacity that depends on the number of carriers and the channel length. This means that if we want to improve the capacity it is necessary to increase the number of carriers. However, in many practical applications this is not possible, or desirable, and we have to consider another solution to increase the system capacity.

4. UTILIZATION OF LONG CODES

In the previous sections we have considered that the codes length is equal to the spreading gain (i.e., to the number of carriers). Thus finding that the system capacity is limited by the number of carriers. In this section we are going to explore the idea of increasing the system capacity by means of increasing the codes length without modifying the number of carriers, i.e., we will consider codes length larger than the spreading gain.

To simplify the notation we consider pL-chip signature codes, where $p \geq 1$ is an integer number (i.e., the code length is a multiple of the spreading gain). This means that the user codes will be repeated after p symbols. The modulator for a block of p consecutive symbols is represented in figure 1 where the j-th symbol of this block is modulated by the j-th part of the user code to obtain the following CDMA signal

$$v_{n,j}^i(k) = s_{np+j}^i c_{i,j}(k); \quad k = 0, \cdots, L-1; \quad j = 0, \cdots, p-1; \quad n = 0, 1, 2, \cdots$$

This equation equala to (1) for $p = 1$. As in section 2, we can obtain a vector of L chips for the j-th symbol of a block as

$$\mathbf{x}_{n,j} = [x_{n,j}(0), \cdots, x_{n,j}(L-1)]^T$$

$$= \sum_{i=1}^{N} s_{np+j}^{i} \mathbf{C}_{i,j} \mathbf{F} \mathbf{h}_i + \mathbf{\Gamma}_{n,j} = \sum_{i=1}^{N} s_{np+j}^{i} \tilde{\mathbf{c}}_{i,j} + \mathbf{\Gamma}_{n,j}$$

for $j = 0, \cdots, p-1$.

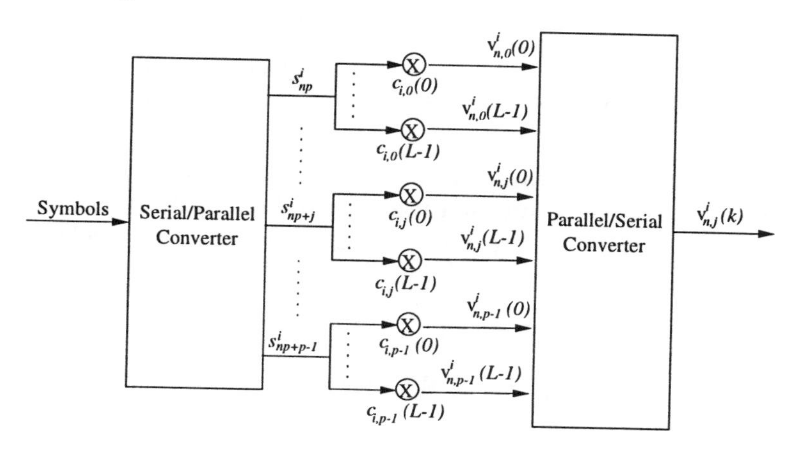

Figure 1 Block diagram of the i-th user modulator for transmiting p symbols with long codes.

Arranging the p consecutive vectors $\mathbf{x}_{n,j}$, corresponding to a single block, into a single vector of length pL, we obtain

$$\mathbf{x}(n) = [\mathbf{x}_{n,0}^T, \cdots, \mathbf{x}_{n,p-1}^T]^T = \sum_{i=1}^{N} \mathbf{C}_i \mathcal{H}_i \mathbf{s}^i(np) + \mathbf{\Gamma}_n \tag{9}$$

where $\mathbf{C}_i$ is a diagonal matrix containing the pL chips of the code associated to the i-th user, $\mathbf{s}^i(np) = [s_{np}^i, \cdots, s_{np+p-1}^i]^T$, $\mathbf{\Gamma}_n = [\mathbf{\Gamma}_{n,0}^T, \cdots, \mathbf{\Gamma}_{n,p-1}^T]^T$, $\mathcal{H}_i = \mathbf{F}\mathbf{h}_i \otimes \mathbf{I}_p$, with $\mathbf{I}_p$ equal to the identity matrix of size $p \times p$ and $\otimes$ represents the kronecker product. Let us consider the $pL \times pL$ autocorrelation matrix of (9)

$$\mathbf{R} = \sum_{i=1}^{N} \sigma_i^2 \mathbf{C}_i \mathcal{H}_i \mathcal{H}_i^H \mathbf{C}_i^H + L\sigma_r^2 \mathbf{I} \tag{10}$$

Denoting $\lambda_0 \geq \lambda_1 \geq \cdots \geq \lambda_{pL-1}$ to the eigenvalues of $\mathbf{R}$, the eigenvectors associated to the pN most significants eigenvalues ($\mathbf{u}_l$, $l = 0, \cdots, pN-1$) span the signal subspace where the different user signals lie. The remaining $p(L-N)$ eigenvectors ($\mathbf{u}_l$, $l = pN, \cdots, pL-1$) span the noise subspace. Taking into account the orthogonality between signal and noise subspace, we obtain the following system of equations

$$(\mathbf{C}_i \mathcal{H}_i)^H \mathbf{u}_l = 0 \quad \Rightarrow \quad \mathbf{h}_i^H \mathbf{F}^H \mathbf{C}_{i,j}^H \mathbf{u}_{l,j} = 0 \tag{11}$$
$$l = pN, \cdots, pL-1 \quad j = 0, \cdots, p-1$$

where $\mathbf{u}_{l,j} = \mathbf{u}_l(jL+1 : jL+L-1)$ and the notation $\mathbf{u}(i:j)$ represents a vector containing the elements of $\mathbf{u}$ between the indexes i and j (both included). Recall that this system of equations still has M unknowns but $p^2(L-N)$ equations. It will be solvable if and only if the number of equations is greater or equal than the number of unknowns, that is,

$$p^2(L-N) \geq M \quad \Rightarrow \quad N \leq \left\lfloor \frac{p^2 L - M}{p^2} \right\rfloor \tag{12}$$

where $\lfloor \cdot \rfloor$ is the floor operation which rounds down to the previous integer value. Therefore, the system capacity can be improved by increasing the p value, i.e., the code length. Similarly, there exists a relationship for the parameter p when there is a fixed number of users: $p \geq \left\lceil \sqrt{\frac{M}{L-N}} \right\rceil$, where $\lceil \cdot \rceil$ is the ceiling operator.

5. CHANNEL IDENTIFICATION ALGORITHM

In order to solve the equations system (11), we can consider the following equivalent system

$$||\mathbf{h}_i^H \mathbf{F}^H \mathbf{C}_{i,j}^H \mathbf{u}_{l,j}||^2 = \mathbf{h}_i^H \mathbf{F}^H \mathbf{C}_{i,j}^H \mathbf{u}_{l,j} \mathbf{u}_{l,j}^H \mathbf{C}_{i,j} \mathbf{F} \mathbf{h}_i = 0 \tag{13}$$

for $l = pN, \cdots, pL-1$ and $j = 1, \cdots, p$. The solution to these equations can be found by solving the following minimization problem

$$\begin{aligned}
\hat{\mathbf{h}}_i &= \arg \min_{||\mathbf{h}_i||^2=1} \left[\sum_{j=1}^{p} \sum_{l=pN}^{pL-1} \mathbf{h}_i^H \mathbf{F}^H \mathbf{C}_{i,j}^H \mathbf{u}_{l,j} \mathbf{u}_{l,j}^H \mathbf{C}_{i,j} \mathbf{F} \mathbf{h}_i \right] \\
&= \arg \min_{||\mathbf{h}_i||^2=1} \mathbf{h}_i^H \left[\sum_{j=1}^{p} \sum_{l=pN}^{pL-1} \mathbf{F}^H \mathbf{C}_{i,j}^H \mathbf{u}_{l,j} \mathbf{u}_{l,j}^H \mathbf{C}_{i,j} \mathbf{F} \right] \mathbf{h}_i \\
&= \arg \min_{||\mathbf{h}_i||^2=1} \mathbf{h}_i^H \left[\sum_{j=1}^{p} \mathbf{F}^H \mathbf{C}_{i,j}^H \mathbf{U}_j \mathbf{U}_j^H \mathbf{C}_{i,j} \mathbf{F} \right] \mathbf{h}_i \\
&= \arg \min_{||\mathbf{h}_i||^2=1} \mathbf{h}_i^H \mathbf{Q} \mathbf{h}_i
\end{aligned} \tag{14}$$

where the solution $\hat{\mathbf{h}}_i$ is an estimation of the channel impulse response vector, $\mathbf{U}_j$ is a $L \times p(L-N)$ matrix whose columns are the subvectors, $\mathbf{u}_{l,j}$, extracted from the eigenvectors associated to the noise subspace (i.e., $\mathbf{u}_l \;\; l = pN, \cdots, pL-1$) and $\mathbf{Q} = \sum_{j=1}^{p} \mathbf{F}^H \mathbf{C}_{i,j}^H \mathbf{U}_j \mathbf{U}_j^H \mathbf{C}_{i,j} \mathbf{F}$. The solution can be obtained by the least squares method and it corresponds to the eigenvector of $\mathbf{Q}$ associated to its minimum eigenvalue [5].

In practice, we do not know *a priori* the autocorrelation matrix (10). However it can be estimated from the sampled averaged matrix as $\hat{\mathbf{R}} = \frac{1}{N_s}\sum_{n=0}^{N_s}\mathbf{x}(n)\mathbf{x}^H(n)$ where N_s is the number of received symbols used to obtain the estimation.

Finally, note that when using second order statistics, the channel impulse response can be obtained up to a complex constant. To remove this ambiguity, and analyze the algorithm performance, it is necessary to compensate it. Towards this aim, we normalize the estimation of the impulse response vector as $\hat{\mathbf{h}}_{i,normalized} = \frac{h_1(0)}{\hat{h}_1(0)}\hat{\mathbf{h}}_i$ where $h_1(0)$ and $\hat{h}_1(0)$ are the first elements of the true and estimated channel impulse response vectors, respectively.

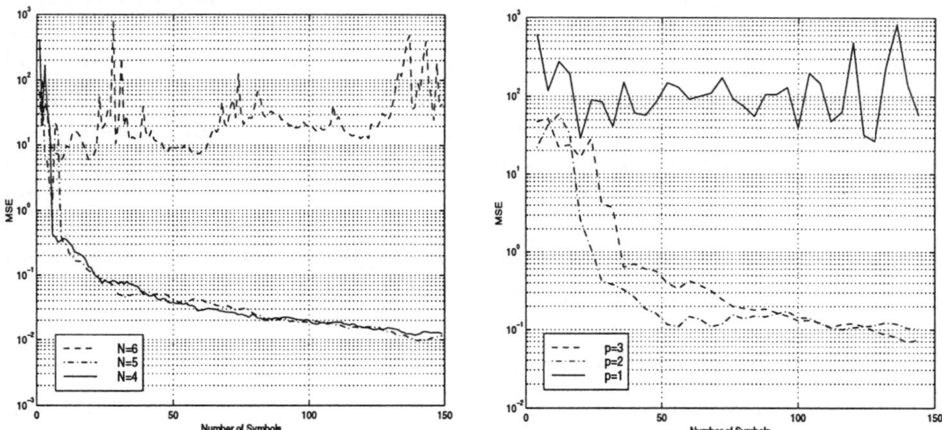

Figure 2 **Left**: Averaged Mean Square Error time evolution for different values of N. **Right:** Averaged Mean Square Error time evolution for different values of N.

6. SIMULATIONS

To illustrate the effectiveness of the proposed identification technique, we have carried out several computer simulations. The performance of the approach is evaluated in terms of the Mean Square Error (MSE)

$$MSE = ||(\hat{\mathbf{h}}_{i,normalized} - \mathbf{h}_i)||^2 \tag{15}$$

Figure 2 (Left) presents the averaged MSE time evolution (mean value for 10 realizations) for different values of N. An environment using short codes, a spreading factor $L = 10$ and $M = 5$ has been considered. The users are received with a $SNR = 10dB$ through a Rice channel [4] with ratio $\frac{A_0}{\sigma} = 1$. According to (8) a maximum value of $N = 5$ is acceptable and it can be seen that for higher values, the channel cannot be estimated.

Figure 2 (Right) presents the averaged MSE time evolution (mean value for 10 realizations) for long codes with different values of p. We have considered the same environment as before with 8 users. In this case the relationship (12) reduces to $p \geq 2$. It can be seen that for values below this condition (i.e., $p = 1$), the channel estimation does not achieve good performance.

7. CONCLUSIONS

We have seen that when using subspace decomposition for blind channel estimation, the capacity of a MC-CDMA system is limited by the number of carriers. As we have shown, to eliminate this dependence and increase the system capacity, it is necessary to raise the users code length. Considering codes larger than a symbol period we have introduced a new channel estimation algorithm that is based on the orthogonality between the subspace spanned by channel matrices (i.e., the signal subspace) and noise subspace. Simulations show the performance in terms of the Mean Square Error of the estimation and they corroborate the theoretical limit obtained for the system capacity.

Acknowledgments

This work has been supported by CICYT (grant TIC96-0500-C10-02) and FEDER (grant 1FD97-0082).

References

[1] Fazel, K., Fettweis, G.P., *Multi-Carrier Spread-Spectrum*, Kluwer Academic Publishers, 1997.

[2] Yee, N., Linnartz, J. P., Fettweis, G., "Multi-Carrier CDMA in Indoor Wireless Radio Networks", *Proc. International Symposium on Personal, Indoor and Mobile Radio Communications, (PIMRC93)*, Yokohama, pp. 109-113, 1993.

[3] E. Moulines, P. Duhamel, J. F. Cardoso and S. Mayrargue, "Subspace Methods for the Blind Identification of Multichannel FIR Filters", *IEEE Transactions on Signal Processing*,vol. 43,no. 2,pp. 516-525,February 1995.

[4] P. Z. Peebles, *Probability, Random Variables and Random Signal Principles*, McGraw-Hill International Editions, Singapore, 1993.

[5] G. Strang, *Linear Algebra and its Applications*, Harcourt Brace Jovanovich, Third Edition, 1988.

Multiuser Detection with Iterated Channel Estimation

Alexander Lampe
Telecommunications Institute II
University of Erlangen, Germany
alampe@nt.e-technik.uni-erlangen.de

Christoph Windpassinger
Telecommunications Institute II
University of Erlangen, Germany
Christoph.Windpassinger@informatik.uni-erlangen.de

Abstract In this paper, we investigate the transmission over time–variant multipath Rayleigh–fading channels employing Direct Sequence Code Division Multiple Access (DS–CDMA). In order to estimate the actual channel state as well as to detect the users' data symbols an iterated channel estimation scheme employing pilot symbols in conjunction with soft decision interference cancellation is applied. We show that in this way the reliability of the channel estimation as well as data symbol detection can be considerably improved in the course of the iterative procedure and that for half and full load of the system nearly the same performance as that of a single user can be achieved.

Keywords: Wideband CDMA, Multiple Access, Channel Estimation

1. INTRODUCTION

In recent works it was shown that multiuser interference can be eliminated almost completely with the help of iterated soft decision interference cancellation

(ISDIC) (Teich and Seidl, 1996, Müller and Huber, 1998, Lampe and Huber, 1999) provided perfect channel knowledge is available at receiver site. In the following we study the question whether these results are still valid if no channel knowledge is a priori given to the receiver but has to be obtained by channel estimation. The main problem we face is how to get sufficiently reliable information about the channel in the presence of not only channel noise but severe multiuser interference. As the complexity of the overall maximum–likelihood multiuser receiver for joint data and channel estimation grows exponentially with the number of users as well as the transmission length (Vasudevan and Varanasi, 1996, Windpassinger, 1999) we suggest the application of an appropriately modified ISDIC scheme for the combined problem of channel estimation and data symbol detection. More explicitly, transmitting data symbols and pilot symbols (being used for channel estimation) in multiplex we propose to carry out the channel estimation iteratively where each channel estimation is followed by a multiuser detection employing the latest, hopefully improved channel estimates and whose soft decisions in turn are used to reduce the multiuser interference distorting the channel estimation.

We show that in this way the performance achievable for system load $K/N = 0.5$ as well as $K/N = 1$ (K and N denote the number of users and processing gain, respectively) is almost as good as that of a single receiver depending on the number of iterations carried out.

The paper is arranged as follows. In Section 2., the transmission model is given and the proposed receiver scheme is introduced. Simulation results are presented for performance evaluations in Section 3.. Finally, Section 4. points out conclusions.

2. ITERATIVE CHANNEL ESTIMATION AND DATA SYMBOL DETECTION

We consider the transmission of K users over independent frequency selective fading channels with CDMA to a single receiver. Assuming that all users transmit symbol asynchronous but chip synchronous[1] the underlying discrete time baseband transmission model is illustrated in Fig. 1a.

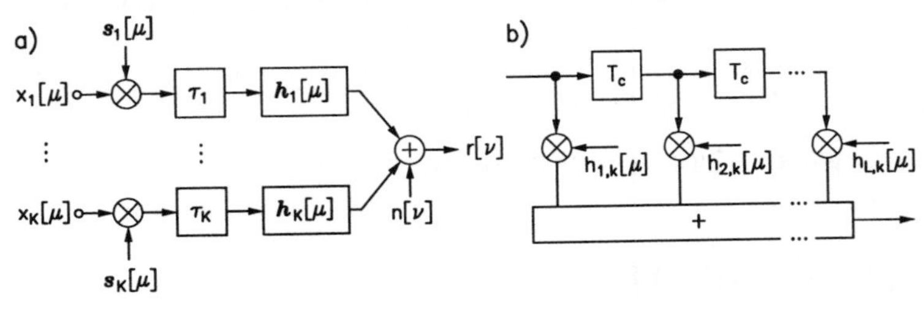

Figure 1 a) Transmission model of CDMA system with K users b) Tapped–delay–line channel model with L paths.

Here, $x_k[\mu]$ and $\boldsymbol{s}_k[\mu] = (s_{0,k}[\mu], \ldots, s_{N-1,k}[\mu])^T$ denote the kth user's transmitted symbol and his random unit energy spreading sequence with $s_{j,k}[\mu] \in \{(\pm 1 \pm \mathrm{j})/\sqrt{2N}\}$ in the μth transmission interval, respectively. The channel symbols $x_k[\mu]$ are chosen from the set $\mathcal{X} = \{\pm 1; \pm \mathrm{j}\}$ and every pth symbol is a pilot symbol, i.e., $x_k[\mu_p = \eta p] = 1, \eta \in \mathbb{Z}$. The asynchronous transmission is modeled by choosing the users' delays as τ_k being multiples of one chip interval with $\tau_k \in \{0, 1, \ldots, N-1\}, \forall k$, where we suppose $\tau_1 = 0$ and $\tau_k \geq \tau_j, k > j$. Next, the kth user's modulated sequence $(x_k[\mu]s_{0,k}[\mu], \ldots, x_k[\mu]s_{N-1,k}[\mu])$ is sent over a channel with impulse response $\boldsymbol{h}_k[\mu]$ representing the well known tapped–delay–line channel model (see Fig. 1b) with L resolvable paths. The tap weights $h_{1,k}[\mu], \ldots, h_{L,k}[\mu], \forall k$, are generated according to the Gaussian WSSUS model with autocorrelation function $\phi_{hh}[\Delta\mu] = \frac{1}{L}J_0(2\pi f_D T_c \Delta\mu), \forall l, k$. [2] Supposing constant tap gains for one symbol interval the received signal sampled with chip rate is

$$r[\nu] = \sum_{\kappa=1}^{K}\sum_{\lambda=1}^{L} x_\kappa\Big[\lfloor \nu, \tau_\kappa, \lambda \rfloor\Big] h_{\lambda,\kappa}\Big[\lfloor \nu, \tau_\kappa, \lambda \rfloor\Big] s_{(\nu,\tau_\kappa,\lambda),\kappa}\Big[\lfloor \nu, \tau_\kappa, \lambda \rfloor\Big] + n[\nu],$$

where $n[.]$ denotes zero mean complex additive white Gaussian channel noise with power σ_n^2 and $(\nu, \tau_\kappa, \lambda) \triangleq (\nu - \tau_\kappa - \lambda)\mathrm{mod}N$ as well as $\lfloor \nu, \tau_\kappa, \lambda \rfloor \triangleq \lfloor (\nu - \tau_\kappa - \lambda)/N \rfloor$[3].

As mentioned above we propose to split up the whole task at hand into two subproblems – channel estimation and multiuser detection – being carried out iteratively several times. In order to explain roughly the algorithm we consider the mth iteration and the kth user.

First, to derive improved estimates for the weights $h_{l,k}[.]$ of the kth user's lth path the estimates $\hat{h}_{\lambda,\kappa}^{m-1}[\mu]$ and $\hat{x}_\kappa^{m-1}[\mu]$ obtained in the $(m-1)$st iteration for path gain $h_{\lambda,\kappa}[\mu]$ and data symbol $x_\kappa[\mu]$, respectively, are used to cancel out multiuser interference due to users $\kappa \neq k$, interpath interference resulting from paths $\lambda \neq l$ and intersymbol interference in the received signal $r[\nu]$ with $\nu \in \Omega_{l,k}^p = \{\mu_p N + \tau_k + l, \ldots, (\mu_p + 1)N + \tau_k + l - 1 | \mu_p = \eta p, \eta \in \mathbb{Z}\}$, i.e, in chip intervals when the kth user's pilot symbols $x_k[\mu_p]$ are received over path l. This leads to $i_{l,k}^m[\nu] = r[\nu] - \left(\mathrm{MUI}_{l,k}^{m-1}[\nu] + \mathrm{ISI}_{l,k}^{m-1}[\nu]\right), \nu \in \Omega_{l,k}^p$, (see Fig. 2). Now, the best way would be to use the samples $i_{l,k}^m[\nu], \nu \in \Omega_{l,k}^p$, at chip rate for channel estimation turning out to be practically infeasible for large N. Instead, passing the sequence $\boldsymbol{i}_{l,k}^m[\mu_p] = (i_{l,k}^m[\mu_p N + \tau_k + l], .., i_{l,k}^m[(\mu_p + 1)N + \tau_k + l - 1])^T$ through a matched filter adapted to the kth user's spreading sequence $\boldsymbol{s}_k[\mu_p]$ and lth path sampled once per pilot symbol interval we get

$$m_{l,k}^m[\mu_p] = \boldsymbol{s}_k^T[\mu_p]\boldsymbol{i}_{l,k}^m[\mu_p] = h_{l,k}[\mu_p]x_k[\mu_p] + \tilde{n}_{l,k}^m[\mu_p],$$

where $\tilde{n}_{l,k}^m[\mu_p]$ denotes the remaining interference distorting the transmission of the pilot symbol $x_k[\mu_p]$.

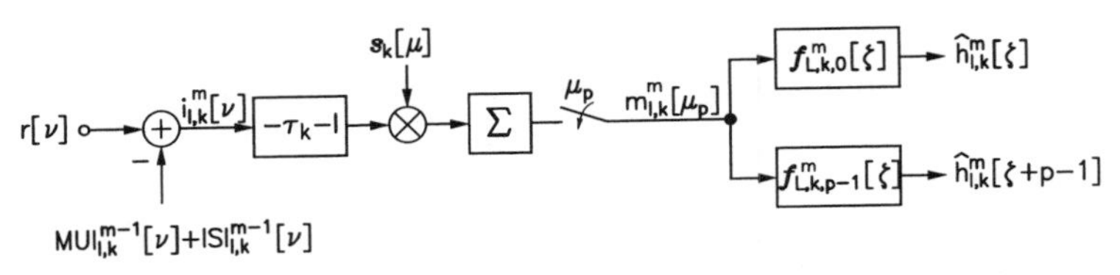

Figure 2 Estimation of path gains $h_{l,k}[\zeta], \ldots, h_{l,k}[\zeta + p - 1]$ of user k in mth iteration.

The samples $\boldsymbol{m}^m_{l,k}[\zeta = \eta p] = (m^m_{l,k}[\zeta - \gamma p], m^m_{l,k}[\zeta - \gamma p + p], \ldots, m^m_{l,k}[\zeta + \gamma p])^T$ belonging to pilot symbols are fed into p channel estimation filters $\boldsymbol{f}^m_{l,k,\pi}[\zeta], \pi = 0, \ldots, p-1$, delivering new estimates $\hat{h}^m_{l,k}[\zeta], \ldots, \hat{h}^m_{l,k}[\zeta+p-1]$. In order to benefit from declining multiuser, interpath as well as intersymbol interference the interpolation filters chosen according to the MMSE–criterion

$$\begin{aligned} \boldsymbol{f}^m_{l,k,\pi}[\zeta] &= \mathcal{E}\{h_{l,k}[\zeta+\pi](\boldsymbol{m}^m_{l,k}[\zeta])^H\} \left(\mathcal{E}\{\boldsymbol{m}^m_{l,k}[\zeta](\boldsymbol{m}^m_{l,k}[\zeta])^H\}\right)^{-1} \\ &= (\boldsymbol{\varphi}^m_{l,k,\pi}[\zeta])^T (\boldsymbol{C}^m_{l,k,\pi}[\zeta])^{-1} \end{aligned}$$

have to be adapted appropriately. This can either be done by incorporating all knowledge on correlations in $\boldsymbol{m}^m_{l,k}[\zeta]$ or by treating the interference $\tilde{n}^m_{l,k}[\zeta - \gamma p], \tilde{n}^m_{l,k}[\zeta - \gamma p + p] \ldots, \tilde{n}^m_{l,k}[\zeta + \gamma p]$ as additional Gaussian noise with power $\sigma^2_{\tilde{n}^m_{l,k}[\eta p]} = \mathcal{E}\{|\tilde{n}^m_{l,k}[\eta p]|^2\}, \forall \eta$, [4], which is of course suboptimum but computationally much more efficient. In fact, we calculate the required covariance matrix as

$$\boldsymbol{C}^m_{l,k,\pi}[\zeta] = \begin{pmatrix} \phi_{hh}(0) + \sigma^2_{\tilde{n}^m_{l,k}[\zeta-\gamma p]} & \phi_{hh}(p) & \cdots & \phi_{hh}(2\gamma p) \\ \phi_{hh}(p) & \ddots & \ddots & \phi_{hh}((2\gamma-1)p) \\ \vdots & \ddots & \ddots & \vdots \\ \phi_{hh}(2\gamma p) & \phi_{hh}((2\gamma-1)p) & \cdots & \phi_{hh}(0) + \sigma^2_{\tilde{n}^m_{l,k}[\zeta+\gamma p]} \end{pmatrix}.$$

This channel estimation procedure is performed in parallel for all paths and all users according to the users' delays.

Next, as to estimate the data symbol $x_k[\mu]$ while suppressing the asynchronous interferers we consider the three consecutive symbol intervals $\mu - 1, \mu, \mu + 1$ of the user of interest. More explicitely, we employ the received signal $r[\nu]$ with $\nu \in [(\mu-1)N + \tau_k + 1, \ldots, (\mu+1)N + \tau_k + L]$. Then, subtracting the assumed interference $\overline{\mathrm{MUI}}^m_k[\nu] + \overline{\mathrm{ISI}}^m_k[\nu]$ we get the sequence $y^m_k[\nu]$ (see Fig. 3)

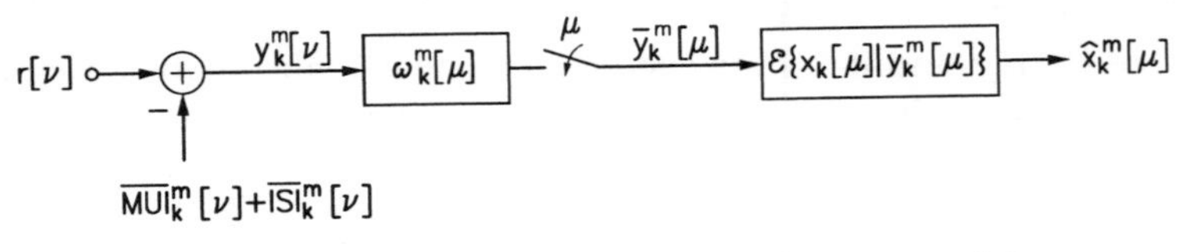

Figure 3 Estimation of the kth user's data symbol $x_k[\mu]$.

$$
\begin{aligned}
y_k^m[\nu] = r[\nu] \quad & - \sum_{i=1}^{2} \sum_{\kappa}^{K} \sum_{\lambda=1}^{L} \hat{x}_{\kappa}^{m}\left[\mu - i\right] \hat{h}_{\lambda,\kappa}^{m}\left[\mu - i\right] s_{(\nu,\tau_\kappa,\lambda),\kappa}\left[\mu - i\right] \\
& - \sum_{\kappa<k} \sum_{\lambda=1}^{L} \hat{x}_{\kappa}^{m}\left[\mu\right] \hat{h}_{\lambda,\kappa}^{m}\left[\mu\right] s_{(\nu,\tau_\kappa,\lambda),\kappa}\left[\mu\right] \\
& - \sum_{\kappa>k} \sum_{\lambda=1}^{L} \hat{x}_{\kappa}^{m-1}\left[\mu\right] \hat{h}_{\lambda,\kappa}^{m}\left[\mu\right] s_{(\nu,\tau_\kappa,\lambda),\kappa}\left[\mu\right] \\
& - \sum_{i=1}^{2} \sum_{\kappa}^{K} \sum_{\lambda=1}^{L} \hat{x}_{\kappa}^{m-1}\left[\mu + i\right] \hat{h}_{\lambda,\kappa}^{m}\left[\mu + i\right] s_{(\nu,\tau_\kappa,\lambda),\kappa}\left[\mu + i\right].
\end{aligned}
$$

In the above equation we took into account that we can make use of new estimates $\hat{x}_{\kappa}^{m}[\mu], \kappa = 1, \ldots, k-1$, of preceding users made in the same iteration as to cancel the MUI due to serial cancellation.

Next, defining $\boldsymbol{y}_k^m[\mu] = (y_k^m[(\mu-1)N+\tau_k+1], \ldots, y_k^m[(\mu+1)N+\tau_k+L])^T$ and passing this sequence through an unbiased MMSE–filter $\boldsymbol{w}_k^m[\mu]$ chosen as

$$
(\boldsymbol{w}_k^m[\mu])^T \quad = \quad \frac{\boldsymbol{s}_{\text{eff},k}^m[\mu] \, (\boldsymbol{C}_k^m[\mu])^{-1}}{\boldsymbol{s}_{\text{eff},k}^m[\mu] \, (\boldsymbol{C}_k^m[\mu])^{-1} \, (\boldsymbol{s}_{\text{eff},k}^m[\mu])^T},
$$

where $\boldsymbol{s}_{\text{eff},k}^m[\mu] = \mathcal{E}\{x_k[\mu](\boldsymbol{y}_k^m[\mu])^H\}$ and $\boldsymbol{C}_k^m[\mu] = \mathcal{E}\{\boldsymbol{y}_k^m[\mu] \, (\boldsymbol{y}_k^m[\mu])^H\}$ we obtain $\bar{y}_k^m[\mu] = (\boldsymbol{w}_k^m[\mu])^T \boldsymbol{y}_k^m[\mu]$. For calculation of $\boldsymbol{C}_k^m[\mu]$ the path gains are modeled as $h_{\lambda,\kappa}^m[\mu] = \hat{h}_{\lambda,\kappa}^m[\mu] + \hat{n}_{\lambda,\kappa}^m[\mu]$. Here, we assume that the filters $\boldsymbol{f}_{l,k,\pi}^m[\zeta]$ deliver virtually uncorrelated estimates $\hat{h}_{\lambda,\kappa}^m[\zeta + \pi]$ and zero mean Gaussian distributed estimation errors $\hat{n}_{\lambda,\kappa}^m[\zeta + \pi]$ due to the principle of orthogonality with variance $\sigma_{\hat{n}_{\lambda,\kappa}^m}^2[\zeta + \pi] = \mathcal{E}\{|\hat{n}_{\lambda,\kappa}^m[\zeta + \pi]|^2\} \approx 1/L - (\boldsymbol{f}_{l,k,\pi}^m[\zeta])^T \boldsymbol{\varphi}_{l,k,\pi}^m[\zeta]$. Furthermore, for sake of computational tractability the correlations between the estimates and/or estimation errors for different users data symbols and path weights are neglected, so that the expectation values required for calculation of $\boldsymbol{w}_k^m[\mu]$ can be solved independently for all index pairs (k, l). So, we get with $\sigma_x^2 = \mathcal{E}_{x \in \mathcal{X}}\{|x|^2\} = 1$

$$
\begin{aligned}
\boldsymbol{C}_k^m[\mu] \quad = \quad & \boldsymbol{S}_{k,\mu,k}[\mu] \boldsymbol{D}_k^m[\mu] (\boldsymbol{S}_{k,\mu,k}[\mu])^H + \sigma_n^2 \boldsymbol{I}_{3N+L} \\
& + \sum_{\alpha=\mu-2}^{\mu} \sum_{\kappa=1}^{\substack{K \ \text{if}\, \alpha < \mu \\ k-1 \, \text{if}\, \alpha = \mu}} (\sigma_{x_\kappa}^m[\alpha])^2 \boldsymbol{S}_{k,\mu,\kappa}[\alpha] \boldsymbol{D}_\kappa^m[\alpha] (\boldsymbol{S}_{k,\mu,\kappa}[\alpha])^H \\
& + \sum_{\alpha=\mu-2}^{\mu+2} \sum_{\substack{\kappa = k+1 \, \text{if}\, \alpha = \mu \\ K = 1 \ \ \text{if}\, \alpha > \mu}}^{K} (\sigma_{x_\kappa}^{m-1}[\alpha])^2 \boldsymbol{S}_{k,\mu,\kappa}[\alpha] \boldsymbol{D}_\kappa^m[\alpha] (\boldsymbol{S}_{k,\mu,\kappa}[\alpha])^H.
\end{aligned}
$$

with matrices $\boldsymbol{D}_\kappa^m[\alpha] = (\hat{\boldsymbol{h}}_\kappa^m[\alpha](\hat{\boldsymbol{h}}_\kappa^m[\alpha])^H + \mathrm{diag}(\sigma^2_{\hat{n}^m_{1,\kappa}}[\alpha], \ldots, \sigma^2_{\hat{n}^m_{L,\kappa}}[\alpha])$, channel gain vectors $\hat{\boldsymbol{h}}_\kappa^m[\alpha] = (\hat{h}^m_{1,\kappa}[\alpha], \ldots, \hat{h}^m_{L,\kappa}[\alpha])^T$ and conditional variance $(\sigma^m_{x_\kappa}[\alpha])^2 = \mathcal{E}\{|x_\kappa[\alpha] - \hat{x}^m_\kappa[\alpha]|^2 | \bar{y}^m_\kappa[\alpha]\} = 1 - |\hat{x}^m_\kappa[\alpha]|^2$. Further, the elements $(\boldsymbol{S}_{k,\mu,\kappa}[\alpha])_{ij}, 1 \le i \le 3N + L, 1 \le j \le L$, are

$$(\boldsymbol{S}_{k,\mu,\kappa}[\alpha])_{ij} = \begin{cases} s_{(\nu_{\mu,k,i}, \tau_\kappa, j), \kappa}[\lfloor \nu_{\mu,k,i}, \tau_\kappa, j \rceil] & \text{if } \lfloor \nu_{\mu,k,i}, \tau_\kappa, j \rfloor = \alpha \\ 0 & \text{otherwise,} \end{cases}$$

where $\nu_{\mu,k,i} \stackrel{\triangle}{=} (\mu - 1)N + \tau_k + i$ and $\boldsymbol{I}_Z$ denotes the $Z \times Z$ identity matrix.

Finally, approximating the noise in $\bar{y}^m_k[\mu]$ resulting from uncancelled other users, channel estimation errors as well as channel noise by a complex Gaussian distribution with power $(\sigma^m_k[\mu])^2 = 1/((\boldsymbol{s}^m_{\mathrm{eff},k}[\mu])^T (\boldsymbol{C}^m_k[\mu])^{-1} \boldsymbol{s}^m_{\mathrm{eff},k}[\mu]) - 1$ the soft decision $\hat{x}^m_k[\mu] = \mathcal{E}\{x_k[\mu] | \bar{y}^m_k[\mu]\}$ is calculated as

$$\hat{x}^m_{k,\mathrm{I}}[\mu] = \tanh\left(\frac{1}{(\sigma^m_k[\mu])^2} \bar{y}^m_{k,\mathrm{I}}[\mu]\right) \qquad \hat{x}^m_{k,\mathrm{Q}}[\mu] = \tanh\left(\frac{1}{(\sigma^m_k[\mu])^2} \bar{y}^m_{k,\mathrm{Q}}[\mu]\right),$$

where I and Q represent the real and imaginary component of the data symbols, respectively. In that way, soft estimates are derived for all K users as well as all time slots μ. Further, in order to improve the estimates several iteration cycles are carried out.

3. NUMERICAL RESULTS

The performance of the proposed receiver scheme has been evaluated for uncoded transmission using QPSK–modulation with Doppler frequency $f_D =$ 455 Hz, a transmission bandwidth $1/T_c = 2.5$ MHz, a pilot symbol distance of $p = 4$ and channel estimation filter order $2\gamma + 1 = 19$.

In Fig. 4 the results for half load, i.e., $K/N = 1/2$ are depicted for $N = 16$ and $L = 1, 2, 3$. Further, for comparison the corresponding curves for a single user having to perform channel estimation for $L = 2$ as well as the single user bound (SUB) for perfect channel knowledge are given (see also Proakis, 1995). To provide a fair comparison in all cases the increase in the energy per bit to noise ratio $10 \log_{10}(E_b/N_0)$ by $10 \log_{10}(p/(p-1))$ due to the use of pilot symbols is taken into account. It can be seen that for half load almost the same symbol error rate (SER) can be achieved as for perfect channel state information. In fact, for each number of propagation paths the gap to the single user bound with perfect channel knowledge is less than 1 dB. Thus, the curves indicate clearly that multiuser systems do not inevitably suffer from increasing interference caused by rising number of users if channel estimation is imperfect. Moreover, studying the results for $L = 2$ we find that for the considered parameters even a single user acquiring channel state information by estimation does not reach the equivalent SUB for perfect channel information.

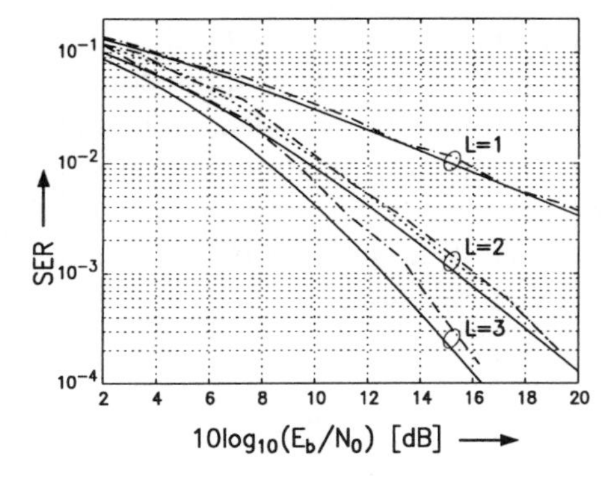

Figure 4 SER vs. $10\log_{10}(E_b/N_0)$ for studied scheme (—·) with $N = 16$ and $K/N = 0.5$ after 5 iterations, single user with channel estimation (· · ·) after 5 iterations and SUB (—) for Rayleigh fading channels with $L = 1, 2, 3$ and $p = 4$.

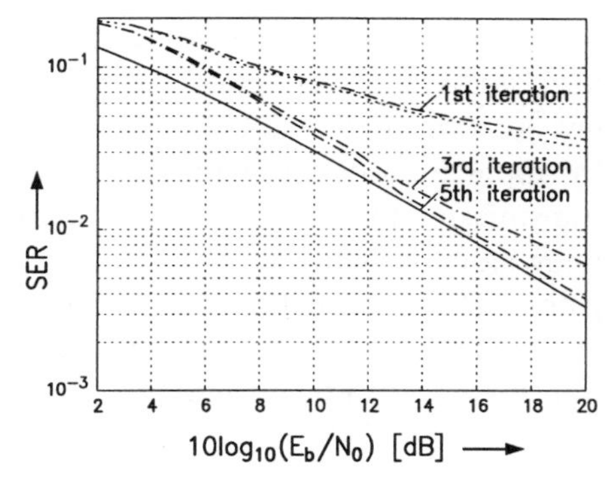

Figure 5 SER vs. $10\log_{10}(E_b/N_0)$ for studied scheme (—·) with $N = 16$ and $K/N = 1$ after 1, 3, 5 iterations, scheme with one channel estimation after 4 multiuser detection cycles (· · ·) and SUB (—) for Rayleigh fading channel with $L = 1$ and $p = 4$.

In order to illustrate the importance of updating the channel estimates let us consider Fig. 5. There, the achievable SER after the 1st, 3rd and 5th iteration for full load, i.e., $K = N$ with $N = 16$ for $L = 1$ is depicted. Further, the SER reachable with only one channel estimation and four consecutive multiuser detection cycles is given.

So, after the first iteration the SER is virtually as bad as that of a linear multiuser receiver employing an MMSE–filter. Moreover, without updating the channel estimates even the application of the iterative soft decision interference cancellation scheme would not yield further gains as can be seen in the plot. In contrast, after the 3rd iteration of joint channel and data estimation nearly minimum SER is already achieved. This underlines again the performance improvement resulting from the significant increase in channel estimation accuracy in course of the iterations and the importance of treating both problems – channel estimation and multiuser detection – together.

4. CONCLUSIONS

We have seen that the so–called "Turbo" principle can be employed successfully for the problem at hand. So, near–optimum multiuser channel estimation and detection can be carried out with a complexity proportional to K^2 instead of exponential order as required by the optimum receiver. Further, taking into account the good performance for uncoded transmission one of the next steps is to combine this algorithm with coding, which is to be expected to yield the same good results. Another one is to involve adaptive channel estimation algorithms into the iterative scheme. More explicitely, we propose to use soft estimates on the data symbols as "pilot" symbols for channel estimation as these soft values are strictly related to the reliability of the estimation. Finally, we find that by application of the iterative algorithm for channel estimation and uncoded transmission the gap to the corresponding single user bound for perfect channel knowledge remains almost constant regardless of the number of propagation paths. Thus, we reckon that by means of iterated channel estimation the performance degradation found in (Schramm, 1996) for increasing L and coded transmission can be overcome.

Notes

1. For sufficiently large spreading gains this is virtually no loss of generality.
2. $J_0(\alpha)$ denotes the bessel function of 1st kind and 0th order, f_D the Doppler frequence and $1/T_c$ the RF bandwidth.
3. The function $\lfloor \alpha \rfloor$ yields the largest integer smaller than or equal to α.
4. All expectation values are conditioned on previously obtained data symbol and channel gain estimates.

References

Lampe, A. and Huber, J. B. (1999). On improved multiuser detection with iterated soft decision interference cancellation. In *Proc. of IEEE International Conference on Communications (ICC)*, pages 172–176, Vancouver, Canada.

Müller, R. R. and Huber, J. B. (1998). Iterated soft–decision interference cancellation for CDMA. In Luise and Pupolin, editors, *Digital Wireless Communications*, pages 110–115. Springer–Verlag, London, U.K.

Proakis, J. G. (1995). *Digital Communications*. McGraw–Hill, New York, 3rd edition.

Schramm, P. (1996). *Modulationsverfahren für CDMA–Mobilkommunikationssysteme unter Berücksichtigung von Kanalcodierung und Kanalschätzung*. Shaker, Aachen.

Teich, W. G. and Seidl, M. (1996). Code division multiple access communication: Multiuser detection based on a recurrent neural network structure. In *Proc. of IEEE Intern. Symp. on Spread Spectrum Techniques and Applications (ISSSTA)*, pages 979–984, Mainz, Germany. Kluwer Academic Publishers.

Vasudevan, S. and Varanasi, M. K. (1996). Achieving near–optimum asymptotic efficiency and fading resistance over the time–varying rayleigh–faded cdma channel. *IEEE Transactions on Communications*, 44(9):1130–1143.

Windpassinger, C. (1999). Gemeinsame Kanalschätzung für Multiuser Übertragung. Studienarbeit (senior project, in German), Lehrstuhl für Nachrichtentechnik, Universität Erlangen–Nürnberg.

MUD Improvement by Using Hexagonal Pilot Distributions for Channel Acquisition and Tracking in MC-CDMA

Faouzi Bader, Santiago Zazo, J. M. Páez Borrallo

Universidad politécnica de Madrid. ETSI Telecomunicación -SSR: C303-Ciudad Universitaria 28040 Madrid (Spain) -Phone: +34 91 3367280 Fax: +34 91 3367350; e-mail. bader gaps.ssr.upm.es

Abstract:

Recently different MC-CDMA (Multicarrier Code Division Multiple Access) systems have been proposed and investigated, with an increasing interest in the field of mobile communications.
An improved technique for MUD (Multi-User Detection) for the uplink concept is proposed. The use of a specific hexagonal pattern for the pilot insertion is the basic support for the distinguishing process of different users for reception. at base station. This particular distribution reduce the MUD complexity at the base station .Based exploiting a free overlapping of user's pilots.
We provide an interference free system to guarantee proper multichannel estimation .

1. INTRODUCTION :

In Mobile Radio Communications Systems much attentions has paid to MC-CDMA, and actually constitute an interest topic of world-wide research. Several methods proposed by [1,2] describing different structure for the downlink transmission process, their main advantage and drawbacks.
The proposal work use the FDM (Frequency Division Multiplexing) concept [1,3,4] where the access methods split the number of users in several groups. This method exploits to the maximum the frequency diversity offered by the mobile radio channel.
In MC-CDMA systems the bandwidth W is splitted in a number of subcarriers for transmitting a sequence S_n $(n = 1......N_C)$,where N_c is the total number of subcarriers. The high rate DS (Direct Spreading) spread data stream is N_C modulations in such away that the L chips of

spread data symbol are transmitted in parallel by the different subcarriers.
The use of pilot symbols is an essential practice for the estimation of the impulse response of the channel and an excellent mean for obtaining information about its variability. The pilot insertion in concretised by a 2-D extension, which offer many advantages and provides a major pilots pattern possibility design and distribution. It also allows to perform channel estimation in two dimensions, in time direction, in addition to the frequency direction, assuming that the sampling rate is sufficient with respect to the channel bandwidth.

The tests and simulation results implemented in [5] demonstrate that the hexagonal pilot distribution in an OFDM-frame offer best performance compared to others distributions like rectangular, linear increase, or frequency sweeping.
Based on this strategy, we proposed using an hexagonal pilot distributions, in addition of some modifications to permit us to improve the process of distinguishing of the pilots locations for the different users in the base station, also decreasing the filtering time process.

2 Transmission Model and Pilot Insertion :

In the uplink concept K mobile users are simultaneously active and transmit their signal to the base station. In our model we assume that each user is affected by an independent channel given by its time impulse response
h(τ;t)

$$h(\tau;t) = \sum_{p=0}^{N_p} h_p(t) e^{j2\pi f_{D,p} t} \delta(\tau - \tau_p) \tag{1}$$

where parametershp are generally complex.

$$h_p = |h_p| e^{j\theta_p} \tag{2}$$

Usually, h_p has a Rayleigh distribution amplitude and θ_p can be modelled as an uniform distribution, $f_{D,p}$ represent the frequency Doppler and τ_p the time delay for the p-th path. The MC transmission can be viewed as a discrete-time and frequency transmission system where each subcarrier is affected by the channel attempt at its corresponding frequency.

Within each OFDM symbol, each user can transmits Q data symbols, $q= 1.....Q,$, each one, with his corresponding spreading word. The received signal in the base station can be expressed as follows.

$$\mathbf{r} = \sum_{k=1}^{K} \mathbf{H}^{(k)} \mathbf{s}^{(k)} + \mathbf{n}^{(k)} \tag{3}$$

where **s** is the transmitted vector symbol corresponding to the k-th user with the spreading code, **H** is a diagonal matrix, which diagonal components of **H** are the complex valued flat fading coefficient H_n , following vector **n** represents the AWGN noise assigned to the subcarrier transmission.

$$\mathbf{H} = \begin{bmatrix} H_1 & 0 & . & . & 0 \\ 0 & H_2 & . & . & . \\ . & . & . & . & . \\ . & . & . & . & 0 \\ 0 & . & . & 0 & H_{N_C} \end{bmatrix} \tag{4}$$

$$\mathbf{n} = (N_1, N_2, ..., N_{Nc}) \tag{5}$$

We analyse the channel estimation aided by a specific hexagonal geometric insertion of pilots within an OFDM-frame duration for different active users. The basic idea in this specificity is that different users exploit different pilot location (therefore avoiding interference between them) but all sharing an hexagonal pattern of pilots, this providing an optimum selection for the uplink multiple access. In this scheme, most of subcarriers for each users are reserved for information transmission, but also several subacarriers do not . This second class is splitted in two parts, one part devoted for channel estimation by the insertion and transmission of pilots of the current user, and the remainder are idle.

The interest of these empty subcarriers is to reserve them for the remaining users pilots. This specific distributions of different user pilots assure a null interference condition between them, because their non overlapping location in time and frequency grid pilots location is guaranteed. Figure 1 shows us the block diagram of the proposed system and the geometry used for the pilot pattern distribution for each user.

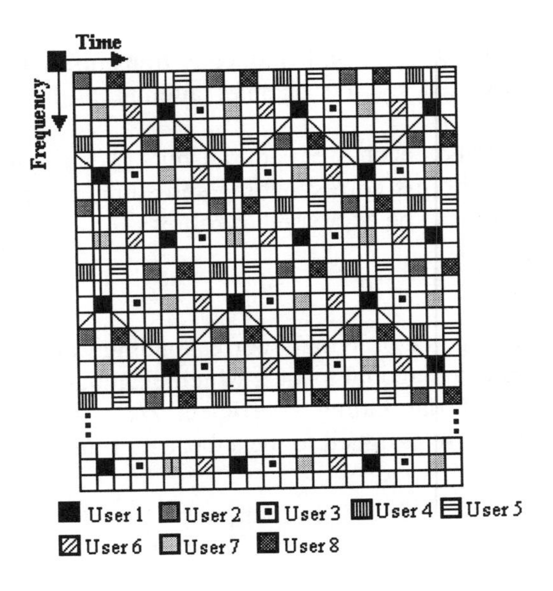

Figure 1. The Hexagonal pilot's distribution representation for eight simultaneously received signals at the base station.

Figure 2, shows this particular scaled scenario, remark the concept of the insertion of hexagonal pilots for different OFDM-frame user's, and conserve the principle of free-overlapping of the different user's pilot pattern.

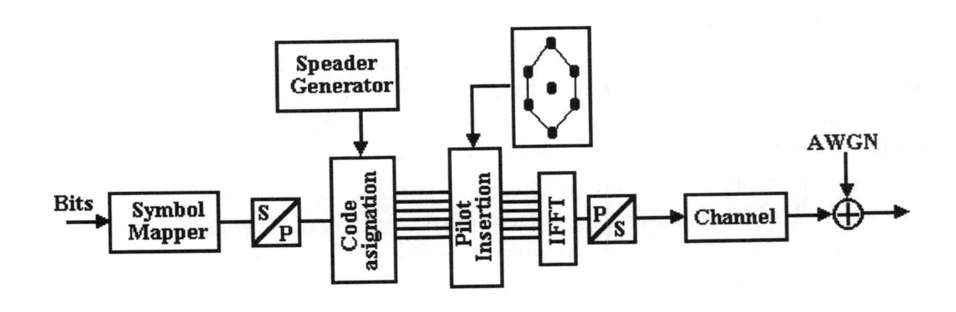

Figure 2 Block diagram of the transmitter for each user

3. FILTERING AND IMPLEMENTATION CONSIDERATIONS

Due the particular insertion of pilots, the transmitted pilot symbol for each user and its reception in the base station can be represented by the expression

$$R_{n,i} = H_{n,i} S_{n,i} + N_{n,i} \tag{6}$$

$n=1.....N_C$, and $i=1,...N_S$, where N_S is the total number of the OFDM-symbols in the OFDM-frame

For each user we proceed the estimation of the channel transfer function at the positions of the pilots symbols, just dividing by the original pilot (note that it is known in advance by the base estation).

$$\hat{H}'_{n',i'} = \frac{R_{n',i}}{S_{n',i'}} = H_{n',i'} + \frac{N_{n',i'}}{S_{n',i'}} \qquad \forall\{n`,i`\} \in \Lambda \tag{7}$$

Λ where represent the set of the pilot positions in an OFDM-frame. The final estimates of the complete channel transfer function belonging to the desired OFDM-frame are obtained from the initial estimates, by using a 2-D Wiener filter. Their coefficient are obtained by applying the orthogonality principle [1,4](noted that the $N_{n',i'}$ has zero mean and is statistically independent from the pilot symbol $S_{n',i'}$).

It is important to indicate that the size of the hexagonal geometry of the pilots pattern depends on the number of users, on the coherence time and on the coherence bandwidth frequency for the expected channel.

The 2-D filtering process is divided in three stages [6], one stage for the starting filtering operation, a second stage for the perfect hexagonal forms in the OFDM-frame, and the third stage deals with at the broken zones the end of OFDM-frame filtering process (if exist).

This treatment of the filtering permitted us to reduce the time of conventional systems for filtering [1,4], an take in consideration the specificity of the hexagonal pilot pattern inserted and used by each user. Our design is able to manage this three stages automatically for the 2-D filtering.

In our proposal, the free overlapping only is concerned for the pilot location, but not for the data symbols which are separable by orthogonal condition of different codes. The MAI may cause a severe degradation of performance even for multiuser detection technique.

This alteration conduct us to implemente a specific MAI cancelator to improve the estimation of the channel transfer function for each user. The specificity concern and take account the special distribution of the pilots pattern for the different users.

Figure 4, represents the block diagram that resume the process of interference cancellation for MAI caused by the multi-user detection where each user has a specific hexagonal pattern for their pilots which are directly detected by base station.

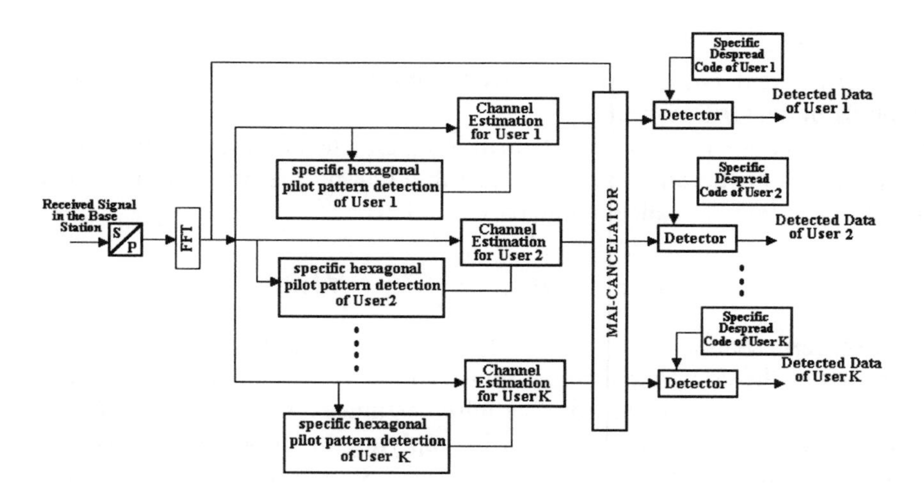

Figure 4 : Block diagram of the receiver at the base station

When, a new user intends to access to the base station, a special resource must be dedicated in his acquisition phase; during this stage the new user captures all the pilots of the rest of users during a determinate number of OFDM symbols time. This fact provides a time-frequency offset correction as an initial channel estimation of the new entering user. In this work, results about the performance degradation when the hexagonal distribution is broken at the edges of the time-frequency grid and, at the acquisition phase of new user are provided.

4. SIMULATION RESULTS:

This proposal use hexagonal pilot patterns in an MC-CDMA system adapted for an uplink transmission process. For the simulation test we have considered K=8 simultaneously active users where each user transmit Q=3 data symbol per OFDM-symbol, the total number of subcarriers including the pilots carriers used in the tests is N_C =32, and a total number of OFDM-symbols equal to N_S= 36. The spreading code is a Walsh-Hadamard of length L=8, and the density of pilots used is the order of dp(%)=2.7. All of these aspects are maintained for all the eight users. Let us remark that we are dealing with an scaled problem where the transmission bandwidth is limited to 3 KHz.

Noted that the percentage of the pilot density used in the simulation tests is lower than the conventional pilot insertions used in [1,4], that no prevent that obtain good estimation, this is relationed with the geometry distribution of the pilots that is well known as the optimum distribution in 2-D sampling, providing better BER (Bit Error Rate) [5], and goods channel estimations in the time and frequency domain [6], the base station can distinguish the different pilot channel response with more fluctuations.

We can show in figure 5 the effect of the brooken hexagonal zones in an OFDM frame]. This fact is obseved in all rest of users, but for each one we have specifics brooked zones.

We can show the good channel estimation in the time domain, this results can be explained by the specifity of the hexagonal pilot distribution in the user's OFDM-frames in adition of free-overlapping condition between them.

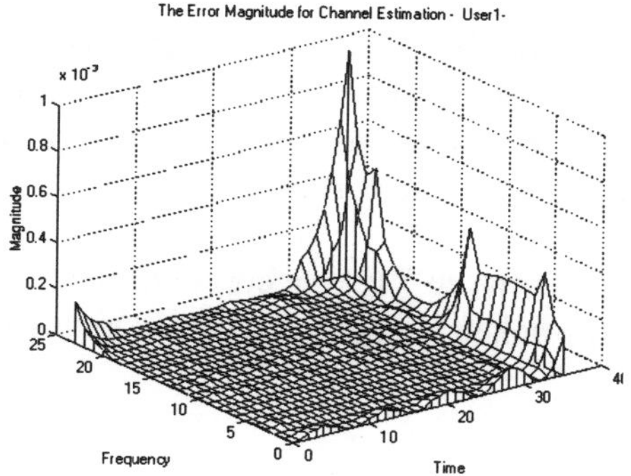

Figure 5: The error magnitude between the received estimated channel and the originally transmitted in the OFDM-frame.

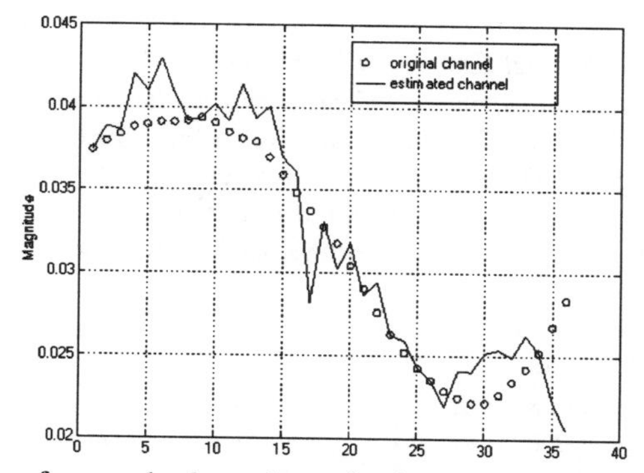

Figure 6: Magnitude of a sample channel transfer function and its estimate attained by time channel estimation (for the user 1).

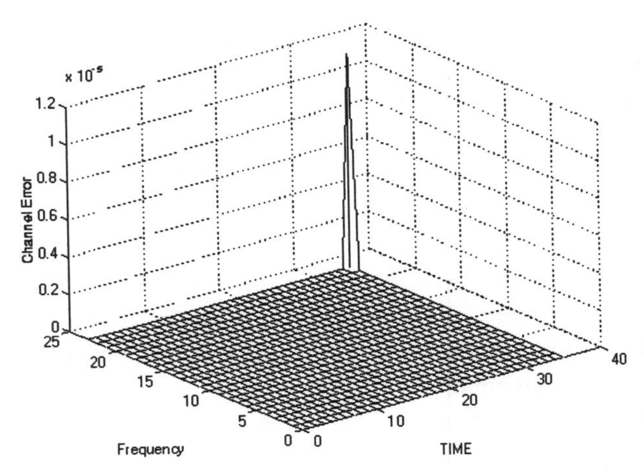

Figure 7: Channel error after MAI cancelation process (for user 1)

The result of the application for the MAI-cancelator shows in figure 4, reduce cosiderabily the effect of interference due to the multi-user detection, remark that all this simulation results do not take account of any kind of coding, which in any case will improve the performance.

The value reflected on the figure 7 (in the corner of the graphic), is relationed with the existance of the boken hexagonal schemes in this zone.

References

[1]: Kaiser, S, " *Multi-Carrier CDMA Mobile Radio System -Analysis and Optimisation of Detection Decoding and Channel Estimation* ", PhD Thesis 1998.

[2]: Nathan Yee and Jean paul Linnartz, " *Controlled Equalization of Multi-Carrier CDMA in an Indoor Rician Faiding Channel* ", In proceeding IEEE Vehicular Technologie Conference, pp1665-1669 ,(VTC'94),Stockholm,Sweden, June 1994.

[3]: Bernd Steiner , " *Uplink Performance of a Multicarrier CDMA Mobile Radio System* ", In proceeding IEEE Vehicular Vehicular Technologie Conference,(VTC'97) pp1902-1906, Phoenix USA, May 1997.

[4]: Khaled Fazel, Gerhald P. Fettweis," *Multi-carrier Spread spectrum* ", Kluwer Academic Publisher 1997,Netherlands.

[5]: Férnandez-Getino García M.Julia, J. M. Páez Borrallo, S Zazo, " *Novel Pilot Pattern for Channel Estimation in OFMD Mobile Systems Over Selective Faiding Channels* ", PIRM'99, Sept 99 Osaka, Japan.

[6]: Faouzi Bader, Santiago Zazo, J. M. Páez Borrallo, " *Optimum Pilot Pattern for Multicarrier-CDMA Systems* ", In proceeding IEEE WPCM'99, Amsterdam, Netherlands.

ADAPTIVE MULTI-SHOT MULTIUSER DETECTION FOR ASYNCHRONOUS MC-CDMA USING BOOTSTRAP ALGORITHM

Pingping Zong[1], Yeheskel Bar-Ness[1] and James K. Chan[2]

[1]Center for Communications and Signal Processing Research
Department of Electrical and Computer Engineering
New Jersey Institute of Technology
University Heights, Newark, NJ 07102
ping@yegal.njit.edu, barness@yegal.njit.edu

[2]RDL
Upalnder way 5800
Culver City, CA 90230
chan@RDL.com

Abstract The performance of an asynchronous multi-carrier code division multiple access (MC-CDMA) system is limited by intersymbol interference (ISI), as well as multiuser interference (MUI) due to the loss of orthogonality among signature codes. Additionally, in the presence of a frequency offset, intercarrier interference (ICI) is introduced due to the loss of orthogonality between adjacent subcarriers. In this paper, we propose an adaptive multi-shot detector using the Bootstrap algorithm that cancels the ISI, MUI and ICI jointly. The detector performs approximately the same as a Wiener filter and does not need a training sequence to converge.

This research was supported by RDL, CA, under the Air Force Research Lab, Rome, NY, SBR program.

1. INTRODUCTION

Multi-carrier code division multiple access (MC-CDMA) is a scheme combining multi-carrier modulation (MCM) and direct-sequence code division multiple access (DS-CDMA). It was proposed in [1] to be a promising candidate to combat the frequency selectivity of a multipath fading channel in wide band wireless digital communication. If the symbol duration in each subcarrier is greater than the delay spread and the separation between each subcarrier is greater than the coherent bandwidth, each subchannel will appear like a flat-fading channel and independent to each other. Frequency diversity is available to be exploited. With additional cyclic prefix, the inter-symbol interference can be avoid.

Using MC-CDMA in the uplink of a communication system results in asynchronous reception of the different users' signals. The output of a coherent detector which is matched to the desired user's signal and sample in synchronous with desired user's symbol consists of partial energy of other users' as well as the desired user's. Hence, in an asynchronous MC-CDMA system, a conventional coherent detector introduces intersymbol interference (ISI). A so-called multi-shot multi-user detector has been used in asynchronous DS-CDMA to eliminate the MUI and ISI [2, 3]. Such multi-shot approach takes into account multiple symbol intervals rather than just the current symbol interval as in the conventional coherent detector.

In this paper we will propose an adaptive multi-shot detector that jointly performs multiuser and intersymbol interference, as well as intercarrier interference cancellation for an asynchronous MC-CDMA system. It is shown that the proposed detector performs close to that of a Wiener filter. Additionally, the proposed detector is a blind algorithm, as it requires no training sequences.

2. SYSTEM MODEL

We consider an asynchronous MC-CDMA system with K active users and M subcarriers. Each user is assigned a unique spreading code. The code length is assumed to be equal to the number of subcarriers, M. Different users are assigned different transmission powers. Before being transmitted, each user's information symbol is replicated into M parallel copies and multiplexed with the spreading code. By applying an M-point IDFT operation, the M coded symbols are modulated onto the M subcarriers. In an asynchronous system, different users transmit symbols at different times, hence the transmitted signal can be described by

$$s(t) = \sum_{k=1}^{K} \sum_{i=-\infty}^{+\infty} \sqrt{a_k} b_k(i) P_{T_b}(t - \tau_k - iT_b) \sum_{m=1}^{M} c_{km} e^{j2\pi \frac{m}{T_b}(t-\tau_k)} \quad (1)$$

where $b_k(i)$ denotes the kth user's ith information symbol generated by a 4-QAM source, a_k corresponds to the transmission power of the kth user, c_{km} is the mth chip of the kth user's spreading code $\mathbf{c}_k$, T_b is the symbol duration, τ_k denotes the delay of the kth user, assuming $\tau_1 \leq \tau_2 \leq \ldots \leq \tau_K$ and

$$P_{T_b}(t) = \begin{cases} 1 & 0 \leq t \leq T_b \\ 0 & otherwise \end{cases} \quad (2)$$

With the IDFT operation, each subcarrier is separated by $\Delta f = \frac{1}{T_b}$.

The channel is assumed to be a frequency-selective channel with $\frac{1}{T_b} \ll BW_c \ll \frac{M}{T_b}$. Each subcarrier's distortion can be uncorrelated or correlated, depending on the subcarrier separation comparing with channel's coherent bandwidth BW_c. The channel's impulse response can be written as:

$$h(t) = \sum_{m=1}^{M} h_m e^{j\theta_m} e^{j2\pi \frac{m}{T_b} t} \quad (3)$$

where θ_m and h_m denote the channel phase and amplitude of the mth subcarrier, respectively. The h_m's are assumed Rayleigh distributed and the θ_m's are assumed uniformly distributed. Supposed that we can track the phase accurately, so $\theta_m = 0$. Then, the received signal is given by:

$$\begin{aligned} r(t) &= s(t) \otimes h(t) \\ &= \sum_{i=-\infty}^{+\infty} \sum_{k=1}^{K} \sqrt{a_k} b_k(i) P_{T_b}(t - \tau_k - iT_b) \sum_{m=1}^{M} c_{km} h_m e^{j2\pi \frac{m}{T_b}(t-\tau_k)} \quad (4) \end{aligned}$$

At the receiver, in order to separate and detect asynchronous MC-CDMA signals, K multi-shot DFT operators are implemented to demodulate the received signal at different users' symbol intervals. The lth DFT operator demodulates the received signal synchronized to the lth user's symbol interval $(iT_b + \tau_l, (i+1)T_b + \tau_l)$. This synchronization can also be done by delaying the received signal by τ_l, then applying the DFT at the interval $(iT_b, (i+1)T_b)$. We use the latter structure here. Additionally, a frequency offset, which is assumed to be same for all users, is considered in our system. Therefore, the output of the desired lth user's DFT operator can be described by:

$$y_l(p) = \text{DFT}\{r(t+\tau_l) e^{j2\pi \frac{\epsilon}{T_b} t} + n(t+\tau_l)\} \quad (5)$$

where $n(t)$ is a complex white Gaussian noise process with zero mean and variance σ^2 and $\epsilon = \frac{\text{offset}}{\Delta f}$ is the normalized frequency offset. The outputs of the lth DFT are despreaded by multiplying with the code sequence $\mathbf{c}_l$:

$$x_l(i) = \sum_{p=1}^{M} y_l(p) c_{lp}$$
$$= \sum_{k=1}^{K} \rho_{lk} \sqrt{a_k} b_k(i) + \sum_{k=1}^{l-1} \rho_{lk}^{L} \sqrt{a_k} b_k(i-1) + \sum_{k=l+1}^{K} \rho_{lk}^{R} \sqrt{a_k} b_k(i+1) + \eta_l \quad (6)$$

where,

$$\rho_{lk}(\epsilon) = \frac{1}{T_b} \sum_{m=1}^{M} \sum_{p=1}^{M} h_m c_{km} c_{lp} \int_0^{T_b} e^{j2\pi \frac{m}{T_b}(t-\tau_{lk})} e^{-j2\pi \frac{(p-\epsilon)}{T_b} t} dt \quad (7)$$

$$\rho_{lk}^{L}(\epsilon) = \begin{cases} \frac{1}{T_b} \sum_{m=1}^{M} \sum_{p=1}^{M} h_m c_{km} c_{lp} \int_0^{T_b} e^{j2\pi \frac{m}{T_b}(t+T_b-\tau_{lk})} e^{-j2\pi \frac{(p-\epsilon)}{T_b} t} dt & l < k \\ 0 & l \geq k \end{cases} \quad (8)$$

$$\rho_{lk}^{R}(\epsilon) = \begin{cases} \frac{1}{T_b} \sum_{m=1}^{M} \sum_{p=1}^{M} h_m c_{km} c_{lp} \int_0^{T_b} e^{j2\pi \frac{m}{T_b}(t-T_b-\tau_{lk})} e^{-j2\pi \frac{(p-\epsilon)}{T_b} t} dt & l > k \\ 0 & l \leq k \end{cases} \quad (9)$$

$$\tau_{lk} = \tau_k - \tau_l \quad (10)$$

$$\eta_l = \frac{1}{T_b} \sum_{p=1}^{M} c_{lp} \int_0^{T_b} n(t+\tau_l) e^{-j2\pi \frac{p}{T_b} t} dt \quad (11)$$

Note that $e^{j2\pi \frac{m}{T_b} t}$ is non-zero only inside the symbol interval $(0, T_b)$ and η_l, $l = 1, \ldots, K$, is a sequence of colored Gaussian noise with zero mean and covariance $\rho_{lk}(0)\sigma^2$.

From equation (6), we notice that besides the multiuser interference (MUI) and intercarrier interference (ICI), the output of the receiver also consists of the intersymbol interference (ISI), which is represented by the second and the third term. If all the users have the same transmission delay (i.e. $\tau_{lk} = 0$), the terms with ρ_{lk}^{L} and ρ_{lk}^{R} vanish and equation (6) reduces to a synchronous model. The proposed multi-shot receiver is shown as Fig 1. We can obtain the ith instant output in matrix notation:

$$\begin{aligned} \mathbf{x}(i) &= \mathbf{PAb}(i) + \mathbf{P}_L \mathbf{Ab}(i-1) + \mathbf{P}_R \mathbf{Ab}(i+1) + \boldsymbol{\eta}(i) \\ &= \begin{bmatrix} \mathbf{P}_L & \mathbf{P} & \mathbf{P}_R \end{bmatrix} \begin{bmatrix} \mathbf{Ab}(i-1) \\ \mathbf{Ab}(i) \\ \mathbf{Ab}(i+1) \end{bmatrix} + \boldsymbol{\eta}(i) \end{aligned} \quad (12)$$

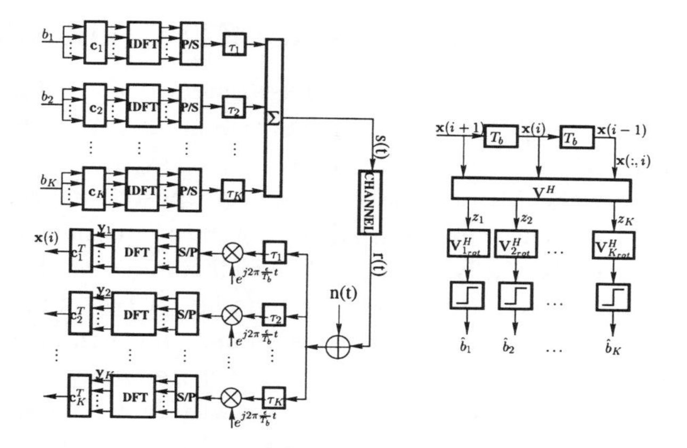

Figure 1 System model of an asynchronous MC-CDMA system.

where,$\mathbf{A} = \text{diag}\left[\sqrt{a_1}, \sqrt{a_2}, \ldots, \sqrt{a_K}\right]$, $\mathbf{b}(i) = [b_1(i), b_2(i), \ldots, b_K(i)]^T$, $\mathbf{P}$ is a matrix defined as $\mathbf{P}(l,k) = \rho_{lk}$, $\mathbf{P}_L$ is an upper triangle matrix defined as $\mathbf{P}_L(k,l) = \rho_{lk}^L$, $\mathbf{P}_R$ is a lower triangle matrix defined as $\mathbf{P}_R(k,l) = \rho_{lk}^R$ and $\boldsymbol{\eta}(i) \sim \mathcal{N}(\mathbf{0}, \sigma^2\mathbf{P}(0))$, where $\mathbf{P}(0)$ denotes matrix $\mathbf{P}$ when $\epsilon = 0$ and $h_m = 1, m = 1, 2, \ldots, M$.

As the structure of equation (12), multi-shot algorithm suggests that $\mathbf{x}(n)(n = i - \frac{N-1}{2}, \ldots, i, \ldots, i + \frac{N-1}{2})$ which includes outputs at N instants can be stacked into an $NK \times 1$ vector to estimate the current bit vector $\mathbf{b}(i)$(see Fig 1). Since most interference to the current bit is contributed by the previous one and the next one bit, $N = 3$ is a reasonable approximation to reduce the computational complexity [2]. Therefore, we rewrite equation (12) as:

$$\begin{aligned} \begin{bmatrix} \mathbf{x}(i-1) \\ \mathbf{x}(i) \\ \mathbf{x}(i+1) \end{bmatrix} &= \begin{bmatrix} \mathbf{P} & \mathbf{P}_R & \mathbf{0} \\ \mathbf{P}_L & \mathbf{P} & \mathbf{P}_R \\ \mathbf{0} & \mathbf{P}_L & \mathbf{P} \end{bmatrix} \begin{bmatrix} \mathbf{Ab}(i-1) \\ \mathbf{Ab}(i) \\ \mathbf{Ab}(i+1) \end{bmatrix} \\ &\quad + \begin{bmatrix} \mathbf{P}_L\mathbf{Ab}(i-2) \\ \mathbf{0} \\ \mathbf{P}_R\mathbf{Ab}(i+2) \end{bmatrix} + \begin{bmatrix} \boldsymbol{\eta}(i-1) \\ \boldsymbol{\eta}(i) \\ \boldsymbol{\eta}(i+1) \end{bmatrix} \\ \mathbf{x}(:,i) &= \mathcal{P}\mathbf{Ab}(:,i) + \mathbf{bias} + \boldsymbol{\eta}(:,i) \end{aligned} \tag{13}$$

where the term of **bias** is due to the truncation approximation, which will be ignored to simplify the analysis without loss of generality, $\boldsymbol{\eta}(:,i) \sim \mathcal{N}(\mathbf{0}, \sigma^2\mathcal{P}(0))$.

3. ADAPTIVE MULTIUSER DETECTOR APPROACH

we will propose an adaptive multiuser detector using bootstrap algorithm in this section. The detector has two stage and will cancel MUI, ICI and ISI jointly for an asynchronous MC-CDMA system. It has been shown that like the MMSE detector, the proposed detector has better performance than the conventional decorrelator in the low interference region and asymptotically delivers the performance of the conventional decorrelator in the high interference region [4]. A two stage adaptive detector has been proposed to correct frequency offset in [5] for an OFDM system and to jointly cancel MUI and correct frequency offset in [6] for a synchronous MC-CDMA system. Here we will show that such a detector can also be adopted in an asynchronous MC-CDMA system.

The structure of the two stage adaptive detector can be described as follows: the first stage is used to minimize the correlation among receiver's outputs $\mathbf{x}(:,i)$ in order to separate symbols of different users, whereas the second stage is used to separate the in-phase and quadrature components of symbols.

The bootstrap implementation of the first stage is described as follows:

$$\mathbf{z}(:,i) = (\mathbf{I} - \mathbf{W})^H \mathbf{x}(:,i) \tag{14}$$

where $\mathbf{W}$ is the weight matrix with zeros on the diagonal. It is chosen such that

$$E[z(k,i)\mathbf{b}(k,i)] = \mathbf{0} \tag{15}$$

where $\mathbf{b}(k,i)$ denotes vector $\mathbf{b}(:,i)$ without the kth element. Equation (15) means that the kth output of the detector is uncorrelated with other users' symbols. $\mathbf{W}$ can be obtained adaptively by a recursive equation:

$$\mathbf{w}_k(i+1) = \mathbf{w}_k(i) + \mu z(k,i)^* \hat{\mathbf{b}}(k,i) \tag{16}$$

where $\hat{\mathbf{b}}(:,i)$ is decisions, $\hat{\mathbf{b}}(:,i) = csign(\mathbf{z}(:,i))$, $\mathbf{w}_k$ is the kth column of $\mathbf{W}$ without the kth element and $\mathbf{z}(k,i)$ denotes vector $\mathbf{z}(:,i)$ without the kth element. Notice that *csign* denotes a complex sign operation:

$$csign(.) = sign(real(.)) + j \times sign(imag(.)) \tag{17}$$

Under the assumption of high interference, $\mathbf{W}$ can be derived from equation (15):

$$\mathbf{w}_k^* = (\mathcal{P}_k)^{-1}\mathbf{p}_k, \tag{18}$$

where $\mathcal{P}_k$ denotes the matrix $\mathcal{P}$ without the kth column and the kth row, $\mathbf{p}_k$ is the kth column of $\mathcal{P}^T$ without the kth element. In this paper, we refer to these $\mathbf{w}_k$ as decorrelating weights. Using the decorrelating weights, the decorrelated signal of the kth user can be written as:

$$z(k,i) = x(k,i) - \mathbf{w}_k^H \mathbf{z}(k,i) = \alpha_k \sqrt{a_k} b(k,i) \tag{19}$$

Notice, $\alpha_k = p_{kk} - \mathbf{p}_k^T \mathcal{P}_k^{-1} \mathbf{q}_k$. Where p_{kk} is the (k,k)th element of $\mathcal{P}$ and $\mathbf{q}_k$ is kth column of $\mathcal{P}$ without the kth element. When a frequency offset presents, α_k will become a complex number and will rotate the constellation of the decision, $\mathbf{b}(k,i)$. The second stage of the adaptive detector is used to cancel the I-Q interference due to α_k. The bootstrap implementation of the second stage is similar to the first stage except that we need $3K$ 2×2 weight matrices $\mathbf{W}_{rot_k}$ to separate $3K$ symbols' in-phase and quadrature components individually. The weight value can be derived as in [6]:

$$w_{rot12_k} = -w_{rot21_k} = \frac{Im\{\alpha_k\}}{Re\{\alpha_k\}} \tag{20}$$

4. SIMULATION RESULTS

Simulation results of the proposed adaptive multi-shot detector over a non-fading AWGN channel is shown in Fig 2 and Fig 3. In our simulations we use an MC-CDMA system with $M = 7$ subcarriers and $K = 3$ active users. The codes are chosen from normalized Gold code sequences with length 7. The transmission delays are chosen as, $\tau_1 = 0$, $\tau_2 = 0.1T_b$ and $\tau_3 = 0.2T_b$. The normalized frequency offset is $\epsilon = 0.1$ for all users. The symbols are generated by a normalized 4-QAM source. We define the SNR as signal power to noise variance ratio(i.e. $\text{SNR}_k = \frac{a_k}{\sigma^2}$). User1 is the desired user.

Fig 2 shows the constellations of the desired user's symbols at different points of the receiver for $\text{SNR}_1 = 10$ dB and $\text{SNR}_2 = 5$ dB, where SNR_2 refers to all other users' signal to noise ratio. Fig 2(b) shows that, at the output of matched filter, the received constellation points are widely scattered and rotated due to the MUI, ISI and ICI. In Fig 2(c), the constellation is much more condensed, since interference from other users is suppressed by the detector after the first stage. After applying the second detector stage the constellation is rotated back, as is shown in Fig 2(d). We can observe that with interference suppression and constellation rotation, the desired user's signal can be well recovered.

Fig 3 shows the comparison of the near-far performances among different detectors as $\text{SNR}_1 = 10$ dB. It can be seen that the performance of the proposed two stage adaptive detector is very close to the performance of the Wiener filter, which is optimal in a MSE sense with respect to both noise and interference.

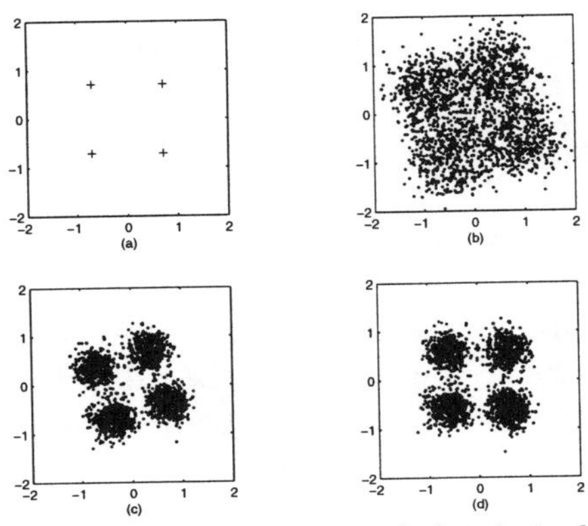

Figure 2 Constellations of the desired symbols for $SNR_1 = 10$ dB and $SNR_2 = 5$ dB, (a) Transmitted constellation, (b) Matched filter output, (c)First stage output, (d)Second stage output.

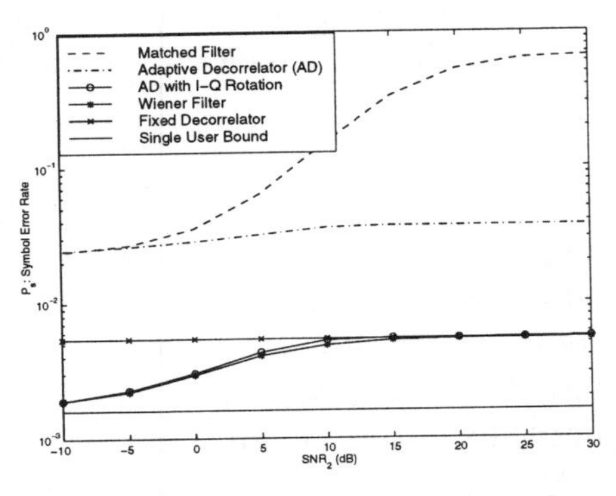

Figure 3 Symbol error rate vs. interference power for SNR1=10dB.

5. CONCLUSION

In this paper, we proposed an multi-shot multiuser detector for an asynchronous MC-CDMA system. The proposed detector achieves similar performance to that of the Wiener filter. Such detector can be implemented adaptively by using bootstrap algorithm.

REFERENCES

[1] N. Yee, J. P. Linnartz, and G. Fettweis, "Multi-carrier CDMA indoor wireless radio networks," in *Proc. of the PIMRC'93*, (Japan), pp. 109–113, Sept. 1993.

[2] H. Ge and Y. Bar-Ness, "Multi-shot approach to multiuser separation and interference suppression in asynchronous CDMA," in *Proc. of the CISS'96*, (U.S.A.), pp. 590–595, Mar. 1996.

[3] N. J. M. van Waes and Y. Bar-Ness, "Adaptive algorithm for the multishot matched filtering multiuser detector in a multipath Rayleigh fading environment," in *Proc. of the VTC'98*, (Canada), pp. 184–188, May 1998.

[4] Y. Bar-Ness and J. B. Punt, "Adaptive bootstrap multi-user CDMA detector," *special issue on "Signal Separation and Interference Cancellation for Personal, Indoor and Mobile Radio Communications," Wireless Personal Communications*, vol. 3, no. 1-2, pp. 55–71, 1996.

[5] M. A. Visser and Y. Bar-Ness, "Frequency offset correction for OFDM using a blind adaptive decorrelator in a time-variant selective Rayleigh fading channel," in *Proc. of the VTC'99*, (U.S.A.), pp. 1281–1284, May 1999.

[6] M. A. Visser and Y. Bar-Ness, "Joint multiuser detection and frequency offset correction for downlink MC-CDMA," in *Proc. of GlobeCom'99*, (Brazil), Dec. 1999.

Section V

INTERFERENCE CANCELLATION

BLIND ADAPTIVE INTERFERENCE CANCELATION FOR MULTICARRIER MODULATED SYSTEMS

Matthijs A. Visser and Yeheskel Bar-Ness
Center for Communications and Signal Processing Research
New Jersey Institute of Technology
Newark, NJ 07102, U.S.A.
e-mail: visser@yegal.njit.edu, barness@yegal.njit.edu

Abstract In this paper we describe various applications of bootstrap based blind adaptive signal separators. Schemes are developed for multiuser detection (MUD) for downlink MC-CDMA, frequency offset correction (FOC) for OFDM and joint MUD and FOC for downlink MC-CDMA.

1. INTRODUCTION

Multicarrier modulation (MCM) is currently being used or proposed as part of many systems and standards. Examples of the use of MCM are the high-rate digital subscriber line (HDSL), digital audio broadcasting (DAB) system and digital terrestrial television broadcasting (dTTb) system. Multicarrier modulation is also being investigated as part of high-speed indoor wireless local area networks (WLAN) and for mobile communications [1, 2, 3, 4, 5].

Advantages of MCM over single carrier modulation (SCM) are: narrowband transmission at each subcarrier, resulting in less complex equalization in the frequency domain; increased immunity to impulse noise due to the longer symbol time; no inter-symbol interference (ISI) with an appropriate cyclic prefix; high spectral efficiency due to overlapping subcarrier spectra; better narrowband interference rejection capability; and greater flexibility in the use of time/frequency slots [1, 6]. Also, MCM may be efficiently implemented using an inverse discrete Fourier transform (IDFT), resulting in orthogonal frequency division multiplexing (OFDM) [7].

Disadvantages of MCM are: increased sensitivity to a frequency offset due to the closely spaced subcarriers, resulting in inter-carrier interference (ICI); increased peak-to-average power ratio (PAPR) and dynamic range, potentially resulting in significant distortion when transmitted through a non-linear device such as a power amplifier [8, 9].

Multicarrier modulation may be combined with spread-spectrum (SS) techniques to create multiple access schemes. Various hybrid MCM/SS schemes have been proposed in the literature [10, 11, 12, 13, 14, 6]. In this paper, multicarrier code division multiple access (MC-CDMA), will be considered [10]. In an MC-CDMA system, the same data symbol is transmitted in parallel over N_c subcarriers, each multiplied by a different element of a users' spreading code. Hence, the code length equals the number of subcarriers.

2. MULTIUSER DETECTION

The performance of an MC-CDMA system—similar to a direct sequence CDMA (DS-CDMA) system—is limited by the presence of multiple access interference (MAI). As a result of the multipath channel effect, downlink (base-to-mobile) communications also suffers from MAI, even if it uses orthogonal code multiplexing. Additionally, transmissions aimed at different mobile users may be assigned different powers, creating a near-far problem for some users. Hence, it is judicious to use a so-called 'near-far resistant' (multiuser) detector to improve the performance of the desired user.

2.1 DOWNLINK MC-CDMA SYSTEM MODEL

A downlink MC-CDMA system with N_c subcarriers and K_a active users ($K_a \leq K$, the maximum number of users) is considered. Multiple access is achieved by assigning a signature sequence from a set of Gold code sequences of length N_c to each user. The received signal $r_m(t)$ is processed by a coherent (matched filter) detector and it is assumed that the orthogonality of the subcarriers is still intact after transmission through the channel, i.e. no significant frequency offset exists. The output of the matched filter in vector notion is:

$$\mathbf{z} = \mathbf{A}\mathbf{C}\sqrt{\mathbf{P}}\,\mathbf{b} + \mathbf{n}, \tag{1}$$

where $\mathbf{P} = \mathrm{diag}\{\mathrm{p}_1, \ldots, \mathrm{p}_{\mathrm{K_a}}\}$ with p_k the power transmitted to user k. The matrix $\mathbf{A} = \mathrm{diag}\{\mathrm{a}_1, \ldots, \mathrm{a}_{\mathrm{N_c}}\}$ contains the subcarrier amplitudes and $\mathbf{n}$ is a vector that contains independent identically distributed (i.i.d.) zero-mean Gaussian noise samples. Additionally, $\mathbf{C} = [\mathbf{c}_1, \mathbf{c}_2, \ldots, \mathbf{c}_{K_a}]$, the $N_c \times K_a$ code matrix, where $\mathbf{c}_k = [c_k(1), c_k(2), \ldots, c_k(N_c)]^{\mathrm{T}}$.

The output of the receiver can be multiplied with either $\mathbf{c}_k$, to get the decision variable using conventional single user detection (SUD) for user k, or with the complete code matrix $\mathbf{C}$, to generate decision variables for all users such that multiuser detection (MUD) can be applied:

$$\mathbf{x} = \mathbf{C}^{\mathrm{T}}\mathbf{z} = \mathbf{C}^{\mathrm{T}}\mathbf{A}\mathbf{C}\sqrt{\mathbf{P}}\,\mathbf{b} + \mathbf{C}^{\mathrm{T}}\mathbf{n} = \mathbf{P}_c'\sqrt{\mathbf{P}}\,\mathbf{b} + \boldsymbol{\eta}, \tag{2}$$

where the conditional (on $\mathbf{A}$) code cross-correlation matrix $\mathbf{P}_c' = \mathbf{C}^{\mathrm{T}}\mathbf{A}\mathbf{C}$.

2.2 CONVENTIONAL DECORRELATOR

The optimal near-far resistant detector performs much better than the conventional matched filter (MF) detector at the expense of a complexity that is exponential in the number of users [15]. A sub-optimal detector, called *conventional decorrelator*, that is linear (in operation as well as in complexity) was presented in [16] and [17].

Equation (2), implies that the signals of the different users are correlated due to MAI. In the case of a DS-CDMA system the *conventional decorrelator* uses the inverse of the code cross-correlation matrix to separate the signals [16]. Now, the inverse of $\mathbf{P}_c'$ is defined as the *conventional decorrelator* for MC-CDMA (henceforth also referred to as the full dimensional decorrelator (FDD)), which completely separates the different users' signals:

$$\mathbf{y} = \mathbf{P}_c'^{-1}\mathbf{x} = \mathbf{P}_c'^{-1}\mathbf{C}^{\mathrm{T}}\mathbf{A}\mathbf{C}\sqrt{\mathbf{P}}\,\mathbf{b} + \mathbf{P}_c'^{-1}\boldsymbol{\eta} = \sqrt{\mathbf{P}}\,\mathbf{b} + \boldsymbol{\xi}'. \tag{3}$$

Due to the signal structure of MC-CDMA, the FDD is dependent on the channel parameters, indicating that adaptive multiuser detection might be preferable, because it may track changes in the channel.

2.3 IMPROVED MULTIUSER DETECTION

In [18] an adaptive, bootstrap based, multiuser detector was proposed for DS-CDMA. This detector does not totally separate the users' signals, but leaves some interference residue in the output. Consequently, the additive noise is less enhanced, resulting in a better output signal-to-interference-plus-noise (SINR) ratio. This detector achieves single-user performance, i.e. outperform the fixed decorrelators, for low interference power, while achieving equal performance in the case of high interference power; as such it behaves similar to MMSE based detectors [19, 20]. The application of bootstrap based detectors to MC-CDMA has also been studied in [21, 22, 23, 24].

The bootstrap signal separator is implemented using a matrix $\mathbf{V}$ to separate the signals, where $\mathbf{V} \neq \mathbf{P}_c'^{-1}$. It was proposed to use $\mathbf{V} = \mathbf{I} - \mathbf{W}$, where $\mathbf{I}$ is the identity matrix and $\mathbf{W}$ is a matrix with zeros on the diagonal (to preserve the desired signals) whose off-diagonal elements will be determined adaptively. The bootstrap algorithm uses the following recursive update equation for the elements of $\mathbf{W}$:

$$\mathbf{w}_k(i+1) = \mathbf{w}_k(i) + \mu\, y_k\, \mathrm{sgn}(\mathbf{y}_k), \tag{4}$$

where $\mathbf{w}_k$ is the kth column of $\mathbf{W}$ without the kth element and μ is the step-size parameter of the adaptation process.

2.4 REDUCED COMPLEXITY MUD

For a large number of users the complexity of the full decorrelator will become prohibitively large as the complexity increases with the number of users. Additionally, the required knowledge about the interfering users—which users are active, their powers and their codes—might inhibit application of the multiuser detector. To reduce the complexity and alleviate these requirements we proposed to use a 'compounded' signature code that represents *all* the possible interfering users. Thus, only the code of the desired user and the compounded code remain. As a result, the dimension of the weight matrix is reduced from $K_a \times K_a$ to 2×2. This also significantly reduces the number of adaptively controlled weights of an adaptive implementation. The codes of all the possible interfering users can be compounded as follows [22, 24]:

$$\mathbf{c}_k^{c^{\mathrm{T}}} = \boldsymbol{\alpha}^{\mathrm{T}} \mathbf{C}_{K_k}^{\mathrm{T}}, \tag{5}$$

where $\boldsymbol{\alpha}$ is a $K - 1 \times 1$ vector that compounds the codes of all the possible interfering users and $\mathbf{C}_{K_k}$ is an $N_c \times K - 1$ matrix containing all K users' signature codes except the code of user k (the desired user).

The reduced complexity *conventional decorrelator* is termed reduced complexity decorrelator (RCD). To derive the RCD it is assumed that $w_{12} = 1$, and $K_a = K$, i.e. the maximum number of users is active. Now, the interference can be completely cancelled by the proper $\boldsymbol{\alpha}$, which defines the RCD [24].

When the maximum number of users (K) is active, all the multiuser interference will be cancelled and the performance will equal that of the full dimensional decorrelator. However, when $K_a < K$ users are active there will be a mismatch in cancelling the interference, because $\boldsymbol{\alpha}$ was calculated based on K active users. This residual interference will be reduced with the adaptive reduced complexity detector(ARCD).

As in the case of the AFDD of section 2.3 the bootstrap algorithm will be used to implement the ARCD. In the case of the compounded codes, however, the weight matrix $\mathbf{W}$ is reduced to a 2×2 matrix. The weight update equation for this $\mathbf{W}$ is the same as previously expressed in equation (4).

2.5 RESULTS

I.i.d. flat Rayleigh fading per subcarrier is assumed, which is characteristic of a system with a subcarrier spacing greater than or equal to the coherence bandwidth. Simulations were performed for $N_c = 31$, the local mean SNR (LSNR) of the desired user is 7 dB, the maximum number of users, $K = 30$. The SUB, FDD and RCD performance have been determined numerically. The

BER curves of the ARCD and AFDD were obtained by bit error counting in the steady state over 10^5 and 5×10^5 bits respectively, and by averaging over 100 Monte-Carlo runs. The various detectors are shown in table 1.

Table 1 The various detectors used for performance comparison.

Detector Type	Transformation	Size
Full Dimensional Decorrelator (FDD)	$\mathbf{P}_c'^{-1}$	$K_a \times K_a$
Adaptive Full Dimensional Detector (AFDD)	$\mathbf{V}$	$K_a \times K_a$
Reduced Complexity Decorrelator (RCD)	$w_{12} = 1$	2×2
Adaptive Reduced Complexity Detector (ARCD)	$\mathbf{V}$	2×2
Single User Detector (SUD)	None	N/A

Figures 1 and 2 compare the BER of the ARCD and AFDD for $K_a = 8$ and $K_a = 20$ respectively. For low ISR the ARCD and AFDD have similar performance. However, for high ISR the AFDD outperforms the ARCD. But, the difference decreases as the number of active users increases. This indicates a possible trade-off between performance and complexity.

Figures 1 and 2 also show that the ARCD always outperforms the SUD and the RCD, by which it is bounded. When the ARCD is compared to the non-adaptive FDD, there exists an ISR level below which the ARCD performs better than the FDD. The ISR level at the cross-over point depends on the number of active users, K_a, and increases when K_a increases (when a maximum number of users is active ($K_a = K$) the FDD and RCD perform the same).

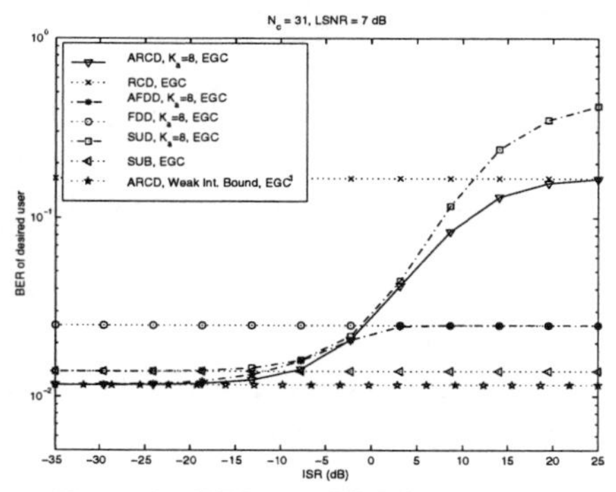

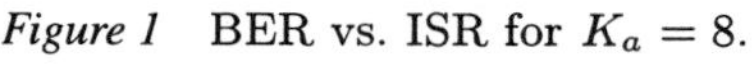

Figure 1 BER vs. ISR for $K_a = 8$.

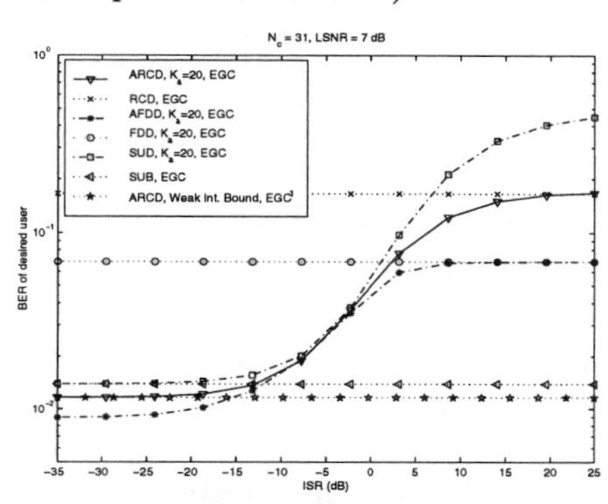

Figure 2 BER vs. ISR for $K_a = 20$.

3. FREQUENCY OFFSET CORRECTION FOR OFDM

For spectral efficiency, an MCM system uses closely spaced subcarriers. A resulting drawback is a high sensitivity of the performance to a frequency offset, which results from a Doppler shift, due to mobile movement, as well as from a mismatch between the carrier frequencies at the transmitter and receiver [8, 25].

The frequency offset causes loss of the subcarriers' orthogonality, resulting in inter-carrier interference (ICI) and reducing the carrier-to-interference ratio (CIR). In addition, the constellation of the desired signal at each subcarrier is rotated, resulting in inter-rail interference (IRI) between the Inphase (I) and Quadrature (Q) rails, which further reduces the carrier-to-interference ratio.

To mitigate the effects of the frequency offset, a blind adaptive decorrelator (bootstrap) based frequency offset correction (FOC) scheme is developed for OFDM. This scheme completely eliminates the ICI as well as the IRI [26, 27].

3.1 SYSTEM MODEL

The considered OFDM system has N_c subcarriers, separated in frequency by $\Delta f = 1/(N_c T_s)$, where T_s is the unmodulated symbol duration, and each subcarrier is M-QAM modulated. The MCM is implemented by an inverse DFT, such that $\mathbf{X} = \text{IDFT}\{\mathbf{b}\}$. $\mathbf{X}$ is called an OFDM frame and consists of N_c samples of duration T_s, i.e. the duration of the OFDM frame is $N_c T_s$. It is instructive to regard the elements of $\mathbf{b}$ as the frequency domain samples and those of $\mathbf{X}$ as the corresponding time domain samples of the DFT pair.

The channel is considered frequency selective with respect to the total bandwidth. The subcarrier bandwidth, however, is assumed to be much smaller than the coherence bandwidth such that each subcarrier is subjected to flat Rayleigh fading. Thus, it is assumed that no ISI occurs such that no guard-interval is needed. Also, it is assumed that the system operates with a frequency offset.

The matched filter can be implemented as a quadrature detector sampled at T_s. The samples are then passed to the N_c-point DFT for demodulation. As a result of the frequency offset the DFT output can be described as:

$$\mathbf{z} = \mathbf{S}^T \mathbf{A} \mathbf{b} + \boldsymbol{\xi}, \tag{6}$$

where the matrix $\mathbf{S}$, which is termed the subcarrier correlation matrix, is defined in [26]. $\mathbf{S}$ is a function of the normalized (w.r.t. Δf) frequency offset ε.

The subcarrier correlation matrix $\mathbf{S}$ has the following three properties: (1) it is non-hermitian, i.e. $\mathbf{S} \neq \mathbf{S}^{\mathrm{H}}$; (2) it is orthonormal, i.e. $\mathbf{S}^{\mathrm{H}}\mathbf{S} = \mathbf{I}$ (or, equivalently, it is unitary as $\mathbf{S}^{\mathrm{H}} = \mathbf{S}^{-1}$); (3) it is a circulant matrix, i.e. the columns (as well as the rows) are shifted versions of each other, as $s(i) = s(i + nN_c)$ for any integer n.

3.2 ADAPTIVE DECORRELATOR BASED FOC

As the ICI introduces correlation between the outputs of the DFT it is proposed to add a decorrelator at the output of the DFT. A *conventional decorrelator* for FOC may be defined as $(\mathbf{S}^{\mathrm{T}})^{-1}$, analogously to the conventional decorrelator previously defined for MUD. From equation 6 it is clear that $(\mathbf{S}^{\mathrm{T}})^{-1} = \mathbf{S}^*$ will completely cancel the ICI and thus correct for the frequency offset.

As ε, and thus $\mathbf{S}$, is unknown at the receiver it is proposed to use the adaptive complex bootstrap algorithm [28], to decorrelate the DFT outputs. Following this first adaptive decorrelator stage, a second adaptive decorrelator stage based on the real-valued bootstrap algorithm is used, which cancels the IRI. Figure 3 shows signal constellations at the outputs of the various stages. It has been shown by simulation that the performance of the adaptive decorrelator equals the performance when using the conventional decorrelator [26, 27].

The decorrelator performance is compared to two other schemes. One, an ICI cancelation scheme based on repetition coding in the frequency domain, which reduces the bandwidth efficiency by half [25]. Two, a scheme based on frequency domain correlative coding with correlation polynomial $F(D) = 1-D$, which does not reduce the bandwidth efficiency [29].

In figure 4 the SINR expressions are plotted as a function of ε for $N_c = 8$ and an SNR of 0 dB, 10 dB and 20 dB. The SINR for the ICI cancelation scheme ($SINR_{ici}$) remains fairly constant and close to the SINR of the decorrelator for low SNRs because the noise power is dominant over the interference power. However, for higher SNRs the decorrelator clearly outperforms the ICI cancelation scheme as now the interference power dominates over the noise power. Additionally, the ICI cancelation scheme has half the bandwidth efficiency due to the repetition coding. The correlative coding scheme ($SINR_{cc}$) performs better than conventional OFDM, however, not as good as the decorrelator or the ICI cancelation scheme.

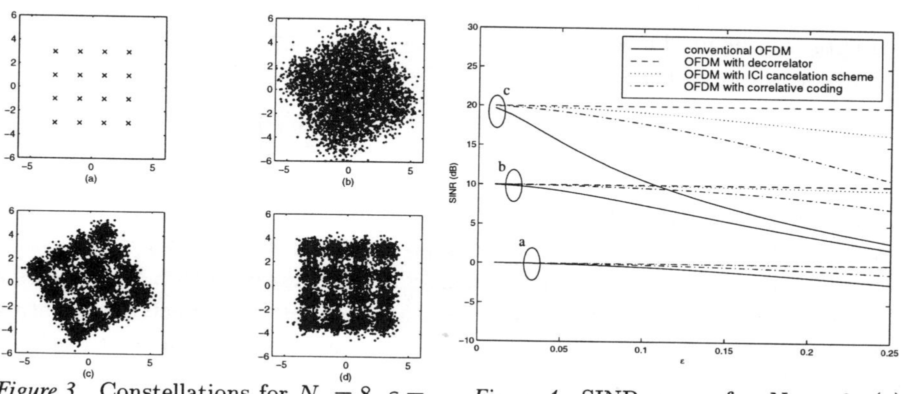

Figure 3 Constellations for $N_c = 8$, $\varepsilon = 0.2$ and SNR = 16 dB: (a) Tx; (b) DFT output; (c) 1st stage; (d) 2nd stage.

Figure 4 SINR vs. ε for $N_c = 8$, (a) SNR = 0 dB; (b) SNR = 10 dB; (c) SNR = 20 dB

4. JOINT MUD & FOC FOR DOWNLINK MC-CDMA

Just as for OFDM, in the case of an MC-CDMA system, a frequency offset results in ICI and IRI. For MC-CDMA, however, the ICI results in additional interference from the desired user, due to non-zero correlation between shifted versions of its code (as a result of the demultiplexing process). For the same reason, additional multiuser interference (MUI) is generated.

4.1 SYSTEM MODEL

A downlink MC-CDMA system is considered with N_c subcarriers and K_a active users (see also section 2.1). Transmissions aimed at different mobile users may be assigned different powers, p_k, creating a near-far problem for some users. The subcarriers are separated in frequency by $\Delta f = 1/T_s$, where T_s is the unmodulated symbol duration, and each subcarrier is M-QAM modulated. The MCM is implemented by an inverse DFT, such that $\mathbf{X} = IDFT\{\mathbf{b}_{ms}\}$. $\mathbf{X}$ is called an MC-CDMA frame and consists of N_c samples of duration T_s/N_c, i.e. the total duration of the MC-CDMA frame is T_s. Additionally, $\mathbf{b}_{ms} = \sum_{k=1}^{K} p_k b(k)\mathbf{c}_k = \mathbf{C}\sqrt{\mathbf{P}}\mathbf{b}$.

The matched filter can be implemented as a quadrature detector, sampled at T_s/N_c. The complex valued samples are then passed to the N_c-point DFT for demodulation.

In section 3, an adaptive decorrelating detector was used for FOC in an OFDM system. This approach was based on the assumption that the data at the different subcarriers are uncorrelated, which is valid for OFDM but not for MC-CDMA. Here, the DFT outputs are, instead, first multiplied with the transpose of the code matrix $\mathbf{C}$ in order to demultiplex the received signal and generate the different users' symbols for detection. The symbols of the different users are assumed uncorrelated and hence it is suggested that a decorrelating detector may be used to separate the different users' signals. The demultiplexed signal may be described as:

$$\mathbf{x} = \mathbf{C}^{\mathrm{T}}\mathbf{S}^{T}\mathbf{A}\mathbf{C}\sqrt{\mathbf{P}}\mathbf{b} + \mathbf{C}^{\mathrm{T}}\mathbf{n} = \mathbf{P}'_{sc}\sqrt{\mathbf{P}}\mathbf{b} + \boldsymbol{\xi} \tag{7}$$

The adaptive joint multiuser detection and frequency offset correction detector is implemented in two stages, similar to the one described in section 3. The first stage decorrelates the users' signals, whereas the second stage performs a decorrelation of the in-phase and quadrature components, effectively rotating the constellation back to its original orientation. Figure 5 shows the signal constellation at the outputs of the various stages.

Figure 6 shows the bit error rate of the detectors under consideration when SIR = -10 dB. It can be seen that the performance of the SUD, which simply performs a hard-decision after the demultiplexing stage, is dominated by the strong multiuser interference. Even the absence of a frequency offset hardly improves the performance of the SUD. Applying the first stage of the joint multiuser detection and frequency offset correction scheme, which decorrelates the different users but does not perform I-Q decorrelation, i.e. it does not rotate the constellation, greatly improves the performance over the SUD. Applying both stages of the joint detector results in a performance that is very close–less than 1 dB difference– to the single user bound (SUB). The SUB is expressed by the probability of bit error (Pe) of 4-QAM.

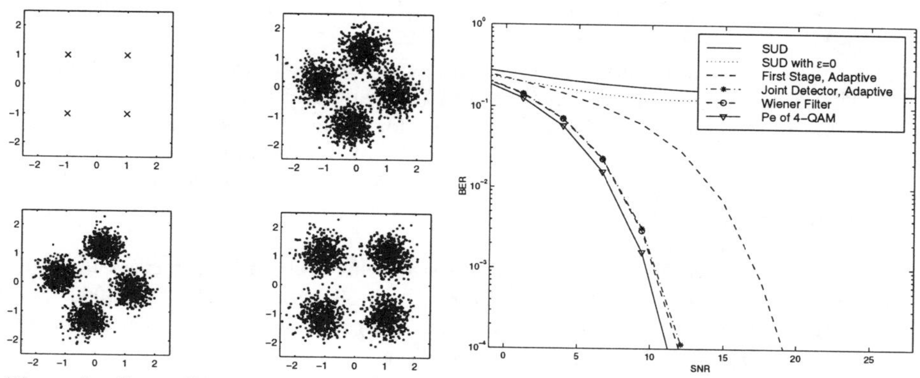

Figure 5 Constellations for $N_c = 7$, $K_a = 4$, $\varepsilon = 0.25$ and SNR = 12 dB: (a) transmitted; (b) demultiplexed; (c) 1st stage; (d) joint detector

Figure 6 BER vs. SNR for $N_c = 7$, $K_a = 4$, $\varepsilon = 0.2$ and SIR = -10 dB

REFERENCES

[1] J. A. C. Bingham, "Multicarrier modulation for data transmission: An idea whose time has come," *IEEE Comm. Mag.*, vol. 28, pp. 5–14, May 1990.

[2] M. Alard and R. Lasalle, "Principles of modulation and channel coding for digital broadcasting for mobile receivers," *EBU Review*, pp. 168–190, Aug. 1987.

[3] L. J. Cimini, Jr., "Performance studies for high-speed indoor wireless communications," *Wireless Personal Communications*, vol. 2, no. 1&2, pp. 67–85, 1995.

[4] J.-P. Linnartz and S. Hara, "Special issue on multi-carrier communications," *Wireless Personal Communications*, vol. 2, no. 1/2, 1995.

[5] K. Fazel, S. Kaiser, P. Robertson, and M. J. Ruf, "A concept of digital terrestrial television broadcasting," *Wireless Pers. Comm.*, vol. 2, no. 1&2, pp. 9–27, 1995.

[6] S. Hara and R. Prasad, "Overview of multicarrier CDMA," *IEEE Communications Magazine*, vol. 35, pp. 126–133, Dec. 1997.

[7] S. B. Weinstein and P. M. Ebert, "Data transmission by frequency-division multiplexing using the discrete fourier transform," *IEEE Transactions on Communication Technology*, vol. 19, pp. 626–634, Oct. 1971.

[8] T. Pollet, M. van Bladel, and M. Moeneclaey, "BER sensitivity of OFDM systems to carrier frequency offset and wiener phase noise," *IEEE Transactions on Communications*, vol. 43, pp. 191–193, Feb. 1995.

[9] X. Li and L. J. Cimini, Jr., "Effects of clipping and filtering on the performance of OFDM," *IEEE Communications Letters*, vol. 2, pp. 131–133, May 1998.

[10] N. Yee, J. P. M. G. Linnartz, and G. Fettweis, "Multi-carrier CDMA indoor wireless radio networks," in *Proc. PIMRC*, (Yokohama), pp. 109–113, Sept. 1993.

[11] K. Fazel, "Performance of CDMA/OFDM for mobile communication system," in *Proc. ICUPC*, (Ottawa, Canada), pp. 975–979, Oct. 1993.

[12] S. Kondo and L. B. Milstein, "On the use of multicarrier direct sequence spread spectrum systems," in *Proc. IEEE Milcom*, (Boston), pp. 52–56, Oct. 1993.

[13] L. Vandendorpe, "Multitone spread spectrum multiple access communications system in a multipath rician fading channel," *IEEE Transactions on Vehicular Technology*, vol. 44, no. 4, pp. 327–337, 1995.

[14] S. Kaiser, "On the performance of different detection techniques for OFDM-CDMA in fading channels," in *Globecom*, (Singapore), pp. 2059–2063, Nov. 1995.

[15] S. Verdú, "Minimum probability of error for asynchronous gaussian multiple-access channels," *IEEE Transactions on IT*, vol. 32, pp. 85–96, Jan. 1986.

[16] R. Lupas and S. Verdú, "Linear multiuser detector for asynchronous code division multiple access channels," *IEEE Transactions on Information Theory*, vol. 35, pp. 123–136, Jan. 1989.

[17] R. Lupas and S. Verdú, "Near-far resistance of multi-user detectors in asynchronous channels," *IEEE Tr. on Comm.*, vol. 38, pp. 496–508, Apr. 1990.

[18] Y. Bar-Ness and J. B. Punt, "Adaptive bootstrap multi-user CDMA detector," *Wireless Personal Communications*, vol. 3, no. 1-2, pp. 55–71, 1996.

[19] U. Madhow and M. L. Honig, "MMSE interference suppression for direct-sequence spread-spectrum CDMA," *IEEE Transactions on Communications*, vol. 42, pp. 3178–3188, Dec. 1994.

[20] H. Ge and Y. Bar-Ness, "Comparative study of linear minimum mean square error (lmmse) and the adaptive bootstrap multiuser detectors for CDMA communications," in *Proc. ICC*, (Dallas, TX, U.S.A), pp. 78–82, June 1996.

[21] Y. Bar-Ness, J. P. Linnartz, and X. Liu, "Synchronous multi-user multi-carrier CDMA communications system with decorrelating interference canceler," in *Proc. PIMRC*, (The Hague, The Netherlands), pp. 184–188, Sept. 1994.

[22] M. A. Visser and Y. Bar-Ness, "Adaptive Multi-Carrier CDMA (MC-CDMA) Structure for Downlink PCS," in *Proc. 9th Tyrrhenian International Workshop on Digital Communication*, (Lerici, Italy), Sept. 1997.

[23] P. Zong, "Signal processing topics in multicarrier modulation: Frequency offset correction for OFDM and multiuser interference cancellation for MC-CDMA," Master's thesis, New Jersey Inst. of Technology, Newark, NJ, U.S.A., May 1998.

[24] M. A. Visser and Y. Bar-Ness, "Adaptive reduced complexity multi-carrier CDMA (MC-CDMA) structure for downlink PCS," *to appear in the special issue on MCSS of the European Transactions on Telecommunications.*

[25] Y. Zhao and S.-G. Häggman, "Sensitivity to doppler shift and carrier frequency errors in OFDM systems – the consequences and solutions," in *Proc. IEEE VTC*, (Atlanta, GA, U.S.A.), pp. 1564–1568, Apr. 1996.

[26] M. A. Visser and Y. Bar-Ness, "OFDM frequency offset correction using an adaptive decorrelator," in *Proc. CISS*, (Princeton), pp. 483–488, Mar. 1998.

[27] M. A. Visser and Y. Bar-Ness, "Frequency offset correction for OFDM using a blind adaptive decorrelator in a time-variant selective Rayleigh fading channel," in *Proc. IEEE VTC*, (Houston, TX, U.S.A.), pp. 1281–1285, May 1999.

[28] N. van Waes and Y. Bar-Ness, "The complex bootstrap algorithm for blind separation of co-channel QAM signals," *Wireless Pers. Comm.*, forthcoming.

[29] Y. Zhao, J.-D. Leclercq, and S.-G. Häggman, "Intercarrier interference compression in OFDM communication systems by using correlative coding," *IEEE Communications Letters*, vol. 2, pp. 214–216, Aug. 1998.

PERFORMANCE OF DOWNLINK MC-CDMA WITH SIMPLE INTERFERENCE CANCELLATION

Hideki Ochiai and Hideki Imai
Institute of Industrial Science, The University of Tokyo,
7-22-1, Roppongi, Minato-ku, Tokyo 106-8558 Japan
ochiai@iis.u-tokyo.ac.jp

Abstract Simple interference cancellation (IC) schemes employing either equal gain combining (EGC) or minimum mean squire error combining (MMSEC) are applied to downlink multi-carrier CDMA and compared by computer simulations. Both fully interleaved and correlated Rayleigh fading channels are considered. The results show that the most significant improvement can be achieved by successive IC with MMSEC, while two-stage reliability-based IC with EGC shows good performance for a correlated fading channel with moderate complexity of the receiver.

1. INTRODUCTION

Recently, a novel application of the orthogonal frequency division multiplexing (OFDM) technique to code-division multiple-access (CDMA) has been proposed [1, 2, 3]. This scheme, often referred to as multi-carrier CDMA (MC-CDMA) or OFDM-CDMA, provides a robustness against multipath fading with certain implementation advantages over direct-sequence (DS) CDMA [2, 4].

In the MC-CDMA system, a detection algorithm of the receiver plays an important role in determining its bit error performance, and various detection schemes for downlink MC-CDMA have been proposed [5, 6]. Most detection schemes, however, cannot avoid a multiple access interference (MAI) even if the orthogonal sequences such as Walsh-Hadamard codes are used for spreading, since the orthogonality cannot be perfectly preserved at the receiver due to a multipath fading. The orthogonality restoring combining (ORC) detector may avoid MAI, but it suffers from a severe noise amplification. Maximum likelihood detection may be computationally too expensive to be applied to the mobile stations especially when the number of active users is large.

In [7], multistage detector is proposed and analyzed where all the undesirable users are canceled. However, appreciable performance improvement may not be observed after the second stage, even though the complexity of the receiver considerably increases. In [8], soft interference cancellation has been proposed where the soft decision of the coded data is fed back for iterative cancellation and decoding. This scheme may achieve near optimal performance over entire coding and modulation processes, but the complexity and the delay required for decoding appear prohibitive for the portable terminals.

In this paper, simple interference cancellation (IC) schemes are applied to the downlink MC-CDMA and their bit error performances are compared by computer simulations.

2. SYSTEM DESCRIPTION

The block diagram of the downlink MC-CDMA system considered in the paper is shown in Fig. 1. For each block, a binary data of each user is spread by user-specific Walsh-Hadamard sequences of length L, denoted by $\mathbf{c}^{(k)} = \{c_0^{(k)}, c_1^{(k)}, \ldots, c_{L-1}^{(k)}\}$ for user k. (For simplicity, BPSK is assumed.) In each OFDM symbol, we assume there are M such blocks. The data of all the users are added, interleaved, and modulated by an N-point inverse fast Fourier transform (IFFT), where $N = ML$. A guard interval is added to avoid intersymbol interference. Note that the number of maximum users supported by each block of the MC-CDMA system is L and we assume that there are K active users where $K \leq L$.

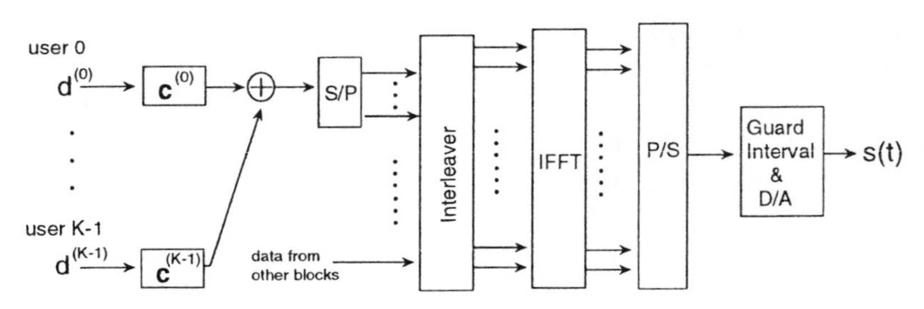

Figure 1 Downlink MC-CDMA Transmitter.

On the assumptions that the guard interval is longer than the maximum delay spread and the fading is slow enough to be considered time-nonselective, the lth subcarrier of each block of the received signal after FFT and deinterleaver can be given by

$$r_l = \sum_{k=0}^{K-1} H_l c_l^{(k)} d^{(k)} + n_l \tag{1}$$

where $d^{(k)}$ denotes the binary data of user k, H_l is a complex channel coefficient, and n_l is an additive white Gaussian noise (AWGN) with zero mean and variance $\sigma_N^2 = N_0/2$.

The receiver of user k combines the received signal components based on the combining algorithm and yields a decision variable:

$$Z^{(k)} = \sum_{l=0}^{L-1} G_l r_l c_l^{(k)} \tag{2}$$

where G_l is given by, for equal gain combining (EGC),

$$G_l = \frac{H_l^*}{|H_l|} \tag{3}$$

and for combining based on the minimum mean square error criterion (MMSEC) [3],

$$G_l = \frac{H_l^*}{|H_l|^2 + \eta} \tag{4}$$

with $\eta = \sigma_N^2/K$. The use of the MMSEC requires the knowledge of the ratio of the noise variance to the total power of the OFDM signal, resulting in the increase of the receiver complexity. In this paper, maximum ratio combing (MRC) or orthogonal restoring combining (ORC) may not be considered, since these combining techniques yield poorer performance than EGC or MMSEC for downlink with large K or low SNR [4]. It should also be noted that ORC may not produce any interference of other users; thus, the interference cancellation technique may not work with ORC.

The decision variable for user k is given by

$$Z^{(k)} = \mathcal{S}^{(k)} + \sum_{\substack{m=0 \\ m \neq k}}^{K-1} \mathcal{I}_m^{(k)} + \mathcal{N}^{(k)} \tag{5}$$

where $\mathcal{S}^{(k)}$ is the desired signal component, $\mathcal{I}_m^{(k)}$ is the interference of user m with user k, and $\mathcal{N}^{(k)}$ is the Gaussian noise term. It can readily be shown that the random variables $\mathcal{N}^{(0)}, \mathcal{N}^{(1)}, \ldots, \mathcal{N}^{(K-1)}$ are i.i.d. Gaussian, since these variables are composed of the orthogonal transformation of i.i.d. Gaussian random variables $n_0, n_1, \ldots, n_{L-1}$, or, $E[\mathcal{N}^{(k)}\mathcal{N}^{(i)}] = 0$ for $i \neq k$. Therefore, given $S^{(k)}$ and $I_m^{(k)}$ for all k and $m \neq k$, $Z^{(k)}$ are uncorrelated Gaussian random variables. Since the interference terms $I_m^{(k)}$ are not known *a priori*, we replace them by their mean, i.e., $I_m^{(k)} = 0$. Consequently, for BPSK modulation, the reliability of the decision of the received symbol $Z^{(k)}$ can be given by its absolute value $|Z^{(k)}|$. Thus, for downlink, the reliabilities of the decisions for each user may differ, even though the received power is the same for all the users.

3. INTERFERENCE CANCELLATION ALGORITHMS

In this paper, the following three interference cancellation algorithms are compared:

Two-stage brute-force IC algorithm

1. Make tentative decisions for all the active users based on the conventional detection scheme (applying either EGC or MMSEC).
2. Based on the tentative decisions, generate the signals of all the other active users, and subtract from the received signal.
3. Make final decision (applying EGC or MMSEC detection) for the desired user.

In [7], threshold ORC (TORC) and MRC were employed for the initial and second stages, respectively. However, in this paper, the identical combining scheme will be employed for both the stages for simplicity of the receiver. Since the receiver tries to cancel all the unwanted users, we refer to this algorithm as *brute force* IC. This approach does not make use of the reliability of the decision variables.

The two-stage brute-force IC may cause error propagation due to the erroneous subtraction of the interferences. In order to mitigate this, we consider the following two-stage reliability-based IC algorithm, where only the interfering users whose reliability is higher than the desired user are subtracted:

Two-stage reliability-based IC algorithm

1. Calculate the reliability $|Z^{(m)}|$ for all the active users (applying EGC or MMSEC detection).
2. For the desired user k, subtract the signals of all the users whose reliability is higher than that of the user k, i.e., $\forall m \neq k$ where $|Z^{(m)}| > |Z^{(k)}|$.

3. Make final decision (applying EGC or MMSEC detection) for the desired user k.

If further increment of the detection stages is acceptable, successive IC [9] may be also considered:

Successive IC algorithm

1. Repeat the following procedure until the desired user is chosen.
   (a) Calculate the reliability (applying EGC or MMSEC detection) for all the remaining users.
   (b) Choose a user with the highest reliability, and subtract the signal component of the user.

In successive IC algorithm, the unwanted user with the highest reliability is canceled one by one, and decision variables for remaining users are updated every cancellation. It should be noted that the maximum delay of the successive IC is proportional to the number of the active users.

4. SIMULATION RESULTS

Computer simulations were performed to examine the performance of the IC schemes. To investigate the ideal performance, a fully interleaved channel is considered; all the $L = 32$ subcarriers experience a statistically independent Rayleigh fading. Perfect estimation of the channel coefficients is assumed. Walsh-Hadamard codes were used for spreading with BPSK modulation.

Figs. 2(a) and (b) show the bit error rate performances of the IC schemes with EGC and MMSEC, respectively, for both stages. Also shown in the figures is the single user bound for optimal (MRC) detection. In both cases, apparently, the most significant gain is observed with the successive IC scheme. It can be also observed that the reliability-based IC outperforms the brute-force IC, even though the improvement is small compared to the successive IC.

As an example of correlated channels, an i.i.d. two-path Rayleigh fading channel is also considered; following [4], we have chosen $M = 8$ and $L = 32$; thus the number of subcarriers is 256. A guard interval normalized by a symbol period is 0.015 [4], with the maximum delay spread equal to the guard interval. The results are shown in Figs. 3, where the lower bound of this channel and bit error rate of MC-FDMA [4] are also shown. From Figs. 3, it is observed that all the cancellation techniques

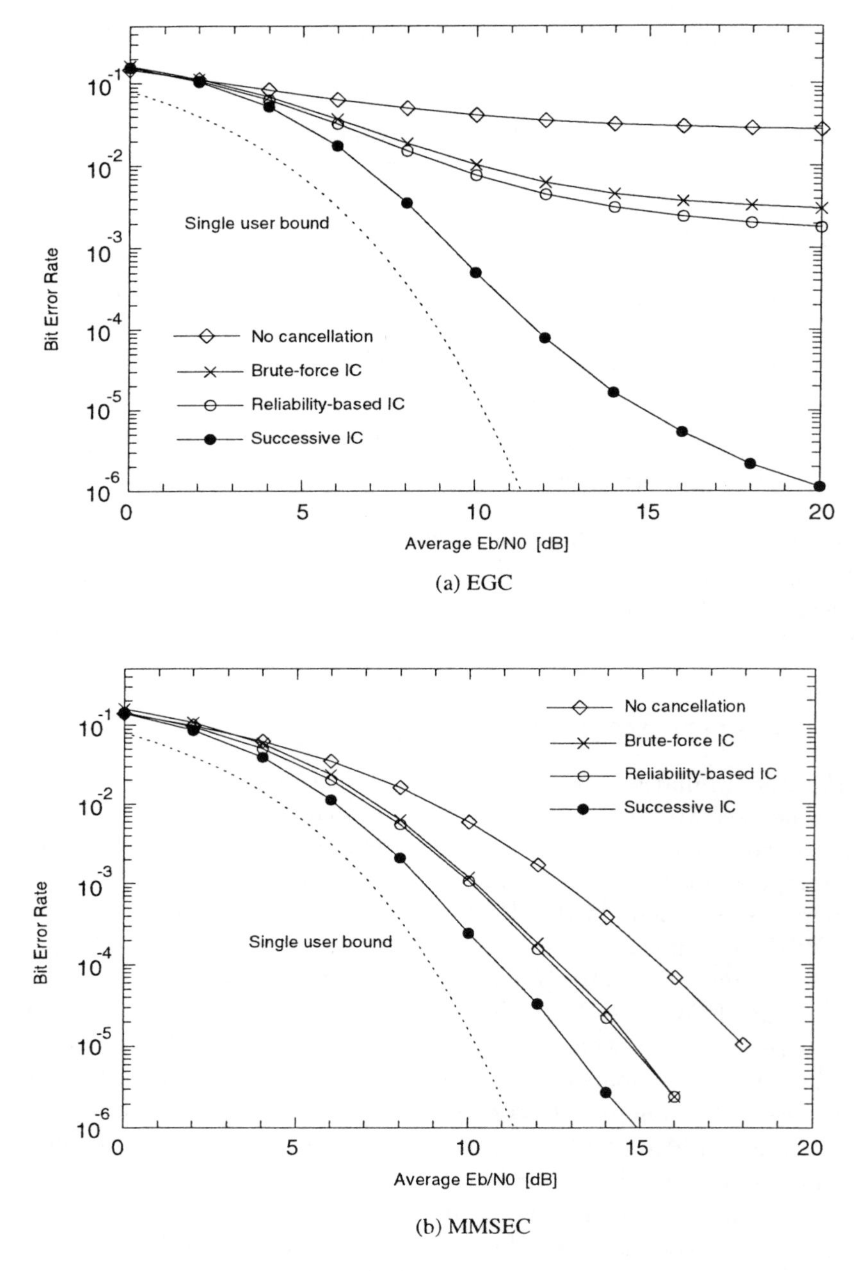

(a) EGC

(b) MMSEC

Figure 2 Bit error rate of MC-CDMA with IC over an ideally interleaved channel; $L = 32$ and $K = 32$.

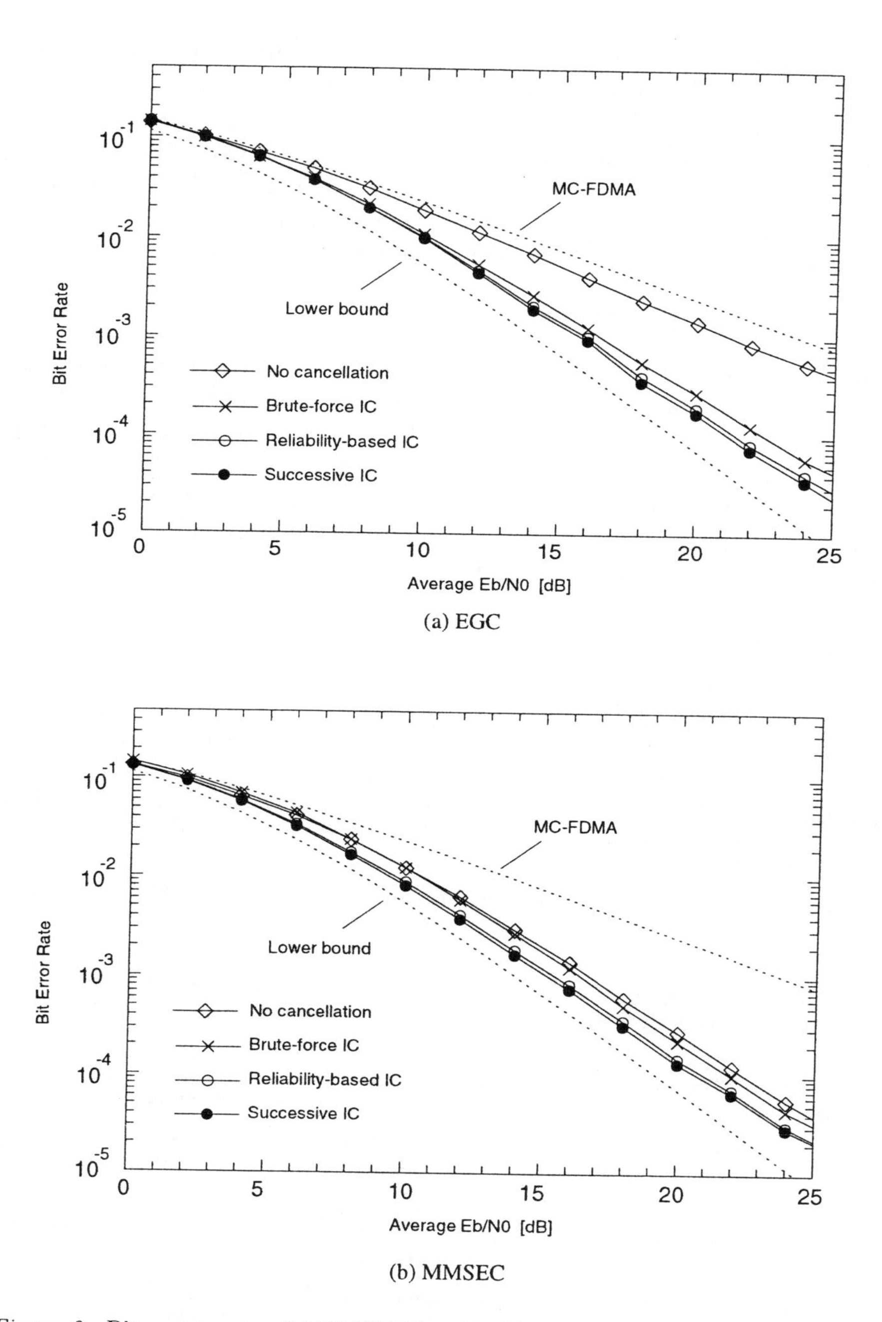

Figure 3 Bit error rate of MC-CDMA with IC over a two-path Rayleigh fading channel; $L = 32$ and $K = 32$.

significantly improve the performance and the difference between the reliability-based IC and successive IC algorithms is small.

It should be noted that even though EGC yields a severe error floor for a fully interleaved Rayleigh fading channel, the error floor will be significantly reduced as the channel becomes correlated. This is because the EGC of the correlated channels somewhat recovers the orthogonality of the users, leading to a considerable reduction in MAI.

5. CONCLUSIONS

The bit error performances of three simple IC techniques (two-stage brute-force IC, two-stage reliability-based IC, and successive IC) were compared by computer simulations. The results show that for a fully interleaved channel, the MMSEC with successive IC gives the best performance of all the others considered in the paper. However, a good performance improvement is observed for the EGC with reliability-based IC over a correlated fading channel.

References

[1] N. Yee, J. P. Linnartz, G. Fettweis, "Multi-carrier CDMA in indoor wireless radio networks," *Proc. PIMRC'93,* pp.109–113, 1993.

[2] K. Fazel, "Performance of CDMA/OFDM for mobile communication system," *Proc. ICUPC'93,* pp.975–979, 1993.

[3] A. Chouly, A. Brajal, S. Jourdan, "Orthogonal multicarrier techniques applied to direct sequence spread spectrum CDMA systems," *Proc. GLOBECOM'93,* pp.1723–1728, 1993.

[4] S. Hara, R. Prasad, "Design and performance of multi-carrier CDMA system in frequency selective Rayleigh fading channels," *IEEE Trans. Veh. Technol., to be published.*

[5] T. Muller, H. Rohling, R. Grunheid, "Comparison of different detection algorithms for OFDM-CDMA in broadband Rayleigh fading," *Proc. VTC'95,* pp.835–838, 1995.

[6] S. Kaiser, "On the performance of different detection techniques for OFDM-CDMA in fading channels," *Proc. GLOBECOM'95,* pp.2059–2063, 1995.

[7] D. N. Kalofonos, J. G. Proakis, "Performance of the multi-stage detector for a MC-CDMA system in a Rayleigh fading channel," *Proc. GLOBECOM'96,* pp. 1784–1788, 1996.

[8] S. Kaiser, J. Hagenauer, "Multi-carrier CDMA with iterative decoding and soft-interference cancellation," *Proc. GLOBECOM'97,* 1997.

[9] A. J. Viterbi, "Very low rate convolutional codes for maximum theoretical performance of spread-spectrum multiple-access channels," *IEEE J. Select. Areas Commun.,* vol. 8, no. 4, May 1990.

WIDEBAND AND NARROWBAND INTERFERENCE CANCELLATION FOR ASYNCHRONOUS MC-CDMA

Andrew C. McCormick, Peter M. Grant and Gordon J. R. Povey
Signals and Systems Group
Department of Electronics and Electrical Engineering
The King's Buildings, Mayfield Road
The University of Edinburgh, Edinburgh, EH9 3JL, Scotland
Andrew.McCormick@ee.ed.ac.uk

Abstract Uplink MC-CDMA is asynchronous, and each user's signal experiences independent fading, therefore the low complexity, linear narrowband receivers used for downlink are not applicable. However, in this case, wideband interference cancellation schemes can provide good multi-user detection. If the asynchronism is constrained to a fraction of the symbol length, a narrowband interference cancellation scheme can be applied. This approach has lower complexity than wideband cancellation, although there may be some degradation in performance. This degradation is estimated for both successive and parallel interference cancellation algorithms, through simulations of their performance in AWGN and indoor wireless channels.

1. INTRODUCTION

Much research into Multi-Carrier CDMA systems has concentrated on synchronous systems, appropriate for a downlink transmission. There has however been comparatively little research on asynchronous systems (appropriate for uplink channels). This is not without justification as the most significant benefit of MC-CDMA, the simple narrowband implementation of linear multi-user detectors/equalisers such as the linear minimum means square error (MMSE) combiner, is lost when an asynchronous approach is taken. Therefore other approaches, such as combining the multi-carrier approach with frequency division between users, have been proposed [Kaiser, 1998].

In the asynchronous case, the required matrix inversions to implement the optimal linear receiver are as complex as those required for direct sequence CDMA. These receivers require accurate channel information to allow stable matrix inversion, even with some degree of synchronisation [Kleer et al., 1999] and the performance degrades severely with imperfect channel information.

Better performance can be achieved in DS-CDMA systems through the use of non-linear interference cancellation (IC) techniques [Varanasi, 1999]. These approaches have been applied in MC-CDMA systems although exploring only the synchronous case [Kalofonos and Proakis, 1996; Teich et al., 1998].

In this paper the application of several interference cancellation (IC), or decision feedback multi-user detection, techniques to asynchronous MC-CDMA is explored. Successive, parallel and partial parallel [Divsalar and Simon, 1995] cancellation techniques are investigated, as wideband techniques applied to the baseband signal. Applying the interference cancellation after the (FFT) demodulation (narrowband IC) is also investigated as this will reduce computational complexity and the resulting deterioration in performance is measured.

2. WIDEBAND AND NARROWBAND INTERFERENCE CANCELLATION ARCHITECTURES

The optimum (maximum likelihood) solution to the multi-user detection problem for both synchronous and asynchronous systems is to compute the distance between the received signal and all possible transmitted signal combinations filtered by the channel model. This detector has exponential complexity $O(2^N)$ (where N denotes the number of users) and is therefore only practical for a small number of users.

Therefore to handle a large number of users, receivers must implement a suboptimal solution for multi-user detection. These can be divided into linear and non-linear receivers. The optimal linear solution can be obtained, by minimising the mean square error. In downlink MC-CDMA, where all users' signals pass through the same channel, this can be implemented with a single tap per carrier. However for uplink systems, each user's signal passes through a different channel and the MMSE solution requires the inversion of the cross-correlation matrix (modified to take acount of the noise) between users' signals. Since this can only be computed from the imperfect channel estimates, and in the case of strong multi-user interference, the cross-correlation estimate will

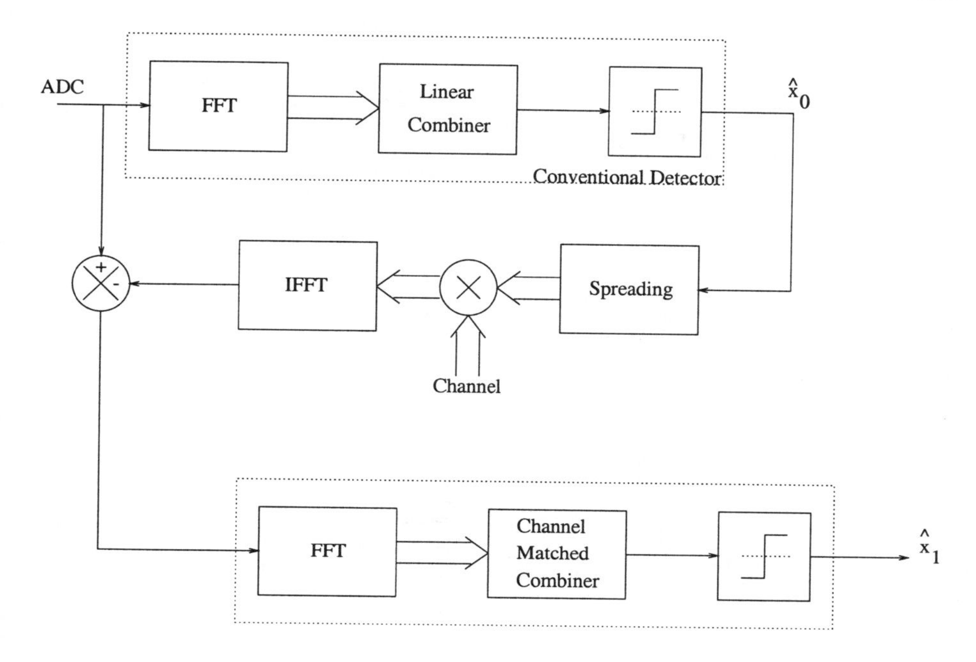

Figure 1 Wideband IC Architecture

contain rows which are almost co-linear, the effect of a small amount of noise may significantly alter the inverse solution.

This paper therefore concentrates on non-linear interference cancellation techniques. These can be divided into successive and parallel algorithms. In successive IC, the user with the largest decision statistic is subtracted from the signal and the decision statistic for the remaining users computed. This is repeated until all the users signals have been cancelled. In parallel IC, a tentative decision using a linear detector is made about all the users transmitted data. For each user, all other users' signals are subtracted and the resulting decision made using a channel matched detector. There is however a significant probability that the data decision made for each user is wrong. This produces 4 times the multi-user interference power of a non-cancelled interfering user. Therefore to reduce this, partial parallel IC can be employed. In this case only a proportion of the interfering signal is cancelled reducing the effect of detection errors at early stages. Both parallel IC algorithms can be applied several times to give better tentative decisions for cancelling at the next stage. With the partial parallel cancellation, the proportion of the signal cancelled can be increased at each iteration.

In MC-CDMA systems, all three of these algorithms can be implemented as either wideband or narrowband systems. With a wideband IC architecture, as

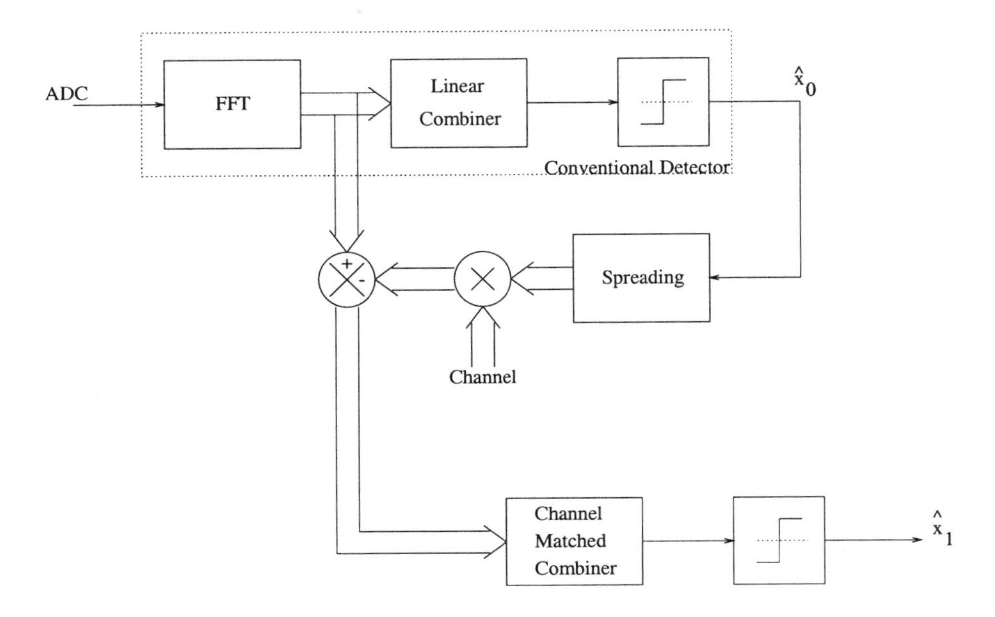

Figure 2 Narrowband IC Architecture

shown in figure 1, the signal subtractions take place in the baseband with the interfering users reconstructed from the tentative data estimates and channel information using an IFFT operation and cyclic-extension. With a narrowband IC architecture, as shown in figure 2, the signal subtractions take place after the demodulation. The advantage of this narrowband approach is that it produces a clear reduction in the computational load as each cancellation does not require an additional IFFT and FFT operation. The time delay of different users is accounted for through a phase shift, but this ignores the inter-carrier interference introduced by the start or end of the users symbol.

For practical operation, both these systems still require some degree of synchronism. Since only one data bit per user is cancelled simultaneously, inter-symbol interference (ISI) due to other users next or previous symbols is not cancelled. While previous data symbols can be easily cancelled, cancelling future symbols requires a block level approach. Alternatively, the ISI can be restricted by limiting the asynchronism to a fraction of the symbol length. The use of cyclic-extensions and guard intervals can be used to eliminate any overlap between with the previous or next symbols of other users.

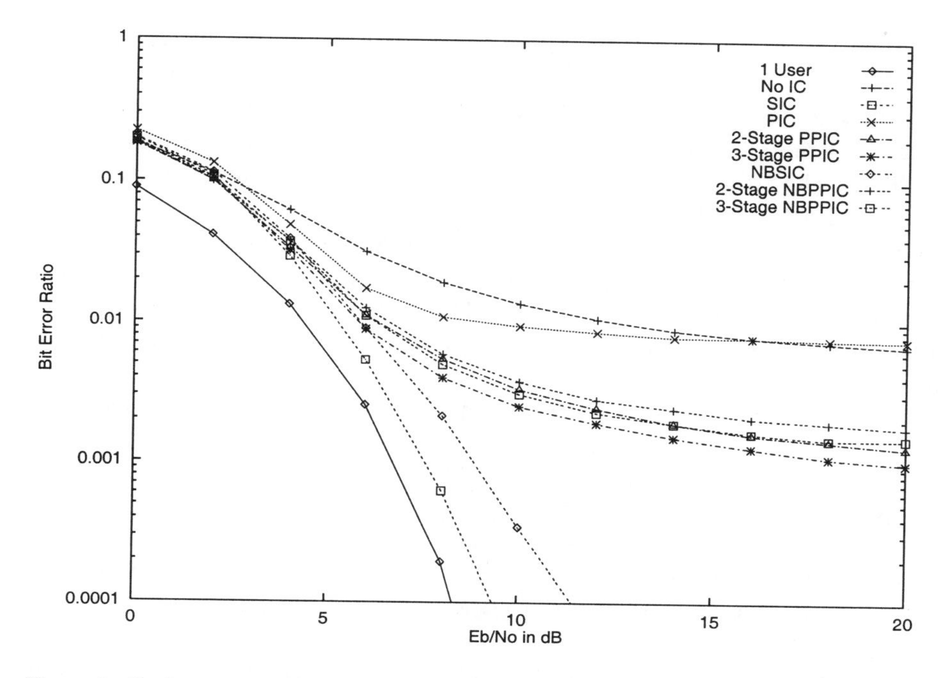

Figure 3 Performance of Interference Cancellation Techniques in AWGN channel with 16 Users

3. SIMULATION RESULTS

Successive, parallel and partial parallel cancellation systems were simulated with wideband cancellation. The successive and partial parallel systems were also simulated using narrowband cancellation. These simulations were performed with a processing gain of 32 in an AWGN channel. However the receivers did not assume an AWGN channel, and channel estimation for each individual user was based on pilot tones transmitted through the channel, giving an imperfect channel estimate. A channel matched (or Maximal Ratio Combiner) linear receiver was employed as the first stage for all cases. The simulations were repeated for 100 randomly chosen delays configurations for the users signals. This asynchronism between users was limited to 25% of the symbol length and a guard interval was also inserted to prevent overlapping symbols. Figure 3 shows the average bit error ratio performance of all the cancellation schemes with 16 users present: wideband and narrowband successive (SIC and NBSIC); wideband parallel (PIC); 2-stage and 3-stage wideband and narrowband partial parallel (PPIC and NBPPIC); and these are compared with the single user case and no interference cancellation (No IC). The partial parameters used were 0.6 for stage 2 and 0.8 for stage 3.

All the cancellation schemes improve the performance, however at high E_b/N_0, the full parallel cancellation does not offer any gain, and struggles to reduce the BER to below 0.01. The partial parallel schemes perform much better with the 3-stage cancellation bringing the BER down to 0.001 at an E_b/N_0 of 20 dB. The 3rd stage provides a small improvement over the 2-stage system of about 2 dB at an E_b/N_0 of 10 dB. The narrowband partial parallel systems show a similar 2 dB degradation over their wideband counterparts. The best performance is obtained from the successive interference cancellation, which performs only 1 dB worse than the single user case. The narrowband system also performs well, and the degradation compared to the wideband system is less than 2 dB.

Simulations were also performed using a fading model of an indoor wireless channel. The data rate was chosen at 256 kbit/s, and the multi-carrier modulation performed using a 32 point IFFT, this was cyclicly extended by 6 samples and a guard interval of 8 samples inserted giving a sample rate of 11.7 MHz. The channel was modelled as 6 taps of exponentially decreasing power down to a tap of -10 dB below the first tap power level, 0.5 μs after the first tap. To overcome the spreading of signal energy, the symbols were cyclicly extended by 6 samples. The taps varied in time to give a Doppler spread of 23 Hz corresponding to motion of 1.4 m/s with a 5 GHz carrier. The system was considered to be uplink and therefore each users channel was simulated independently.

To identify the user capacity of the system, the E_b/N_0 required to obtain a bit error ratio of 0.01 was estimated for 1 to 32 users. The required E_b/N_0 obtained for each system are shown in figure 4. This shows that with no interference cancellation, for more than 12 users, the required E_b/N_0 is 10 dB more than that for the single user case. The parallel interference cancellation outperforms this significantly, increasing the capacity to 21 users. The two-stage partial parallel cancellation improves on this giving a capacity of 25 users. With three-stage partial parallel cancellation, the capacity is increased to 31 users. The best performance is achieved with the successive interference cancellation; this allows full capacity, with less than 4 dB increase in signal strength.

When comparing the wideband and narrowband systems, little difference is observed. With the successive interference cancellation, the narrowband system performs at worst 0.5 dB worse than the wideband system. With the partial parallel systems, the narrowband systems outperform the wideband systems marginally providing a capacity increase of 1 user. These small differences may be due to the partial parameter not being optimal for both systems or the use of narrowband channel estimation.

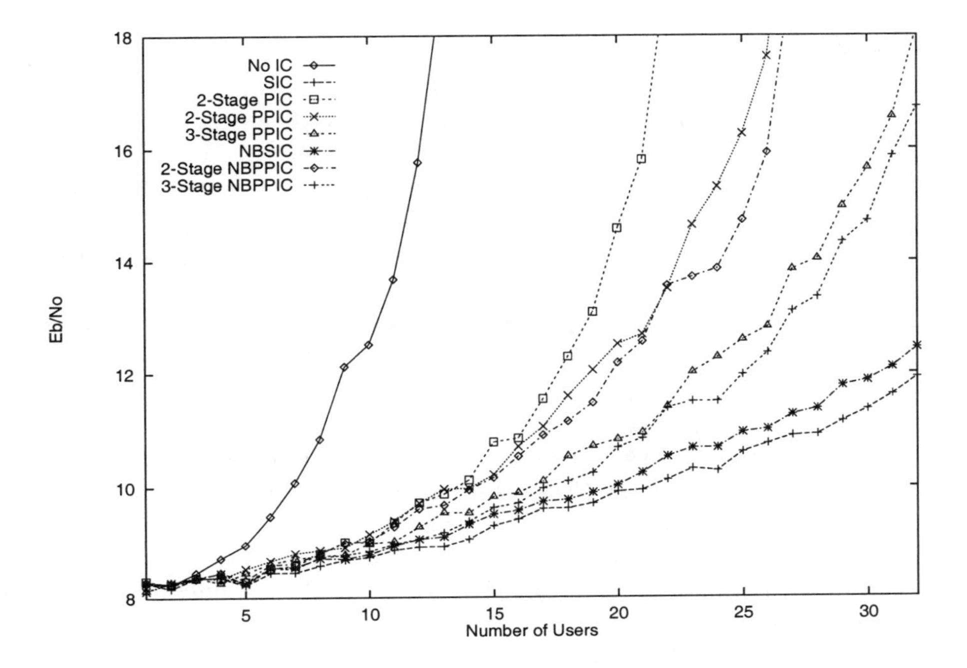

Figure 4 E_b/N_0 in dB Required by each IC System for BER of 0.01

4. CONCLUSION

Interference cancellation techniques allow multi-user detection in uplink MC-CDMA when the users are transmitting asynchronously and each user's signal experiences independent fading. If the asynchronism is constrained, guard intervals can be inserted to prevent overlap between successive symbols. In such situations, a lower complexity narrowband approach to interference cancellation can be applied. This narrowband approach can be applied to either successive and parallel interference cancellation schemes. Simulations comparing narrowband and wideband, successive and parallel, system show that the differences between the wideband and narrowband performance are small. The best performance was achieved with the successive interference cancellation which could support a fully loaded system at a BER of 0.01 with only an increase of 4 dB above the required single user power level.

Acknowledgement

This work is funded by the U.K. Engineering and Physical Sciences Research Council grant no. GR/L/98091 for research into Multi-Carrier CDMA with Multi-User Detection.

References

Divsalar, D. and Simon, M. (1995). Improved CDMA Performance Using Parallel Interference Cancellation. Technical report, Jet Propulsion Laboratory, California Institute of Technology. JPL Publication 95-21.

Kaiser, S. (1998). MC-FDMA and MC-TDMA versus MC-CDMA and SS-MC-MA: Performance evaluation for fading channels. In *Proceedings of the IEEE Int. Symposium on Spread Spectrum Techniques and Applications*, pages 200–204, Sun City, South Africa.

Kalofonos, D. N. and Proakis, J. G. (1996). Performance of the Multi-stage Detector for a MC-CDMA System in a Rayleigh Fading Channel. *IEEE Globecom*, pages 1784–1788, London, U.K.

Kleer, F., Hara, S., and Prasad, R. (1999). Detection Strategies and Cancellation Schemes in a MC-CDMA System. In Swarts, F., van Rooyan, P., Opperman, I., and Lötter, M. P., editors, *CDMA Techniques for Third Generation Mobile Systems*, chapter 7, pages 185–215. Kluwer.

Teich, W. G., Bury, A., Egle, J., Nold, M., and Lindner, J. (1998). Iterative Detection Algorithms for Extended MC-CDMA. In *Proceedings of the IEEE Int. Symposium on Spread Spectrum Techniques and Applications*, pages 184–188, Sun City, South Africa.

Varanasi, M. K. (1999). Decision Feedback Multiuser Detection: A Systematic Approach. *IEEE Transactions on Information Theory*, 45(1):219–240.

MULTIPLE ACCESS INTERFERENCE CANCELLATION IN MULTICARRIER CDMA SYSTEMS USING A SUBSPACE PROJECTION TECHNIQUE

Daniel I. Iglesia, Carlos J. Escudero, Luis Castedo
Departamento de Electrónica y Sistemas. Universidad de La Coruña
Campus de Elviña s/n, 15.071 La Coruña, SPAIN
Tel: ++ 34-981-167150, e-mail: escudero@des.fi.udc.es

Abstract Multicarrier CDMA systems are efficient methods to combat the ISI introduced by dispersive channels without channel equalization. The performance of these systems is degraded by the Multiple Access Interference (MAI). This paper introduces a method based on signal subspace projection for MAI cancellation that overcomes the limitations of the blind adaptive receivers when the received user code is not perfectly known. It is demonstrated that this method yields to a considerable improvement of the output SINR.

1. INTRODUCTION

Multicarrier (MC) transmission methods for Code Division Multiple Access (CDMA) communication systems have been recently proposed [1] as an efficient technique to combat multipath propagation. MC systems split the transmitted digital information in multiple carriers and achieve high transmission speeds using large symbol periods. As a consequence, MC systems can simply remove the Inter-Symbol Interference (ISI) introducing a short guard time between symbols and avoid the utilization of a channel equalizer. Although MC-CDMA systems do not suffer from ISI, channel distortion causes loss of orthogonality between codes and Multiple Access Interference (MAI) arises due to interfering users transmitting simultaneously with the desired one.

Because of the time-varying nature of mobile radio channels, adaptive filtering techniques have to be used in order to suppress the MAI. Conventional adaptive filtering techniques are based on the Minimum Mean Square Error (MMSE) criterion [2], where it is necessary to transmit a training sequence to set up the filter coefficients. To avoid the training sequences, blind techniques based on the Linearly Constrained Minimum Variance (LCMV) criterion [3] have been proposed for MAI cancellation in DS-CDMA. However, these techniques are extremely sensitive to innacuracies in the acquisition of the desired user timing and spreading code.

This paper introduces a new method to improve the performance of the blind LCMV criterion in a MC-CDMA system where we cannot obtain a perfect knowledge of the desired user code. This technique tries to approximate the transmitted code with the received one by means of a projection onto the signal-interference subspace of the input autocorrelation matrix.

The paper is organized as follows: Section 2 describes the signal model of a MC-CDMA system. Section 3 states the problem and shows the performance degradation in a simple environment. Section 4 presents the projection technique and its adaptive implementation. Section 5 contains the results of computer simulations and, finally, Section 6 is devoted to the conclusions.

2. SIGNAL MODEL

Let us consider a discrete-time baseband equivalent model of a synchronous MC-CDMA system with N users using L-chip spreading codes. Each user i multiplies the transmitted symbols s_n^i with a L-chips spreading code, $c_i(k), \quad k = 0, \cdots, L-1$, to produce a sequence

$$v_n^i(k) = s_n^i c_i(k), \qquad k = 0, \cdots, L-1 \tag{1}$$

The transmitter computes the IDFT (Inverse Discrete Fourier Transform) of (1) to yield

$$V_n^i(m) = IDFT[v_n^i(k)] = \frac{1}{L}\sum_{l=0}^{L-1} v_n^i(l) e^{j\frac{2\pi}{L}lm} \tag{2}$$

This signal is transmitted through a dispersive channel with impulse response, $h_i(m)$, that introduces Additive White Gaussian Noise (AWGN) and is shared by the other $N-1$ users. The received signal is

$$X_n(m) \quad = \quad \sum_{i=1}^{N} V_n^i(m) * h_i(m) + r_n(m) \tag{3}$$

where $*$ denotes discrete convolution and $r_n(m)$ is a white noise sequence with variance σ_r^2.

To recover the transmitted symbols, the receiver applies a DFT (Discrete Fourier Transform) to the received signal (3). Assuming that the receiver is perfectly synchronized to the users and that there exists a sufficiently large guard times between symbols, the resulting signal is

$$x_n(k) \quad = \quad DFT[X_n(m)] = \sum_{i=1}^{N} s_n^i \tilde{c}_i(k) + \Gamma_n(k) \tag{4}$$

for $k = 0, \cdots, L-1$ and where $\tilde{c}_i(k) = c_i(k)H_i(k)$ is a perturbed version of the spreading code for the i-th user and $H_i(k)$ and $\Gamma_n(k)$ are the DFT's of $h_i(m)$ and $r_n(m)$, respectively.

Note that the dispersive effects of the channel, $h_i(n)$, are reduced to a multiplication in (4) if we assume a sufficiently large guard time between adjacent symbols. If we consider an unknown channel, both the amplitude and phase of $H_i(k)$ can be modeled as random variables. In the simulations section we will consider that $H_i(k)$ is a vector of Rician random variables.

Finally, if we use matrix notation we can define an observations vector as follows

$$\mathbf{x}_n = [x_n(0), \cdots, x_n(L-1)]^T = \sum_{i=1}^{N} s_n^i \tilde{\mathbf{c}}_i + \mathbf{\Gamma}_n \tag{5}$$

where $\tilde{\mathbf{c}}_i = [\tilde{c}_i(0), \cdots, \tilde{c}_i(L-1)]^T$ is the received user code and $\mathbf{\Gamma}_n = [\Gamma_n(0), \cdots, \Gamma_n(L-1)]^T$ is a vector of Gaussian noise components. Assuming statistical independence between users and noise, the autocorrelation matrix of the observations vector is

$$\mathbf{R}_x \quad = \quad \sigma_d^2 \tilde{\mathbf{c}}_d \tilde{\mathbf{c}}_d^H + \sum_{\substack{i=1 \\ i \neq d}}^{N} \sigma_i^2 \tilde{\mathbf{c}}_i \tilde{\mathbf{c}}_i^H + L\sigma_r^2 \mathbf{I} = \sigma_d^2 \tilde{\mathbf{c}}_d \tilde{\mathbf{c}}_d^H + \mathbf{R}_{i+r} \tag{6}$$

where superindex ^{H} denotes conjugate transpose, σ_i^2 and σ_r^2 are the i-th user signal and noise power, respectively, the subindex $_d$ identifies the desired user and $\mathbf{R}_{i+r}$ is the autocorrelation matrix for the interferences plus the noise.

To produce the receiver output, the observations vector (5) is processed by a linear combiner $\mathbf{w}$ as follows

$$y_n = \mathbf{w}^H \mathbf{x}_n \tag{7}$$

and it is obtained a Signal to Interference plus Noise Ratio (SINR) at the output given by

$$SINR = \frac{\sigma_d^2 |\mathbf{w}^H \tilde{\mathbf{c}}_d|^2}{\mathbf{w}^H \mathbf{R}_{i+r} \mathbf{w}} \tag{8}$$

3. PROBLEM STATEMENT

When considering blind adaptive MAI cancellation, the weight vector $\mathbf{w}$ in (7) is chosen according to the LCMV criterion [3]

$$\min_{\mathbf{w}} E[|y_n|^2] \quad \text{subject to} \quad \mathbf{w}^H \tilde{\mathbf{c}}_d = 1 \tag{9}$$

It is well known [3] that the solution to this optimization problem is

$$\mathbf{w}_{LCMV} = \mu \mathbf{R}_x^{-1} \tilde{\mathbf{c}}_d \quad \Rightarrow \quad SINR_{opt} = \sigma_d^2 \tilde{\mathbf{c}}_d^H \mathbf{R}_{i+r}^{-1} \tilde{\mathbf{c}}_d \tag{10}$$

where μ is an arbitrary constant and $SINR_{opt}$ is the maximum SINR attainable.
However, the weight vector (10) can only be obtained under ideal conditions where we have a perfect knowledge of the received code $\tilde{\mathbf{c}}_d$. In practical situations, code acquisition systems provide an estimation that we will term $\mathbf{c}_d = [c_d(0), \cdots, c_d(L-1)]^T$. In this case, the weight vector is given by

$$\mathbf{w}_{LCMV} = \mu \mathbf{R}_x^{-1} \mathbf{c}_d \tag{11}$$

that is no longer optimum since $\mathbf{c}_d \neq \tilde{\mathbf{c}}_d$. The resulting SINR is extremely low because the receiver interprets the desired signal as an interference and cancels it from the output.

In order to show the degradation of the SINR when using (11), let us consider a simple environment with a single user (the desired one) and Gaussian noise. In this case the autocorrelation matrix for the interferences plus noise is reduced to a diagonal matrix $\mathbf{R}_{i+r} = \sigma_r^2 \mathbf{I}$. Therefore, substituting (11) in (8) and using the matrix inversion lemma, the SINR can be rewritten as

$$SINR = \frac{SINR_{opt} \ \ cos^2\theta}{1 + SINR_{opt} \ (SINR_{opt} + 2) \ \ sin^2\theta} \tag{12}$$

where θ represents the angle between the vectors $\tilde{\mathbf{c}}_d$ and $\mathbf{c}_d$. Note that θ depends on the perturbation introduced by the channel. In the absence of channel distortion, $\theta = 0$ and equation in (12) corresponds with the $SINR_{opt}$. However, when $\theta \neq 0$ the $SINR$ decreases when $SINR_{opt}$ increases.

This degradation of the SINR can be avoided if we use in (11) any vector parallel to $\tilde{\mathbf{c}}_d$ instead of $\mathbf{c}_d$. One of these vectors can be obtained, even without the knowledge of the perturbed code, by projecting the transmitted code $\mathbf{c}_d$ onto the signal subspace since this subspace is the line spanned by the vector $\tilde{\mathbf{c}}_d$. Next section presents the projection technique for a general environment.

4. PROJECTION TECHNIQUE

When we consider a general environment with N users and Gaussian noise it is easy to determine the signal-interference subspace but not the desired signal subspace alone. Nevertheless, projecting $\mathbf{c}_d$ onto the signal-interference subspace results in a better performance as will be shown in the sequel. This projection does not guarantee a vector parallel to the true desired user code, $\tilde{\mathbf{c}}_d$, but we still obtain an approximation close to it. This technique has already been used in [5] for adaptive beamforming with uncalibrated arrays.

Indeed, the transmitted user code can be decomposed as the sum

$$\mathbf{c}_d = \mathbf{c}_d^s + \mathbf{c}_d^r \tag{13}$$

where $\mathbf{c}_d^s$ and $\mathbf{c}_d^r$ are two orthogonal components that lie in the signal-interference and noise subspace, respectively. These two subspaces can be determined from the eigenvectors and eigenvalues of the autocorrelation matrix $\mathbf{R}_x$ [4]. Substituting (13) and (11) in (8), the SINR for the LCMV technique can be expressed as follows

$$SINR_{LCMV} = \left(\frac{\mathbf{c}_d^{s^H} \mathbf{R}_x^{-1} \mathbf{c}_d^s + \mathbf{c}_d^{r^H} \mathbf{R}_x^{-1} \mathbf{c}_d^r}{\sigma_d^2 |\mathbf{c}_d^{s^H} \mathbf{R}_x^{-1} \tilde{\mathbf{c}}_d|^2} - 1 \right)^{-1} \tag{14}$$

This expression can be interpreted as the generalization of (12) to a N users environment and, in general, yields to values which are considerably smaller than the optimum SINR.

However, projecting the transmitted user code onto the signal plus interference subspace is equivalent to eliminating the component $\mathbf{c}_d^r$ and the resulting output SINR is given by

$$SINR_{proj} = \left(\frac{\mathbf{c}_d^{s^H} \mathbf{R}_x^{-1} \mathbf{c}_d^s}{\sigma_d^2 |\mathbf{c}_d^{s^H} \mathbf{R}_x^{-1} \tilde{\mathbf{c}}_d|^2} - 1 \right)^{-1} \tag{15}$$

It is straightforward to see that this value is higher than (14). This indicates that an improvement in performance is obtained when using the projected vector $\mathbf{c}_d^s$ instead of $\mathbf{c}_d$.

4.1 ADAPTIVE IMPLEMENTATION

It is well know that the signal-interference subspace is spanned by the eigenvectors corresponding to the N most significant eigenvalues of the autocorrelation matrix $\mathbf{R}_x$ [4]. Considering the $L \times N$ unitary matrix $\mathbf{U}$ containing these significant eigenvectors, we can build a projection matrix, $\mathbf{P} = \mathbf{U}(\mathbf{U}^H\mathbf{U})^{-1}\mathbf{U}^H = \mathbf{U}\mathbf{U}^H$. Therefore, the projection of $\mathbf{c}_d$ onto the signal-interference subspace is obtained by multiplying the projection matrix with the code, i.e., $\mathbf{P}\mathbf{c}_d = \mathbf{c}_d^s$. If we use this projection, the weight vector (11) can be modified as follows

$$\mathbf{w}_{proj} = \mathbf{R}_x^{-1}\mathbf{P}\mathbf{c}_d = \mathbf{R}_x^{-1}\mathbf{c}_d^s \tag{16}$$

In a practical implementation, the autocorrelation matrix is unknown and is estimated as $\hat{\mathbf{R}}_x = \frac{1}{N_r}\sum_{k=1}^{N_r}\mathbf{x}_k\mathbf{x}_k^H$, where N_r is the number of received symbols used to estimate the matrix. This finite number of data observations produce errors in the estimation that reduce the SINR similarly to situations where we do not have a perfect knowledge of the received codes. Estimation errors can be assumed as an error in the knowledge of the received codes. Therefore, for a fixed number of data observations, the projection technique will obtain a higher output SINR. The proposed projection algorithm exhibits faster convergence even when there is no channel distortion.

5. SIMULATIONS

To illustrate the effectiveness of the proposed algorithm, this section presents the results of some computer simulations. In the simulations we assume a Rician fading model [6] where the channel factors are statistically independent between subcarriers. We also consider Hadamard codes with length $L = 8$.

In the first simulation we compare the performance of the projection technique with respect to the optimum (perfect knowledge of desired user code) and the LCMV techniques. Figure 1 plots the average of the output SINR for 10 channel realizations with respect to the input SNR. We consider three users transmitting with the same input SNR through two different channels with ratios $\frac{A_0}{\sigma} = 1$ and $\frac{A_0}{\sigma} = 3$. Note that the output SINR corresponding to the LCMV reduces as the optimum increases whereas the projection technique improves the performance by reducing the degradation of the SINR with respect to the optimum value.

To illustrate the performance of the proposed algorithm, Symbol Error Rate (SER) with respect to the input SNR for the same three users environment as before and $\frac{A_0}{\sigma} = 3$ is plotted in figure 2 (left). Note that the curve corresponding to the projection method decreases with the input SNR and is close to the SER obtained when we have a perfect knowledge of the received code.

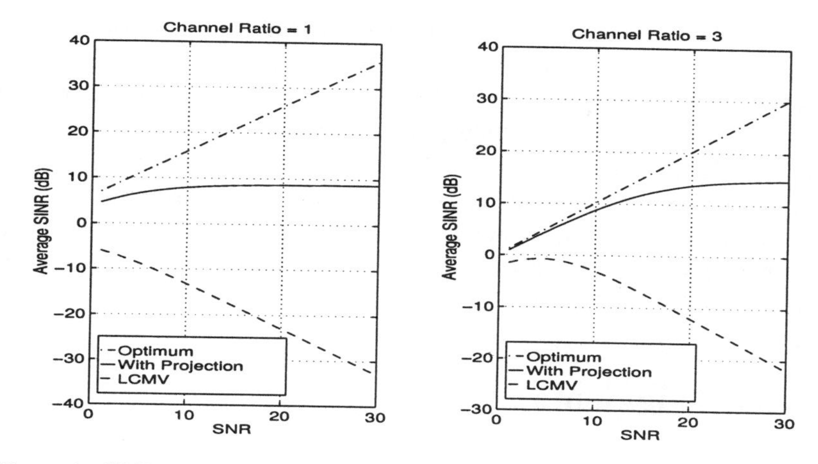

Figure 1 SINR vs. SNR for optimum, projection and LCMV techniques.

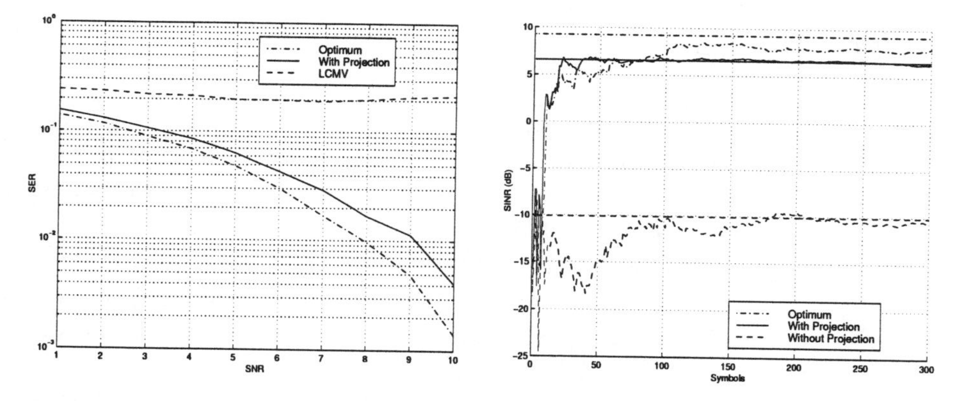

Figure 2 **Left**: SER vs. SNR for optimum, projection and LCMV techniques. **Right:** SINR time evolution for receiver with the optimum, projection and LCMV techniques.

Figure 2 (right) represents the output SINR time evolution of the algorithm. In this simulation we have considered three users transmitting with the same input $SNR = 6dB$ through a channel with $\frac{A_0}{\sigma} = 1$. Note that the projection technique presents an asymptotic convergence higher than the LCMV technique. The difference between the convergence value of the projection technique and the optimum value depends on the channel features. Moreover, the receiver with projection converges faster than the others.

6. CONCLUSIONS

MC-CDMA systems effectively combat the ISI introduced by dispersive channels. Nevertheless, they are still limited by the MAI due to code non orthogonality. The performance of blind adaptive receivers based on the LCMV criterion is very sensitive to inaccuracies in the acquisition of the received user code and typically result in extremely low output SINR. This paper shows that this output SINR is considerably increased if we use a projection of the transmitted user code onto the signal-interference subspace to obtain a new code closer to the received one. This technique also accounts for the errors in the estimation of the autocorrelation matrix and thus exhibits faster convergence. Computer simulations illustrate the performance of this technique in a Rician channel.

Acknowledgments

This work has been supported by CICYT (grant TIC96-0500-C10-02) and FEDER (grant 1FD97-0082).

References

[1] Yee, N., Linnartz, J. P., Fettweis, G., "Multi-Carrier CDMA in Indoor Wireless Radio Networks", *Proc. International Symposium on Personal, Indoor and Mobile Radio Communications, (PIMRC93)*, Yokohama, pp. 109-113, 1993.

[2] U. Madhow, M. L. Honig, "MMSE Interference Suppression for Direct-Sequence Spread-Spectrum CDMA", *IEEE Transactions on Communications*, vol. 42, no. 12, pp. 3178-3188, December 1994.

[3] M. L. Honig, U. Madhow, S. Verdú, "Blind Adaptive Multiuser Detection", *IEEE Transactions on Information Theory*, vol. 41, no.4, pp. 944-960, July 1995.

[4] E. Moulines, P. Duhamel, J. F. Cardoso and S. Mayrargue, "Subspace Methods for the Blind Identification of Multichannel FIR Filters", *IEEE Transactions on Signal Processing*, vol. 43, no. 2, pp. 516-525, February 1995.

[5] D. D. Feldman and L. J. Griffiths, "A Projection Approach for Robust Adaptive Beamforming", *IEEE Transactions on Signal Processing*, vol. 42, no. 4, pp. 867-876, April 1994.

[6] P. Z. Peebles, *Probability, Random Variables and Random Signal Principles*, McGraw-Hill International Editions, Singapore, 1993.

INTERFERENCE CANCELLATION FOR A MULTI-CARRIER SPREAD-SPECTRUM FCDMA SYSTEM[†]

Achim Nahler, Jörg Kühne, and Gerhard P. Fettweis

Dresden University of Technology, Mobile Communications Systems
D-01062 Dresden, Germany, Tel.: +49 351-463 5521, Fax.: +49 351-463 7255
e-mail: nahler@ifn.et.tu-dresden.de

ABSTRACT

This paper deals with investigations of interference cancellation technique for a multi-carrier spread-spectrum system in that all subscribers use the same but frequency-shifted signature sequence. Firstly, the cross-correlation behavior resulting from this special multiple access scheme will be examined for different code signals. Then, bit error rates will be evaluated for successive and parallel interference cancellation, respectively. Furthermore, two proposals considering the code-dependent cross-correlation function will be given to modify this multiple access scheme so that the uplink performance of such a system is enhanced without additional hardware cost.

I. INTRODUCTION

In [1] and [2], the new modulation techniques called *multi-carrier spread-spectrum* (MC-SS) was introduced. To generate the wideband code signal using a weighted trigonometric sum is the basic idea of MC-SS. This kind of spread-spectrum signal has more degrees of freedom to influence system characteristics such as spectral properties, correlation behavior or signal's dynamic range compared to conventional *direct-sequence spread-spectrum* (DS-SS) based on *pseudo-noise* (PN) sequences. In [3], it was shown that MC-SS has better spectral properties then DS-SS and also a good auto-correlation behavior. The latter allows time-domain equalization like a *matched filter* (MF) or a RAKE receiver instead of frequency-domain equalization that was proposed in first publications about MC-SS [1] and [2]. In [4], it is proposed to de-spread the incoming signal with an fast analog *surface acoustic wave* (SAW) device to reduce the complexity of the MC-SS receiver. In a multi-user system, every user should work with an identical SAW filter to lower the costs. That means that all users have the same spreading sequence. Therefore, a special *frequency code division multiple access* (FCDMA) scheme is required. The distinction of the different users is achieved by assigning to each user a main carrier frequency

[†]This work is partially supported by the German National Science Foundation (Deutsche Forschungsgemeinschaft contract Fe 423/2)

offset. To guarantee orthogonal signaling, the frequency offset should be equal to a multiple of the sub-carrier spacing [4].
For such an MC-SS FCDMA system, the cross-correlation function between the different users is determined by the time-frequency correlation of the code signal, the so-called ambiguity function. In [5], it is shown that different MC-SS code sequences lead to different ambiguity functions. In this paper, it will be investigated how the multi-user performance degradation depends on the chosen code, what are the reasons for the different performance and how to combat the degradation without additional implementation cost. This paper also concentrates on successive and parallel interference cancellation techniques because of their low complexity compared to optimum multi-user detection developed in [6] or other sub-optimum algorithms like solutions based on the minimum mean square error criterion or the decorrelation matched filter.
The paper is organized as follows. In section II, the transmitter and the multiple access scheme are described and the resulting cross-correlation behavior is examined. Section III deals with different interference cancellation techniques and their uplink performance. In section IV, two modifications of the multiple access scheme are proposed and their performance enhancement is investigated.

II. THE MC-SS FCDMA TRANSMITTER

The uth user transmit signal $s_u(t)$ is obtained by multiplying the narrowband data signal $d_u(t)$ with the symbol duration T_S by the MC-SS code signal $c(t)$ and then shifting the wideband signal by u sub-carriers in the frequency domain, hence

$$s_u(t) = d_u(t) \cdot c_u(t) \text{ with } c_u(t) = c(t) \cdot e^{j2\pi f_u t} \text{ and } c(t) = \sum_{k=0}^{K-1} C_k e^{j2\pi \frac{k}{T_S} t} \quad (1)$$

and $f_u = \frac{u}{T_S}$ as proposed in [4]. The spreading signal $c(t)$ is composed of K weighted and superimposed sinusoids. The complex-valued weighting coefficients C_k have the same magnitude $|C_k| = 1/\sqrt{K}$ and differ only in their phases. These coefficients determine the system properties such as spectral shape, dynamic range and correlation behavior. The multiple access is realized as mentioned above through the frequency shift by u sub-carriers for the uth user, hence $c_u(t) = c(t) \cdot e^{j2\pi \frac{u}{T_S} t}$. This scheme leads to the even cross-correlation function between user m and n

$$\varphi_{m,n}(t) = \frac{1}{T_S} \int_0^{T_S} c_m^*(\tau) c_n(\tau + t)\, d\tau, \quad (2)$$

thus

$$\varphi_{m,n}(t) = \frac{1}{K} \begin{cases} \sum\limits_{k=m-n}^{K-1} C_k^* C_{k+m-n} e^{j2\pi \frac{k}{T_S} t} & \text{for } n > m \\ \sum\limits_{k=0}^{K-1-(m+n)} C_k^* C_{k+m-n} e^{j2\pi \frac{k}{T_S} t} & \text{for } m > n \end{cases} \tag{3}$$

and it is also determined by the chosen coefficients C_k. It is worth noting that the odd cross-correlations behave very similar to the even cross-correlations. Two typical examples [5], which show the influence of the code selection on the cross-correlation behavior, are given in the Fig. 1. Figure 1(a) shows the cross-correlation magnitude of a so-called chirp-like code sequence with a quadratic phase. The main interference is concentrated in a narrow time interval with duration of $\frac{2T_S}{K}$. Outside if this area, the interference is very low. In the Fig. 1(b), the cross-correlation magnitude is shown for a PN-like code sequence C_k with pseudo-random chosen phases of -1 or $+1$. Its interference is distributed much more evenly in the time-frequency (time-user) plane then for chirp like codes. The influence of such very different cross-correlation behavior on the performance of *interference cancellation* (IC) algorithms will be investigated in the next section.

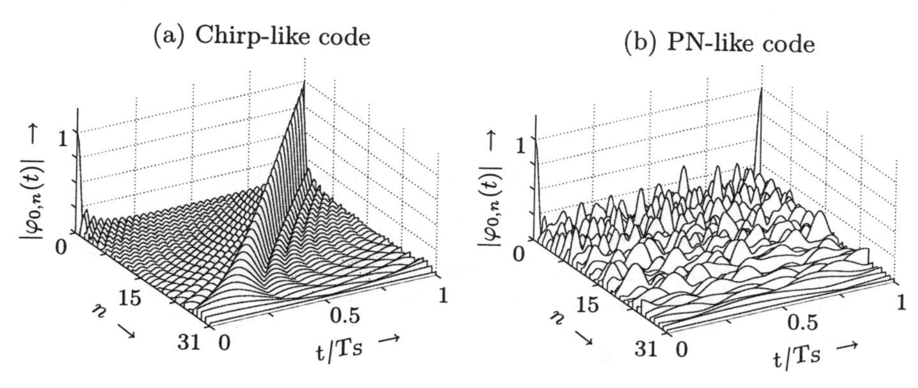

Fig. 1: Cross-correlation behavior $|\varphi_{0,n}(t)|$ of the 0th user ($u = 0$) with the nth user ($u = n$) for different codes

III. PERFORMANCE OF THE INTERFERENCE CANCELLATION TECHNIQUES

Decision-driven IC algorithms are interesting sub-optimum multi-user detection techniques because of their low additional hardware requirements and nevertheless significant performance improvement and near-far resistance. Therefore, the serial approach of *successive interference cancellation* (SIC) [7] and the parallel multi-stage approach of *parallel interference cancellation* (PIC) [8] are investigated. Throughout this paper, both schemes work internally with hard-decision IC.

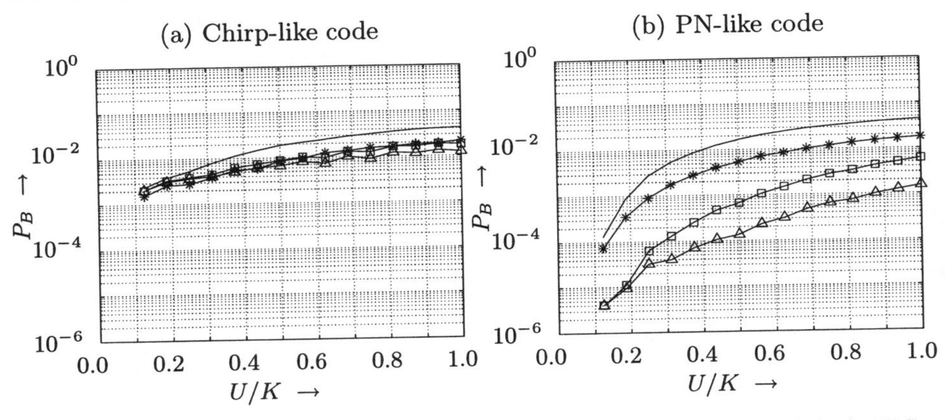

Fig. 2: BER for different codes and different receiver structures: MF (—), SIC (-⋆-), PIC (1st stage: -□-, 2nd stage: -△-) (SNR =10 dB)

The simulation scenario is a time and frequency non-selective uplink channel with U users, hence the received signal is

$$r(t) = \sum_{u=0}^{U-1} s_u(t - \tau_u) + n(t), \tag{4}$$

where the delays τ_u are uniform distributed over one symbol duration and $n(t)$ is additive white Gaussian noise with a two-sided power spectral density of $\frac{N_0}{2}$. In Fig. 2, the performance is shown for two codes (with chirp-like and PN-like phases, respectively, according to the distinction of section II and $K = 16$ sub-carriers) and for different receiver architectures. For a *signal-to-noise ratio* (SNR) equal to 10 dB, the uncoded *bit error rate* (BER) P_B is plotted versus the load ratio $\frac{U}{K}$ for the MF (—), the SIC receiver (-⋆-), the PIC receiver after its first stage (-□-) and after its second stage (-△-). From Fig. 2(a) it can be seen that neither SIC nor PIC lead to a substantial performance enhancement for chirp-like codes. In contrast to this, the performance is significant better for all receiver structures in the case of using PN-like codes (Fig. 2(b)), whereas more then two PIC stages do not lead to any further performance improvement. What are the reasons for those quite different performance results? A worst case scenario for two users is constructed. It means that the difference between the delays of both users leads to maximum cross-correlation $\varphi_{0,1}(\tau_0 - \tau_1) = \varphi_{\max}$. If the 0th user transmits a "+1" and the 1st one a "-1" then the output of the matched filter is $\tilde{d}_0 = 1 - \varphi_{\max} + \eta$ for the 0th user and $\tilde{d}_1 = -1 + \varphi_{\max} + \eta$ for the 1st user, where η is the filtered noise. For the chirp-like code used for the previous simulations, $\varphi_{\max}$ is 0.73 so that the SNR decreases dramatically (Fig. 3). In such a case the BER increases and also the probability of error propagation to the next SIC or PIC stage increases and the interference cancellation schemes fail, hence. The resulting BER, that is obtained by averaging the BER's over all possible cross-correlation constellations, is dominated by that worst case.

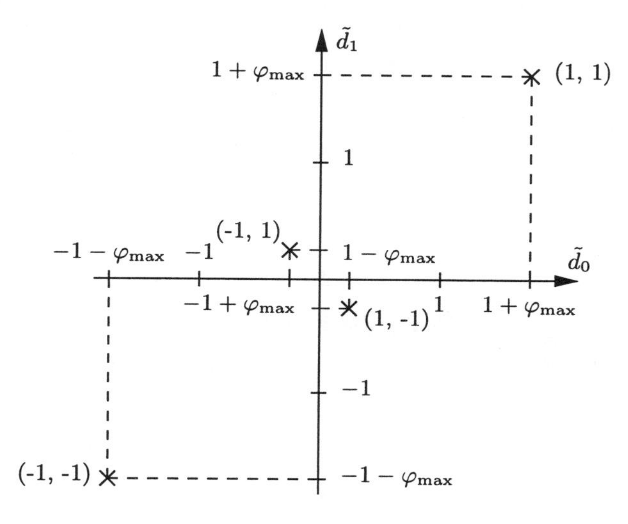

Fig. 3: Decision region of matched filter outputs in a worst case scenario for two users

Considering this cross-correlation behavior described in section II (see Fig. 1), it is obvious that the error distribution over the time-frequency plane is similar: Looking along the frequency (user) axis, the interference (BER degradation) is the lower the higher the frequency distance is for both kinds of codes. Looking along the time axis, both codes differ significantly. Using a chirp-like code, the BER is strongly decreased if the delay difference is inside the area with the cross-correlation peak and outside of this interval the BER is only very slight decreased. In contrast to this, the BER is degraded in about the same order over the symbol duration for the PN-like code. The consequences of such error distributions over the time-frequency plane are discussed in the next section.

IV. PERFORMANCE IMPROVEMENT THROUGH ADAPTIVE FREQUENCY ALLOCATION AND TIMING ADVANCE

In this section, opportunities are investigated to exploit the behavior of the cross-correlation for performance improvements without additional hardware costs. The goal of the optimization is to minimize the overall interference for each additional user step-by-step inside the correlation window of interest given through the maximum delay of the channel. For that purpose, two proposals are presented.

The first one considers that the interference amount between two users depends on the carrier spacing of both users. A user, that is only one sub-carrier spaced from another user, interferes with this other user more than a user, that ist e. g. 10 sub-carriers spaced from the user of interest. This behavior leads to a modified multiple access scheme that assigns to each additional user such a frequency offset that the frequency distance between the users is maximum within the available bandwidth, hence $f_0 = 0$, $f_1 = (K-1)/T_S$, $f_2 = K/(2 \cdot T_S)$ and so on. As it can be seen from Fig. 4, the system working

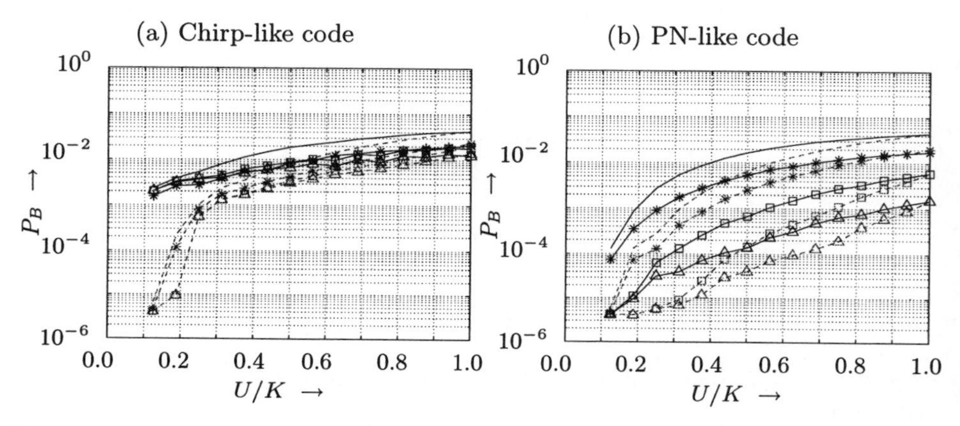

Fig. 4: BER comparision of the system using adaptive frequency allocation (MF: – –, SIC: – –★– –, PIC 1st stage: – –□– –, 2nd stage: – –△– –) with the conventional system (MF: —, SIC: –★–, PIC 1st stage: –□–, 2nd stage: –△–) (SNR = 10 dB)

with the modified multiple access scheme performs better than the original system. High performance gains are obtained in the low-load case because of the high sub-carrier distance between the individual users.

For the second approach, it is assumed that the maximum delay of the channel $\tau_{\max}$ is much lower than the symbol duration. This assumption is justified for applications in pico- or micro-cellular environments. For those radio channels it is proposed to send the signals of the different users with such a user-specific time offset ΔT_u, so that the cross-correlations are very low within the time intervall of interest given through the maximum delay of the channel. A worst case scenario has been simulated to gain insight into the improvement capabilities of that approach. The user and channel-specific delays τ_u are within the interval $[0, 2 \cdot T_S/K]$, so that the maximum cross-correlation

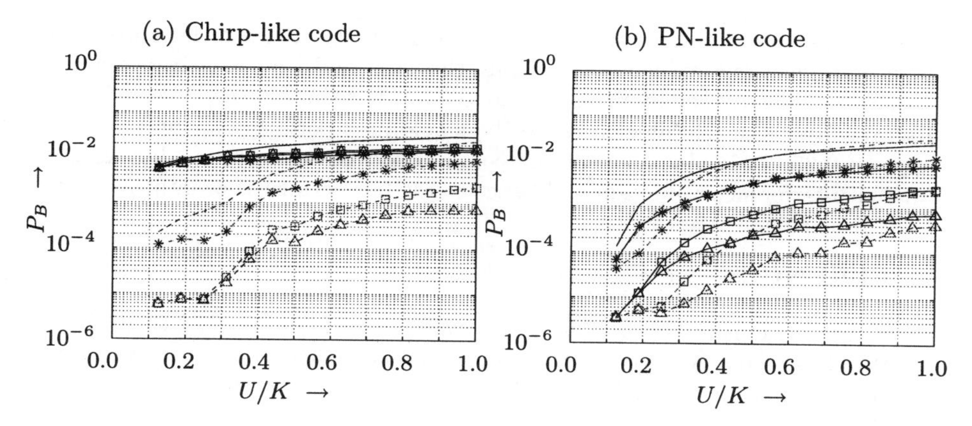

Fig. 5: BER comparision of the system using adaptive timing advance (MF: – –, SIC: – –⋆– –, PIC 1st stage: – –□– –, 2nd stage: – –△– –) with the conventional system (MF: —, SIC: –⋆–, PIC 1st stage: –□–, 2nd stage: –△–) (SNR = 10 dB)

value $\varphi_{\max}$ is ever within the correlation window and the SNR is substantially determined by $\varphi_{\max}$. To overcome this obstacle, the time offsets are assigned to $\Delta T_u = 2(u-1)T_S/K$. Figure 5 shows the corresponding simulation results. As expected, the MAI is substantially reduced and for both codes the 2nd PIC-stage is capable to cancel out nearly all MAI for low load-ratios.

V. CONCLUSIONS

For an MC-SS FCDMA system in that all users are spread with the same but frequency-shifted code, the performance of successive and parallel interference cancellation schemes has been investigated in a time and frequency non-selective uplink scenario for codes with chirp-like and PN-like phases, respectively. The completely different cross-correlation behavior arising from both code types leads to quite different performance. PN-like codes are much better suited than chirp-like codes because the interference is more equally distributed over time. That allows the IC schemes to improve the performance much more than in the case using a code with quadratic phases. Considering the interference distribution over the time-user plane, the multiple access scheme is modified in two ways to enhance the performance without additional hardware cost. 1. Adaptive frequency allocation: Each user's frequency offset is chosen so that the frequency distance of all users is maximal because the cross-correlation is the lower the higher the frequency distance is between the users. 2. Adaptive timing advance: If the channel maximum delay is much shorter than the symbol duration, the signals are sent with an user-specific time offset, so that the maximum cross-correlation value is outside the correlation window. Both modifications reduce dramatically the MAI and lead to substantial performance gains, especially in the low-load case. For the PN-like code, a 2-stage PIC receiver is nearly able to reach the single user performance up to a load-ratio of 0.25. It is obvious that the combination of both approaches leads to further interference reduction and hence to further performance improvements.

REFERENCES

[1] N. Yee, J.-P. Linnartz, and G. P. Fettweis, "Multi-Carrier CDMA in Indoor Wireless Radio Networks," in *Proc. of the 4th IEEE International Symposium on Personal, Indoor and Mobile Radio Communications*, pp. 109–113, 1993.

[2] K. Fazel, "Performance of CDMA/OFDM for Mobile Communication Systems," in *Proc. of the 2nd IEEE International Conference on Universal Personal Communications*, pp. 975–979, 1993.

[3] A. Nahler and G. P. Fettweis, "An Approach for a Multi-Carrier Spread Spectrum System with RAKE-Receiver," in *Proc. of the 1st International Workshop on Multi-Carrier Spread-Spectrum*, pp. 97–104, Kluwer Academic Press, 1997.

[4] V. Aue and G. P. Fettweis, "Higher-Level Multi-Carrier Modulation and Its Implementation," in *Proc. of the 4th International Symposium on Spread Spectrum Techniques and Applications*, pp. 126–130, 1996.

[5] J. Kühne, A. Nahler, and G. P. Fettweis, "Multi User Interference Evaluation for a One-Code Multi-Carrier Spread-Spectrum CDMA System with Imperfect Time and Frequency Synchronization," in *Proc. of the 5th IEEE International Symposium on Spread Spectrum Techniques and Applications*, pp. 479–483, 1998.

[6] S. Verdú, "Minimum Probability of Error for Asynchronous Gaussian Multiple-Access Channels," *IEEE Transactions on Information Theory*, vol. IT-32, pp. 85–96, January 1986.

[7] A. J. Viterbi, "Very Low Rate Convolutional Codes for Maximum Theoretical Performance of Spread-Spectrum Multiple-Access Channels," *IEEE Journal on Selected Areas in Communications*, vol. 8, pp. 641–649, May 1990.

[8] M. K. Varanasi and B. Aazhang, "Multistage Detection in Asynchronous Code-Division Multiple-Access Communications," *IEEE Transactions on Communications*, vol. 38, pp. 509–519, April 1990.

NARROWBAND INTERFERENCE MITIGATION IN OFDM-BASED WLANS

Reto Ness, Steven Thoen*, Liesbet Van der Perre,
Bert Gyselinckx and Marc Engels
Interuniversity Micro Electronics Center (IMEC)
Kapeldreef 75, 3001 Heverlee, Belgium

Key words: OFDM, Narrowband Interference

Abstract: We propose two narrowband interference cancellation algorithms for OFDM-based Wireless Local Networks. The first consists of adapting the constellation sizes of the carriers to their Signal-to-Interference-plus-Noise Ratio (SINR). The second relies on a spreading sequence to exploit frequency diversity. The BER performance in the indoor multipath channel is evaluated by simulations using a very flexible model of typical narrowband interference. The channel is modeled based on a ray tracing approach. Both interference cancellation algorithms provide an excellent robustness against the frequency selectivity of the multipath fading channel and strong narrowband interference.

1. INTRODUCTION

The interest in indoor wireless communications has grown rapidly in recent years because of the clear advantages over cabled networks, such as mobility of users, reduced cabling and increased setup flexibility. This trend does not only affect the professional office, but also the domestic environment. The ISM band at 2.4 GHz has been standardized globally for unlicensed usage, making it especially attractive for such in-home applications. From activities of industry consortia like HomeRF and Bluetooth, we expect that 2.4 GHz technology will soon become available for consumer applications.

The design of in-home communication systems is impaired by severe multipath and shadowing effects in the indoor environment. Also, strong

* PhD student at KULeuven and supported by a F.W.O.-scholarship

narrowband interference, due for instance to the leakage of microwave ovens or interfering wireless communication systems operating in the same band, severely hampers the performance of these systems. Therefore, interference mitigation is a crucial aspect for wireless indoor links.

We focus on OFDM-based systems using a cyclic prefix [1] for ISI mitigation, as proposed in the Hiperlan II standard. These systems reduce the complexity of the equalizer to a one-tap operation per carrier. Moreover, OFDM allows for easy signal processing in the frequency domain and therefore has advantages with respect to narrowband interference mitigation.

In this paper, we propose two narrowband interference cancellation algorithms and evaluate their performance. In Section 2 the system model is introduced. The channel model and the interference model used in our analysis are presented in Sections 3 and 4 respectively. In Section 5 we focus on a first algorithm, based on Adaptive Loading. A second algorithm that exploits frequency diversity by applying Direct Sequence Spreading (DSS) is described in Section 6. Finally, some conclusions are drawn.

2. SYSTEM MODEL

The baseband system model is schematized in *Figure 1*. The subscripts k and m denote the carrier and the signalling interval index respectively. $n(t)$ is AWGN and $i(t)$ is additive interference.

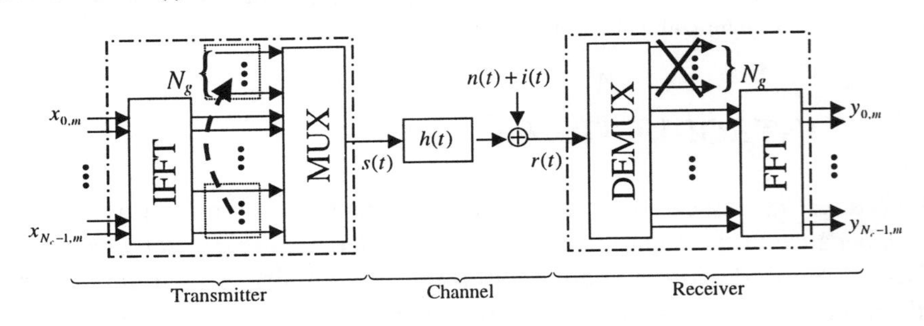

Figure 1. System Model

For a sufficiently long cyclic prefix, the received symbols are given by

$$y_{k,m} = h_k x_{k,m} + n_{k,m} + i_{k,m}, \tag{1}$$

where h_k is the Fourier Transform of the channel response $h(t)$ sampled at the kth carrier frequency, $n_{k,m}$ is AWGN and $i_{k,m}$ is a term due to the interference. In all simulations presented in this paper the net data rate is 20 Mbps at a net data symbol rate of 20 MHz, so that each OFDM symbol contains N_c information bits. The data is transmitted in burst of 6400 bits. We use a fixed number of carriers (N_c=64) and a cyclic prefix with a fixed length (N_g=8). The choice for N_g is motivated by an ISI-free transmission, since for the considered sample rate, typically the number of significant taps in the impulse response of the indoor channel does not exceed this value. The channel is considered static during the transmission of one burst. Also, we assume that transmitter and receiver are perfectly synchronized in time and frequency and that the receiver has perfect knowledge of the channel. We evaluate the performance of the proposed algorithms with reference to a non-adaptive OFDM system that uses BPSK on all carriers.

3. CHANNEL MODELING

We model the complex baseband impulse response $h(t)$ of the indoor multipath channel as [2]

$$h(t) = \sum_{l=0}^{N-1} \beta_l e^{j\theta_l} \delta(t - \tau_l), \tag{2}$$

where l is the path index, β_l is the path gain, θ_l is the path phase shift and τ_l is the path delay. We derive the parameters from a propagation study based on 2D ray tracing [3]. Input data is the floorplan of the setup including reflection and transmission coefficients. Only paths having an attenuation <30 dB are considered. *Figure 2* shows an example of the multipath propagation. The Power Delay Profile (PDP) and the frequency response are illustrated for a center frequency at 2.45 GHz.

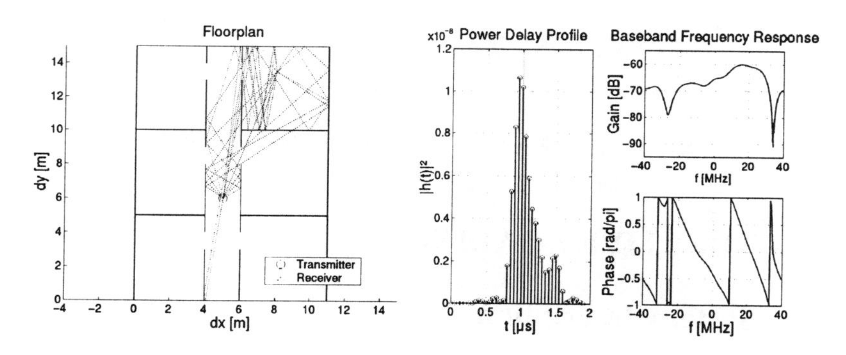

Figure 2. Example of indoor multipath channel modeling based on 2D ray tracing

To average out over different channels, we iterate over 36 responses.

4. INTERFERENCE MODELING

To simulate two different types of narrowband interference we developed a flexible complex baseband model, given by [4]

$$i(t) = A(t)e^{j(2\pi f_c(t)t+\phi(t))}. \tag{3}$$

Type A - Microwave oven emission: The carrier frequency is set to the design frequency of the magnetron in the baseband (f_c=0 Hz for a baseband center frequency of 2.45 GHz). $A(t)$ is a rectangular waveform with a frequency of 50 Hz to model the switching of the magnetron due to its power supply [5]. We constructed $\phi(t)$ to match the results of broadband measurements [6] in the time and frequency domain. The spectrogram of a typical interfering signal generated by this model is given in *Figure 3a*

Type B - Frequency hopping interference: The amplitude being constant ($A(t)$=1), we apply continuous phase frequency shift keying to modulate the carrier using a pseudo-random BPSK sequence with a symbol period of 1 μs. $f_c(t)$ implements a pseudo-random hopping sequence of frequencies spaced by 1 MHz within a bandwidth ranging from –10 to 10 MHz. The hopping rate is 1600 Hz with reference to the Bluetooth system. *Figure 3b* shows the spectrogram of a typical interfering signal generated by this model.

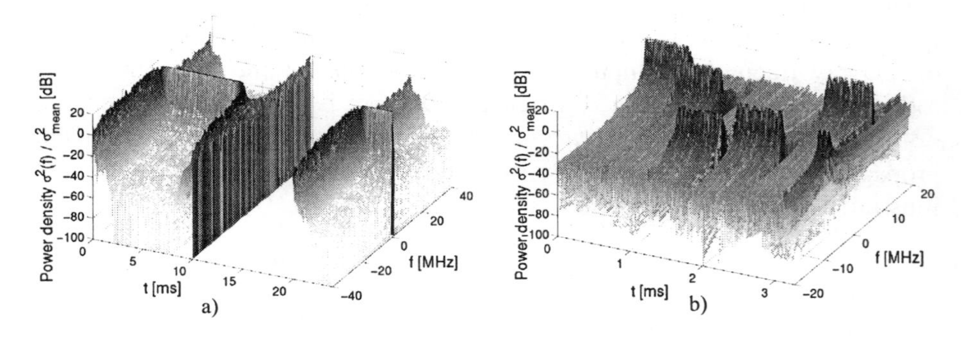

Figure 3. Spectrograms of the interference generated by the model

5. ADAPTIVE LOADING

5.1 System description

A high channel attenuation and strong interference on some of the sub-carriers cause high carrier bit error rates which dominate the average performance. Consequently, it is beneficial not to use these carriers and to perform a joint optimization of the data rates and the transmit power on the residual carriers. Since we consider a constant bit rate per OFDM symbol R_{tot} and a total transmit power S_{tot}, we start from the bit loading algorithm by Fischer et al. [7]. This algorithm equalizes the symbol error probability of the used carriers assuming rectangular QAM constellations and AWGN. To account for the interference, we replace the noise power by an estimate of the interference-plus-noise power on the carriers. In general, carriers affected by narrowband interference are avoided and the residual interference on the used carriers can be neglected with respect to the noise. Thus, the assumption of AWGN remains valid. The optimization problem is given by

$$\Pr\{Symbol\ Error\} = const. \rightarrow \min., \ \forall k \in D, \tag{4}$$

$$\text{with } \sum_{k \in D} R_k = const. = R_{tot} \text{ and } \sum_{k \in D} S_k = const. = S_{tot}, \tag{5}$$

where D is the set of used carriers, R_k is the constellation size (bits/QAM symbol) and S_k is the transmit power of the kth carrier. Introducing interference-plus-noise, the non-integer rate distribution is

$$R_k = \frac{R_{tot}}{\mathrm{M}\{D\}} + \frac{1}{\mathrm{M}\{D\}} \mathrm{ld}\left(\frac{|h_k|^2}{NI_k} \prod_{l \in D} \frac{NI_l}{|h_l|^2} \right), \quad \forall k \in D, \tag{6}$$

where NI_k is the average interference-plus-noise power of the kth carrier and $\mathrm{M}\{D\}$ is the number of elements in set D. After quantization to integer rates Rq_k, the transmit power distribution becomes

$$S_k = \frac{S_{tot} \cdot NI_k / |h_k|^2 \cdot 2^{Rq_k}}{\sum_{l \in D} NI_l / |h_l|^2 \cdot 2^{Rq_l}}, \quad \forall k \in D. \tag{7}$$

5.2 Simulation results

We assume perfect channel knowledge and estimate the NI_k at the receiver during a silent phase of the duration of 10 OFDM symbols initially to the transmission of each burst (100 OFDM symbols). Thus, the loading scheme is updated every 312 μs by the receiver and fed back to the transmitter. In *Figure 4* we compare the performance of a non-adaptive system using BPSK constellations and identical transmit power on all carriers to Adaptive Loading with the same number of bits per OFDM symbol for different *SIR* per bit (E_b/I). For both considered types of interference a significant gain in the BER performance is achieved.

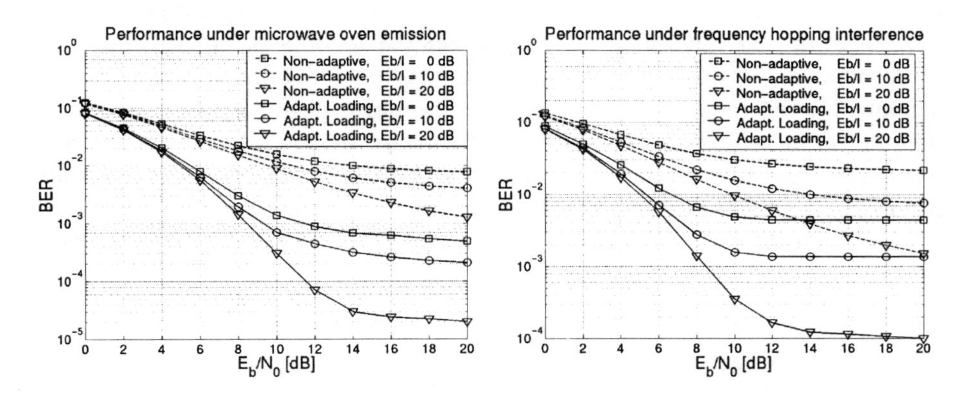

Figure 4. BER performance of Adaptive Loading

6. DIRECT SEQUENCE SPREADING

6.1 System description

Another strategy to mitigate interference is to exploit frequency diversity, inspired on an algorithm proposed by Fazel [8]. The idea consists of applying DSS with a code length L to spread each data symbol over a set of carriers that are maximally apart. Thus, by combining the received chips with respect to their reliability the data symbols can be recovered, even if some carriers have a low *SINR*. We impose $|c_k|=1$, $\forall k$ for the elements in the spreading code. At the receiver we apply weighting coefficients w_k to the chips prior to despreading, so that a non-biased estimate of x is given by

$$\hat{x} = x\sum_{k=1}^{L} w_k h_k c_k + \sum_{k=1}^{L} w_k ni_k, \quad \text{with} \quad \sum_{k=1}^{L} w_k h_k c_k = 1, \tag{8}$$

where ni_k is interference-plus-noise. We use Maximal Ratio Combining to determine the w_k. Assuming that the ni_k on different carriers are not correlated with each other, Lagrange optimization yields

$$w_k = \frac{h_k^* c_k^* / NI_k}{\sum_{l=1}^{L} |h_l|^2 / NI_l}. \tag{9}$$

In [8] the insertion of "zero-pilots" into the OFDM symbols is proposed to track the NI_k. However, this reduces the bandwidth efficiency. Instead we use a decision feedback estimator that gives a comparable performance without loss of bandwidth efficiency. It consists of a one-tap IIR-filter per carrier implementing the transfer function $(1-\lambda)/(1-\lambda Z^{-1})$, $0<\lambda<1$. The filter is initialized by an interference-plus noise estimation during a silent interval of 10 OFDM symbols prior to the transmission. *Figure 5* shows an example of DSS/OFDM for a code length L=2.

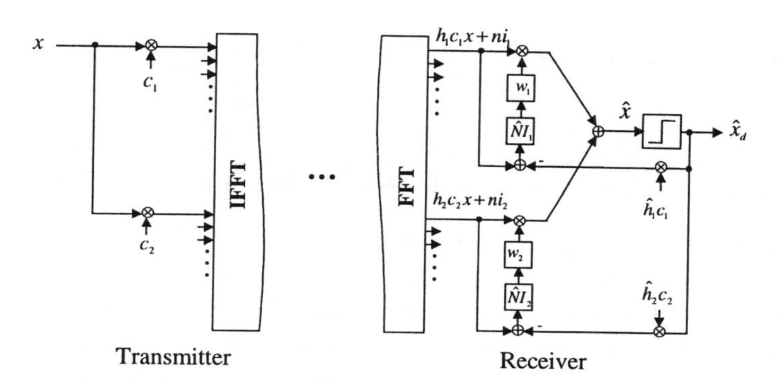

Figure 5. Example of DSS with chip weighting for a code length L=2

6.2 Simulation results

In *Figure 6* we compare the performance of a non-adaptive system using BPSK constellations to DSS/OFDM with λ=0.3 and a pseudo-random BPSK code of length 2 using QPSK constellations to maintain the same bandwidth. An excellent robustness against both types of interference is achieved.

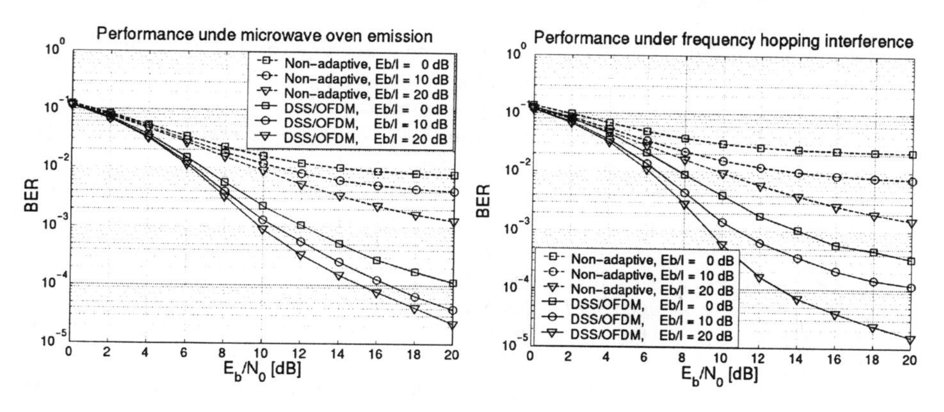

Figure 6. BER performance of DSS with chip weighting

7. CONCLUSION

We proposed two narrowband interference mitigation algorithms for OFDM-based WLANs. Two typical interference sources were modelled for performance evaluations in an indoor multipath propagation environment in the 2.4 GHz band. It was shown that both algorithms provide a high robustness against both interference and frequency selective fading. The gain in E_b/N_0 at a bit error probability of 10^{-2} is given in *Table 1.*

Table 1. Performance gain in terms of E_b/N_0 at BER = 10^{-2}

Algorithm	Adaptive Loading						DSS/OFDM					
Interference	Type A			Type B			Type A			Type B		
E_b/I [dB]	0	10	20	0	10	20	0	10	20	0	10	20
Gain [dB]	8.2	5.3	4.5	>13	8.6	4.9	7.2	4.5	3.4	>12	7.4	3.8

Adaptive Loading avoids carriers with very low *SINRs* and optimizes the bandwidth usage. However, broadcast applications are not supported, because of the adaptation to the channel and the interference at the transmit side. DSS/OFDM with chip weighting mitigates narrowband interference by exploiting frequency diversity. The decision feedback estimator allows for interference tracking on the signal. Also, the adaptation takes place only at the receiver side, so that broadcast applications can be supported.

REFERENCES

[1] J.A.C. Bingham, "Multicarrier Modulation for Data Transmission: An Idea Whose Time has Come", *IEEE Comm. Magazine, vol. 28, pp. 5-14, May 1990.*

[2] A.M. Saleh et al., "A Statistical Model for Indoor Multipath Propagation", *IEEE Selected Areas of Comm., vol. CSA-5, no. 2, pp. 1384-87, July 1991.*

[3] Liesbet Van der Perre et al., "Adaptive Loading Strategy for a High Speed OFDM-based WLAN", *IEEE Globecom '98, Sydney, Australia, pp. 1936-40, November 1998.*

[4] S. Vasudevan et al.,"Reliable Wireless Telephony using the 2.4 GHz ISM Band: Issues and Solutions", *IEEE 4th Int. Symposium on Spread Spectrum Techniques and Applications, New York, vol. 2, pp. 790-94, 1996.*

[5] S. Miyamoto et al., "Performance of Radio Communication Systems under Microwave Oven Interference Environment", *IEEE Int. Symposium on Electromagnetic Compatibility, New York, Cat. no. 97TH821, pp. 308-11, 1997.*

[6] L. Leyten and H. Visser, "Radiation Properties of the Microwave Oven", *Philips, Nat. Lab. Technical Note no. 203/93, RWR-537-LL-93083-11, 1993.*

[7] R. Fischer and J. Huber, "A New Loading Algorithm for Discrete Multitone Transmission", *IEEE Globecom '96, London, England, pp. 724-28, November 1996.*

[8] K. Fazel, "Narrowband Interference Rejection in Orthogonal Multicarrier Spread-Spectrum Communications", *IEEE 3rd Annual Int. Conference on Universal Personal Communications, New York, pp. Xvii+674, 1994.*

MULTI-CARRIER CDMA USING INTERFERENCE CANCELLATION

J-Y. Baudais, J-F. Hélard, J. Citerne
Institut National des Sciences Appliquées, INSA/LCST, 20 avenues ds Buttes de Coëmes, CS 14315, 35043 RENNES CEDEX, France.
E-mail: jean-yves.baudais@insa-rennes.fr, jean-francois.helard@insa-rennes.fr

Key words: *MMSE, Interference Cancellation, Iterative Detection, Downlink.*

Abstract: *In this paper, various detection techniques in the synchronous case of a multi-user MC-CDMA system operating in frequency selective Rayleigh channel are compared with respect to the achievable spectral efficiency. It is shown that a parallel interference cancellation MMSE scheme with two stages could be a good compromise between performance and complexity.*

1. Introduction

Since 1993, many researchers have investigated the suitability of the Multi-Carrier Code Division Multiple Access (MC-CDMA) for cellular systems [1], [2]. This promising multiple access scheme with high bandwidth efficiency is based on a serial concatenation of Direct Sequence (DS) spreading with Multi-Carrier (MC) modulation: The MC-CDMA transmitter spreads the original data stream over different subcarriers using a given spreading code in the frequency domain. For a synchronous system as the downlink mobile radio communication channel, the application of orthogonal codes such as Walsh-Hadamard codes guarantees the absence of Multiple Access Interference (MAI) in an ideal channel. However, through a frequency selective fading channel, all the subcarriers have different amplitude levels and different phase shifts, which results in a loss of the orthogonality among users and then generates MAI. So, after direct FFT and frequency desinterleaving, the received sequence must be "equalized" by employing a bank of adaptive one tap equalizers to combat the phase and amplitude distorsions caused by the mobile radio channel on the subcarriers. The channel estimation is usually derived from the FFT of the channel impulse response which can be estimated using pilot inserted between the data.

To combat the MAI, one may use various Single-user Detection (SD) techniques which do not take into account any information about this MAI. When the receiver does not know either the number of users or the received SNR, conventional and simple detection methods are Maximum Ratio Combining (MRC), Equal Gain Combining (EGC) or Orthogonal Restoring Combining (ORC). For more sophisticated receivers structures, the Minimum Mean Square Error (MMSE) algorithm [3] can provide performance improvements.

In order to improve the performance of the receiver still further, Multi-user Detection (MD) techniques can be processed. Multi-user detection is based on the important assumption that the codes of the different users are known to the receiver a priori. Most of the proposed multi-user detectors can be classified in one of the two categories: Maximum Likelihood (ML) detectors and Interference Cancellation (IC) detectors.

In this paper, various equalization strategies are studied in the case of a downlink transmission, *i.e.*, from the base station to the mobiles. First of all, the promising potential of MC-CDMA receivers with SD based on minimum mean square error (MMSE) detection is confirmed for full load and non full load systems. Then a comparison of the performance of various interference cancellation receivers with two stages and even three stages is presented.

2. System description

The block diagram of the considered MC-CDMA transmitter and receiver is depicted in figure 1. Each data symbol x^n_j assigned to user j, $j = 1,\ldots,N_u$ and transmitted during the symbol interval n is multiplied bit-synchronously with its user specific Walsh-Hadamard spreading code $\boldsymbol{C}_j = [c_j^1, c_j^2, \ldots, c_j^{Lc}]^T$ of length L_C, where $[.]^T$ denotes matrix transposition. L_C corresponds to the bandwidth expansion factor and is equal to the maximum number of simultaneously active users. The vector of the data symbols transmitted during the nth OFDM symbol by all the users can be written $\boldsymbol{X}^n = [x^n{}_1, x^n{}_2, \ldots, x^n{}_j, \ldots, x^n{}_{Lc}]^T$ with $x^n{}_j = 0$ when user j is inactive. The code matrix $\boldsymbol{C}$ is defined to be:

$$\boldsymbol{C} = \begin{bmatrix} c_1^1 & c_2^1 & \ldots & c_{Lc}^1 \\ c_1^2 & c_2^2 & \ldots & c_{Lc}^2 \\ \ldots & \ldots & \ldots & \ldots \\ c_1^{Lc} & c_2^{Lc} & \ldots & c_{Lc}^{Lc} \end{bmatrix} \tag{1}$$

where the jth column vector of $\boldsymbol{C}$ corresponds to the spreading code $\boldsymbol{C}_j$ of the user j.

Since we consider the synchronous downlink of an MC-CDMA system, the different data modulated spreading codes of the N_u users can be added before Serial-to-Parallel (S/P) conversion. Furthermore, the N_u user signals are supposed to be transmitted with the same power. The number N_p of subcarriers which are 4PSK modulated is chosen equal to the spreading code length L_C. Thus, each of the N_p subcarrier In phase and Quadrature waveforms is modulated by a single chip belonging to a spreading Walsh-Hadamard code.

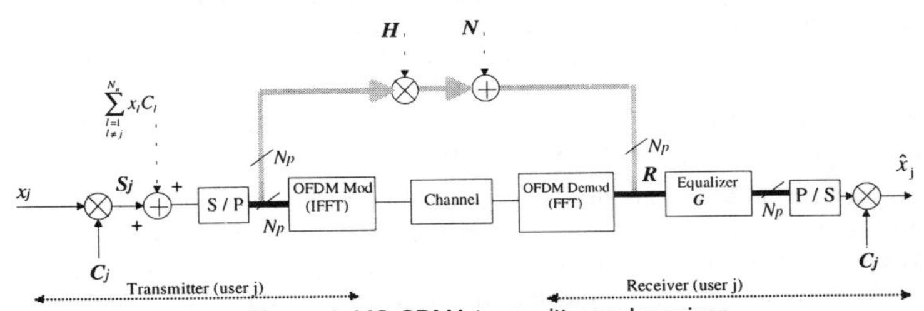

Figure 1. MC-CDMA transmitter and receiver.

For this study, frequency non-selective fading per subcarrier and time invariance during one OFDM symbol are assumed. Furthermore, the absence of Intersymbol Interference and Intercarrier Interference is guaranteed by the use of a guard interval longer than the maximum excess delay of the impulse response of the channel. Under theses assumptions and considering an ideal interleaving, the channel can be modelised in the frequency domain as represented in grey lines in figure 1. The complex channel fading coefficients are considered as independent for each subcarrier and constant during each OFDM symbol. Using this assumption, the temporel index n can be suppressed and the channel response can be estimated for the subcarrier k by $h_k = \rho_k\, e^{i\theta_k}$. Due to the absence of ICI, the channel matrix is diagonal and equal to:

$$\boldsymbol{H} = \begin{bmatrix} h_1 & \dots & 0 & 0 \\ \dots & h_2 & \dots & 0 \\ 0 & \dots & \dots & \dots \\ 0 & 0 & \dots & h_{Np} \end{bmatrix} \tag{2}$$

$\boldsymbol{N} = [n_1, n_2, ..., n_{Np}]^T$ is the vector containing the AWGN terms with n_k representing the noise term at the subcarrier k with variance given by $\sigma_N^2 = E\{|n_k|^2\}$, $k = 1,...,N_p$. The received vector is:

$$\boldsymbol{R} = [r_1, r_2, ..., r_{Np}]^T = \boldsymbol{H.C.X} + \boldsymbol{N} \tag{3}$$

The N_pxN_p matrix $\boldsymbol{G}$ represents the complex equalization coefficients obtained from the channel estimation which can be based on known transmitted pilot symbols inserted between the data carriers.

3. Single-user detection

3.1 Basic detection techniques

The complex channel coefficients h_k are supposed to be estimated by the receiver and are consequently treated as deterministic constants. Various detection techniques can be carried out [1], [5]. For those well known basic techniques, the assigned equalization coefficient is equal to:

$$g_k = h_k^* \qquad \text{Maximum ratio combining (MRC)} \tag{4}$$

$$g_k = h_k^* / |h_k| \qquad \text{Equal Gain Combining (EGC)} \tag{5}$$

$$g_k = 1/h_k \qquad \text{Orthogonality Restoring Combining (ORC)} \tag{6}$$

3.2 MMSE equalization

The MMSE equalization minimises the mean square error between the transmitted symbol and the estimated one. This equalization technique is the application of Wiener filtering. Let $\boldsymbol{W}_j = [w_j^0, w_j^1, ..., w_j^{Np}]$ be the weighting optimal vector. The estimated symbol of the jth user can be written:

$$\hat{x}_j = \boldsymbol{W}_j^T . \boldsymbol{R} = \boldsymbol{C}_j^T . \boldsymbol{G} . \boldsymbol{R} \tag{7}$$

According to the Wiener filtering, the optimal weighting vector is equal to:

$$\boldsymbol{W}_j = \boldsymbol{\Gamma}_{R,R}^{-1} . \boldsymbol{\Gamma}_{R,x_j} \tag{8}$$

where $\Gamma_{R,R}$ is the autocorrelation matrix of the received vector $\boldsymbol{R}$ and $\Gamma_{R,xj}$ is the cross-correlation vector between the desired symbol, x_j and the received signal vector, $\boldsymbol{R}$. Those quantities are equal to:

$$\begin{aligned} \boldsymbol{\Gamma}_{R,R} &= E\{\boldsymbol{R}^*.\boldsymbol{R}^T\} = \boldsymbol{H}^*.C.E\{\boldsymbol{X}^*.\boldsymbol{X}^T\}.C^T.\boldsymbol{H} + E\{\boldsymbol{N}^*.\boldsymbol{N}^T\} \\ \boldsymbol{\Gamma}_{R,x_j} &= E\{\boldsymbol{R}^*.x_j\} = \boldsymbol{H}^*.C.E\{\boldsymbol{X}^*.x_j\} \end{aligned} \tag{9}$$

where (.)* denotes complex conjugation. Then, the optimal weighting vector can be written:

$$\boldsymbol{W}_j^T = E\{x_j.\boldsymbol{X}^{*T}\}.\boldsymbol{C}^T\boldsymbol{H}^*.\left(\boldsymbol{H}.\boldsymbol{C}.E\{\boldsymbol{X}.\boldsymbol{X}^{*T}\}.\boldsymbol{C}^T.\boldsymbol{H}^* + E\{\boldsymbol{N}.\boldsymbol{N}^{*T}\}\right)^{-1} \quad (10)$$

The subcarrier noises have the same variance and are independent. Thus, $E\{\boldsymbol{N}.\boldsymbol{N}^{*T}\} = \sigma_N^2.\boldsymbol{I}$ where $\boldsymbol{I}$ is the identity matrix. Since the user signals have the same power ($E\{x_j^2\} = E_s$) and are independant, we can write $E\{\boldsymbol{X}.\boldsymbol{X}^{*T}\} = E_s.\boldsymbol{A}$, where $\boldsymbol{A} = \{a_{ij}\}$ is a diagonal matrix with the term $a_{jj} = 1$ if the user j is active and $a_{jj} = 0$ if the user j is inactive. Then, the equalization coefficient matrix can be expressed as:

$$\boldsymbol{G} = \boldsymbol{H}^*.\left(\boldsymbol{H}.\boldsymbol{C}.\boldsymbol{A}.\boldsymbol{C}^T.\boldsymbol{H}^* + \frac{\sigma_N^2}{E_s}.\boldsymbol{I}\right)^{-1} \quad (11)$$

In the full load case ($N_u = L_c$), the quantity $\boldsymbol{C}.\boldsymbol{A}.\boldsymbol{C}^T$ is equal to the identity matrix and the equalization coefficients matrix $\boldsymbol{G}$ is a diagonal matrix with the kth subcarrier equalization coefficient equal to [6]:

$$g_k = \frac{h_k^*}{|h_k|^2 + \frac{1}{\gamma_C}} \quad (12)$$

where γ_C is the subcarrier signal to noise ratio which is in this case equal to γ_X the signal to noise ratio of the received data symbol x.

In the non full load case, the equalization coefficient matrix is not diagonal. Then, to avoid matrix inversion, a suboptimal solution obtained when the wiener filter is optimised independently on each carrier, is to choose the kth subcarrier equalization coefficient equal to [6]:

$$g_k = \frac{h_k^*}{|h_k|^2 + \frac{N_p}{N_u}.\frac{1}{\gamma_X}} \quad (13)$$

4. Multi-user detection.

Based on the exploitation of the maximum likelihood criterion, the ML detector is the optimum detector. Theoretically, this method is applicable to both uplink and downlink channels. However, since its complexity grows exponentially with the number of users and with the code length, this method appears to be applicable only when the spreading sequences of all users are relatively short.

The Parallel Interference Cancellation (PIC) detector estimates the interference due to the simultaneous other users in order to remove this multiple user interference component from the received signal. Interference Cancellation can be carried out iteratively in multiple detection stages. To cope with the MAI, various combinations of single detection techniques have been studied, as for example EGC in all stages [4] which appears to be less efficient than a solution with MMSE in all stages presented in [5] and [6]. An other combination with ORC (or Zero Forcing) or ORC with threshold at the first stage followed by MRC is presented in [7], [8] and [9]. In this case, the multiuser interference is eliminated using the orthogonality restoring detection and then the user detects its own information applying the maximum ratio combining method. The block diagram of the considered MC-

CDMA receiver with parallel interference cancellation is illustrated in figure 2.

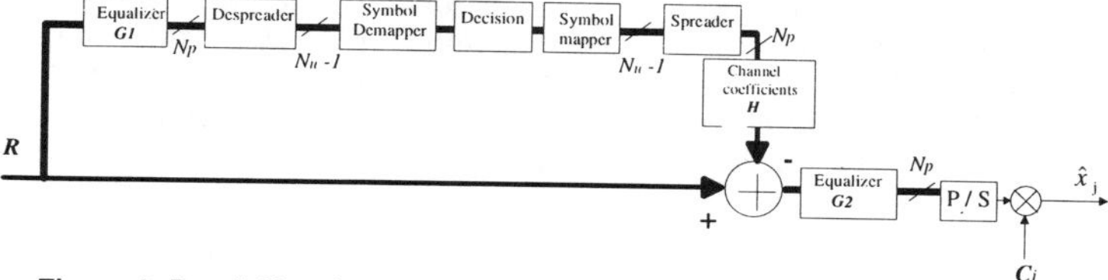

Figure 2. Parallel interference cancellation MC-CDMA receiver with two stages.

In the initial detection stage, the data symbols of all N_u *-1* active users are detected in parallel by the first SD equalizer with a gain ***G1***. After the despreading and the demapping, the decisions of this initial stage are used to reconstruct the interfering contribution in the received signal ***R***. The resulting interference is then subtracted from the received signal and the data detection is performed again on the signal with reduced MAI. Thus, the second and further detection stages work iteratively by using the decisions of the previous stage which yields the estimated data symbols at the mth iteration:

$$\hat{x}_j^{(m)} = \boldsymbol{C}_j^T.\boldsymbol{G}^{(m)}.\left(\boldsymbol{R} - \boldsymbol{H}\sum_{\substack{l=1 \\ l \neq j}}^{N_u} \hat{x}_j^{(m-1)}.\boldsymbol{C}_j\right) \quad ,m = 1,...,M_{it} \tag{14}$$

where M_{it} is the total number of iterations.

5. Simulation results.

The simulation results are presented without channel coding for various detection techniques and for various numbers N_u of active users and number N_p of subcarriers. Each of the independent subcarriers is QPSK modulated at the transmitter side and then multiplied by a decorrelated Rayleigh fading. The diversity N_D offered by the channel is then equal to the number N_p of subcarriers. Furthermore, it is assumed that accurate estimates of the frequency channel response for each subcarrier is available and that all the signals are received with equal power.

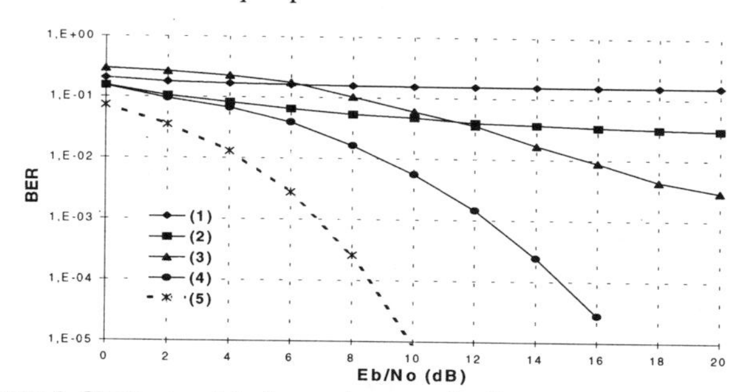

Figure 3. Single user detection performance for $N_u = L_c = N_p = 64$ with various receiver techniques: MRC (1), EGC (2), ORC (3), MMSE (4), MF bound (5).

The single detection based on minimum mean square error (MMSE) equalization offers the best results (figure 3) with full load systems as it has already been demonstrated in [5] and [6]. In figure 3, the number ($N_u = 64$) of active users is equal to the length ($L_C = 64$) of the Walsh-Hadamard code. The Matched Filter (MF) bound is given as reference (curve 5). The MF bound for an uncoded MC-CDMA system corresponds to the BER obtained in the case of data transmissions over N_p statistically independent Rayleigh fading channels with MRC detection and without MAI. The MMSE (curve 4) outperforms the other techniques avoiding an excessive noise amplification for low signal to noise ratios while restoring the orthogonality among users for large signal to noise ratios.

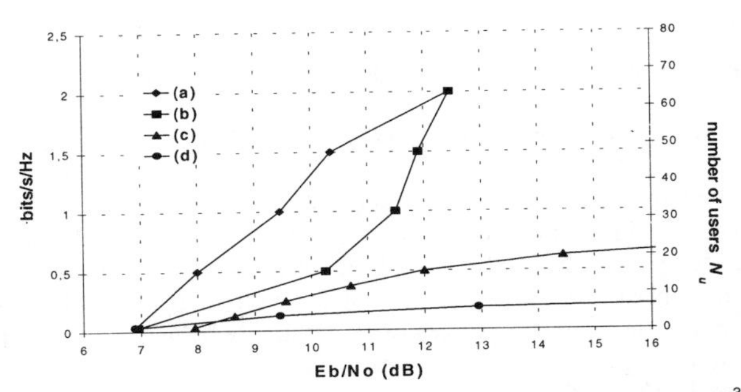

Figure 4. Spectral efficiency for various receiver techniques for BER=10^{-3} and $N_p = 64$: matrix-MMSE (a), MMSE (b), EGC (c), MRC (d).

In figure 4, the performance of various single detection systems with $L_c = N_p = 64$ are compared, taking into account the resulting spectral efficiency in terms of the necessary E_b/N_0 to achieve a bit error probability of $P_b = 10^{-3}$. The maximum achievable spectral efficiency is 2 bit/s/Hz, corresponding to full load system ($N_u = L_c = N_p$), because the loss due to the guard interval, the synchronisation and the channel estimation is not taken into account. In any cases, MRC and EGC perform poorly. The curves (b) gives suboptimal MMSE system performance with the equalization coefficients optimised independently on each subcarrier and equal to the expression (13). The curve (a) corresponds to MMSE system performance according to matrix approach with the equalization coefficient matrix $\boldsymbol{G}$ equal to the expression (11). For full load systems ($N_u = 64$), the performance of the two MMSE approaches are the same. On the other hand, comparing them for non full load systems shows a gain of more than 2 dB in the matrix MMSE case with $N_u = 32$ or 16, which corresponds to a rough spectral efficiency respectively equal to 1 or 0.5 bit/s/Hz.

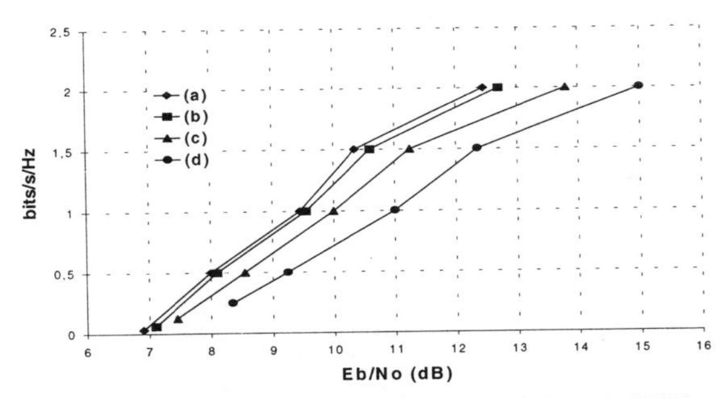

Figure 5. Spectral efficiency of QPSK matrix-MMSE SD MC-CDMA systems for various lengths of the spreading code for BER=10^{-3}: $N_p = 64$ (a), $N_p = 32$ (b), $N_p = 16$ (c), $N_p = 8$ (d).

In figure 5, the performance of matrix-MMSE MC-CDMA systems with various code lengths, are compared taking into account the resulting rough spectral efficiency in terms of the necessary E_b/N_0 to achieve a bit error probability of $P_b = 10^{-3}$. As expected, when the channel diversity (equal to the number N_p of independent subcarriers) increases, the results are better. For example, the matrix-MMSE system with $N_p = L_c = 64$ can handle a full user capacity ($N_u = 64$ in this case) at $E_b/N_0 = 12.4$ dB. However, the performance improvement due to the increase of the number of subcarriers (and then the channel diversity) from 32 to 64, is inferior to 0.3 dB for full load systems.

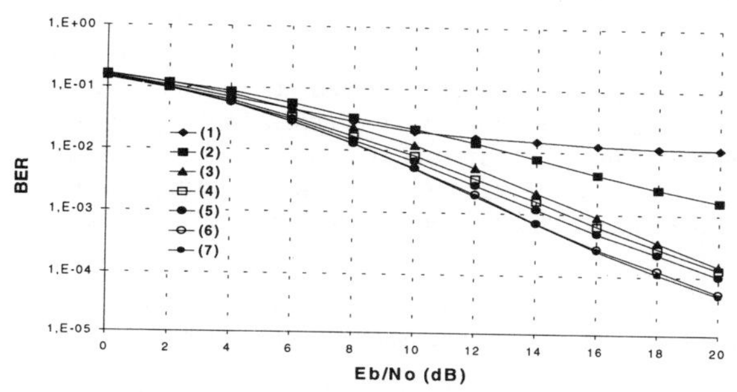

Figure 6. Multiuser detection performance for $N_u = N_p = 8$ with various receiver techniques: EGC1/EGC2 (1), ZF1/MRC2 (2), MMSE1/MRC2 (3), EGC1/MMSE2 (4), MMSE1/EGC2 (5), MMSE1/MMSE2 (6), MMSE1/MMSE2/MMSE3 (7)

A performance comparison of full load multi-stage parallel interference cancellation systems with various receiver techniques for the first and the second stage and, if necessary for the third stage, is presented in figure 6. Obviously, the systems with MMSE equalisation at the first stage (curves 3, 5, 6, 7) outperforms other schemes. The combination with Zero Forcing for the first stage followed by MRC (curve 2, ZF1/MRC2) requires a 8 dB higher E_b/N_0 to achieve a BER equal to 10^{-3} compared to a system with MMSE for the two stages (curve 6, MMSE1/MMSE2). Finally, the performance improvement obtained by a third MMSE stage is inferior to 0.1 dB (comparison of the curves 6 and 7), which does not justify the additional complexity.

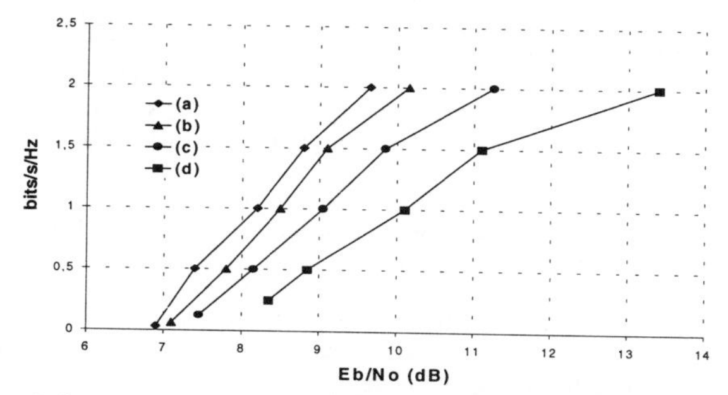

Figure 7. Spectral efficiency of 2 stage IC QPSK matrix-MMSE MC-CDMA systems for various lengths of the spreading code and for BER = 10^{-3}: $N_p = 64$ (a), $N_p = 32$ (b), $N_p = 16$(c), $N_p = 8$ (d).

The resulting spectral efficiency in terms of the necessary E_b/N_0 to achieve a bit error probability of $P_b = 10^{-3}$ is given in figure 7 for 2 stage interference cancellation matrix-MMSE systems with various code lengths L_c = 8, 16, 32, 64 equal to the number N_p of the subcarriers. The comparison of the curve (b) of the figure 5 with the curve (b) of the figure 7 shows that the full user capacity can be obtained with a 2.6 dB lower E_b/N_0 with the 2 stage interference cancellation 32 subcarrier scheme compared to the SD MMSE 32 subcarrier system. Furthermore, it appears that a system which will take advantage of a channel diversity equal to 32 will offer rather good performance.

6. Conclusion.

These simulation results have confirmed the potential of MMSE detection techniques in the synchronous case of a multiuser MC-CDMA system operating in frequency selective Rayleigh channel. For non full load systems, the novel matrix MMSE approach presented in this paper, offers a gain of more than 2 dB for a BER equal to 10^{-3} compared to a system applying the MMSE algorithm independently on each subcarrier. Besides, it has been shown that a two stage interference cancellation matrix MMSE scheme could be a good compromise between performance and complexity.

Remark: *Concerning the novel matrix approach, presented in this paper to implement the MMSE detection technique, a patent application has been filed.*

References:

[1] S. Hara, R. Prasad, "Overview of multicarrier CDMA", IEEE Communications Magazine, December 1997, pp 126-133.

[2] K. Fazel and L. Papke, "On the performance of convolutionnally-coded CDMA/OFDM for mobile communication system", Proceedings of IEEE PIMRC'93, Yokohama, Japan, September 1993, pp 468-472

[3] N. Yee, J.P. Linnartz, "Wiener filtering of multi-carrier CDMA in a Rayleigh fading channel", Proceedings of IEEE PIMRC'94, September 1994, vol.4, pp 1344-1347.

[4] K. Fazel, "Performance of CDMA/OFDM for mobile communication system", Proceedings of IEEE International Conference on Universal Personal Communications (ICUPC'93), Ottawa, Canada, October 1993, pp 975-979.

[5] S. Kaiser, "On the performance of different detection techniques for OFDM - CDMA in fading channels", Proceedings of IEEE Global Telecommunications Conference (GLOBECOM'95), Singapore, November 1995, pp 2059-2063.

[6] S. Kaiser, "Analytical performance evaluation of OFDM-CDMA mobile radio systems", Proceedings First European Personal and Mobile Communications Conference (EPMCC'95), Bologna, Italy, November 1995, pp 215-220.

[7] S. Hara, T.H. Lee and R. Prasad, "BER comparison of DS-CDMA and MC-CDMA for frequency selective fading channels", Proceedings 7th Tyrrhenian International Workshop on Digital Communications, Tirrenia, Italy, September 1995, pp 3-14.

[8] D.N. Kalofonos and J.G. Proakis, "Performance of the multi-stage detector for a MC-CDMA system in a Rayleigh fading channel", Proceedings of IEEE Global Telecommunications Conference (GLOBECOM'96), London, United Kingdom, November 1996, pp 1784-1788.

[9] J.J. Maxey, R.F. Ormondroyd, "Multi-carrier CDMA using convolutional coding and interference cancellation over fading channels", First Workshop on Multi-Carrier Spread-Spectrum, Oberpfaffenhofen, Germany, April 1997, pp 89-96.

Section VI

SYNCHRONIZATION AND CHANNEL ESTIMATION

An Overview of MC-CDMA Synchronisation Sensitivity

Heidi Steendam and Marc Moeneclaey
Department of Telecommunications and Information Processing, University of Ghent, B-9000 GENT, BELGIUM

Key words: MC-CDMA, synchronisation errors

Abstract: This paper presents an overview of the effect of synchronisation errors on MC-CDMA performance in downlink communications. We distinguish two types of synchronisation errors: carrier phase errors and timing errors. We show that the MC-CDMA system is very sensitive to a carrier frequency offset or a clock frequency offset. For a maximal load, carrier phase jitter and timing jitter give rise to a degradation that is independent of the spectral content of the jitter; moreover, the degradation caused by carrier phase jitter and timing jitter is (essentially) independent of the number of carriers. A constant carrier phase offset and a constant timing offset cause no degradation of the MC-CDMA system performance.

1. INTRODUCTION

The enormous growth of interest for multicarrier (MC) systems can be ascribed to its high bandwidth efficiency and its immunity to channel dispersion. Recently, different combinations of orthogonal frequency division multiplexing (OFDM) and code division multiple access (CDMA) have been investigated in the context of high data rate communication over dispersive channels [1]-[9]. One of these systems is multicarrier CDMA (MC-CDMA), which has been proposed for downlink communication in mobile radio. In MC-CDMA the data symbols are multiplied with a higher rate chip sequence and then modulated on orthogonal carriers.

This paper presents an overview of the effect of synchronisation errors on MC-CDMA performance in downlink communications. We can distinguish mainly two levels of synchronisation: carrier synchronisation and timing recovery. In carrier synchronisation, a local reference carrier with a phase and frequency as closely matching to that of the carrier used for upconverting the transmitted signal, must be generated for the downconversion of the signal to a baseband signal. The effect of the carrier phase errors, caused by the error between the carrier used for the upconversion of the data signal and the local reference carrier has been investigated in [10]-[15]. The effect of a frequency offset was studied in [10], [12]-[14], while the sensitivity of MC-CDMA to carrier phase jitter was described in [11], [13]-[14].

The next problem is the recovery of the timing instants, as the sampling clock oscillator of the receiver has a phase and frequency drift against that of the transmitter. The influence of the timing errors, made in the process of extracting the sampling instants, was studied in [13]-[15]. The sensitivity of MC-CDMA to a clock frequency offset between the transmitter clock and the receiver sampling clock and the effect of timing jitter resulting from a phase-locked sampling clock have been studied in [13]-[14].

2. SYSTEM DESCRIPTION

In this paper, we consider the MC-CDMA system that is shown in figure 1 for one user. The data symbols $\{a_{i,m}\}$, transmitted at a rate R_s, where $a_{i,m}$ denotes the i-th symbol belonging to user m, is first multiplied by a higher rate chip sequence of length N, $\{c_{n,m}|n=0,\ldots,N-1\}$, $c_{n,m}$ denoting the n-th chip of the sequence belonging to user m. Sequences belonging to different users are assumed to be orthogonal. The resulting samples are mapped on the orthogonal carriers and modulated using the inverse fast Fourier transform (IFFT). We insert a guard interval, consisting of a cyclic prefix of the transmitted samples, to avoid interframe interference. The resulting samples are applied to a transmit filter, which is a unit-energy square-root Nyquist filter (e.g. a cosine rolloff filter with rolloff α) and transmitted over a (possibly dispersive) channel. The channel output signal is disturbed by additive white Gaussian noise (AWGN) with power spectral density N_o and affected by a carrier phase error. The resulting signal is then fed to the receiver filter, which is matched to the transmit filter and sampled at the instants $t_{i,k}=kT+i(N+\nu)T+\varepsilon_{i,k}T$, where $\varepsilon_{i,k}$ is the normalised timing error at the k-th instant of the i-th transmitted frame.

When the phase error is slowly varying as compared to $T=1/((N+\nu)R_s)$, it was shown in [13] that the synchronisation errors can be included in an equivalent time-varying impulse response with Fourier transform $H_{eq}(f;t_{i,k})=H(f)e^{j\phi(t_{i,k})}e^{j2\pi f\varepsilon_{i,k}T}$, where $\phi(t_{i,k})$ and $\varepsilon_{i,k}$ are respectively the carrier phase error and the timing error at the instant $t_{i,k}$ and H(f) consists of the cascade of the transmit filter, the channel and the receiver filter. We assume that the duration of the equivalent time-varying impulse response does not exceed the duration of the guard interval. We select the N samples outside the guard interval for further processing and demodulate the signal using the FFT. Then each FFT output is multiplied with the corresponding chip of the considered user and applied to a one-tap MMSE equaliser. The outputs of the equalisers are summed to obtain the samples at the input of the decision device. The signal-to-noise ratio (SNR) is defined as the ratio of the useful power to the sum of the interference power and the noise power. In the case of an ideal channel and in the absence of synchronisation errors, the SNR yields E_s/N_0, where E_s is the energy per symbol transmitted to each user and N_0 is the noise power spectral density. The SNR will degrade in the presence of synchronisation errors. We define the degradation as Deg=10log(E_s/N_0)-10log(SNR).

In the following, we consider the case of downstream communication. As in downstream communication, the signals sent to the different users are synchronised at the basestation, the timing errors are the same. In addition, as all transmitted carriers are generated by the same oscillator, they exhibit the same carrier phase errors. To clearly isolate the effect of the synchronisation errors, we consider the case of an ideal channel.

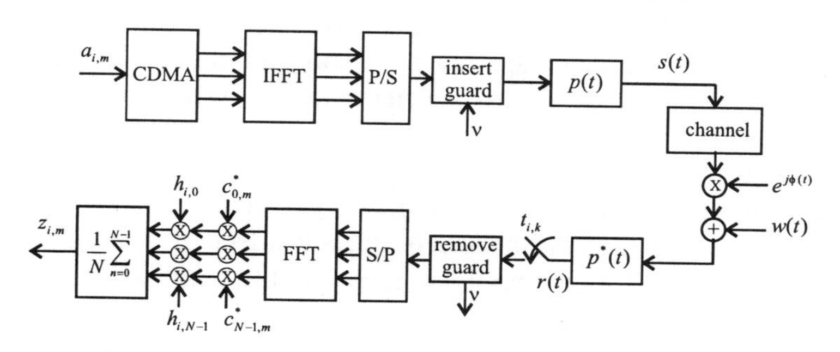

Figure 1. Conceptual block diagram of the MC-CDMA system for one user

3. CARRIER PHASE ERRORS

3.1 Constant Phase Error

In the case of a constant phase offset, the orthogonality between the different users is not affected, i.e. no multiuser interference (MUI) is introduced, as a constant phase offset only causes a carrier independent phase rotation of the FFT outputs. This means that, when no correction was applied by the equaliser, the samples at the input of the decision device are rotated over an angle ϕ. To avoid the reduction of the noise margins, this phase rotation of the useful component is corrected by the equaliser: the equaliser rotates the FFT outputs over (an estimate of) the angle $-\phi$, i.e. $h_{i,k}=e^{-j\phi}$. As a phase rotation of the FFT outputs has no influence on the noise power level, this constant phase offset is compensated by the equaliser without loss of performance.

3.2 Carrier Frequency Offset

If the downconversion of the signal is performed by means of a free running local oscillator, a carrier frequency offset can occur. The carrier frequency offset ΔF causes a shift of the frequency band of the transmitted signal. When we focus on the n-th transmitted carrier, we observe in figure 2 that the frequency shift of this transmitted carrier gives rise to an attenuation of the n-th observed carrier, thus an attenuation of the useful component. Furthermore, all other observed carriers are disturbed by a non-zero interference caused by the n-th transmitted carrier. In [11] it is shown that a one-tap equaliser is not able to eliminate this MUI, i.e. a carrier frequency offset will introduce a performance degradation that depends on the product of the number of carriers N and the frequency offset ΔF. In figure 3, this degradation is shown for the maximum load (i.e. the number of users equals N). We observe a high sensitivity of the MC-CDMA system to the carrier frequency offset. From figure 2 it follows that a frequency offset equal to the carrier spacing ($\Delta F=1/NT$) gives rise to a severe performance degradation, as the spread data $\{a_o c_n\}$ is shifted over one carrier and is not correlated with the corresponding chips. Therefore, the frequency offset must be limited, i.e. $\Delta FT \ll 1/N$. The carrier frequency offset therefore must be compensated before demodulation, i.e. in front of the FFT.

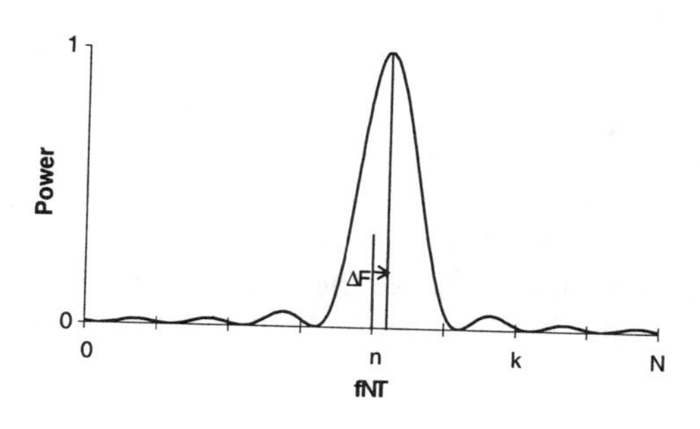

Figure 2. Contribution of the n-th transmitted carrier to the k-th FFT output (N=8, ΔFT=0.1)

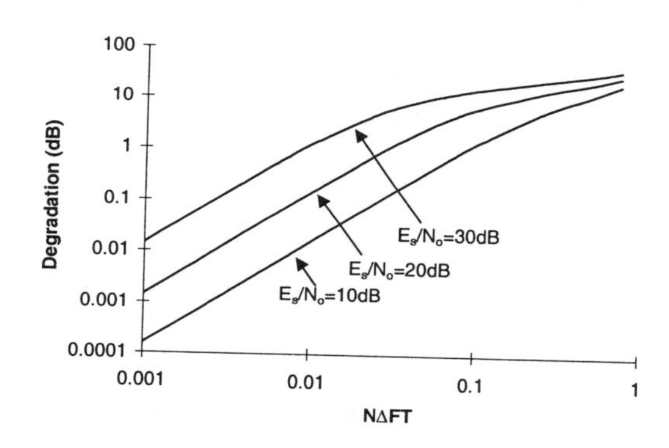

Figure 3. Influence of carrier frequency offset

3.3 Carrier Phase Jitter

To avoid the strong degradation caused by the carrier frequency offset, a phase-locked local oscillator can be used for downconverting the RF signal. The phase error resulting from this PLL can be modelled as a zero-mean stationary random process with jitter spectrum $S_\phi(f)$ and jitter variance σ_ϕ^2. Assuming slowly varying phase errors and small jitter variances, the equaliser coefficients are essentially the same as in the absence of carrier phase jitter. In [12] it is shown that the fluctuation of the useful component, caused by the random character of the jitter, mainly consists of the low frequency components (<1/NT) of the jitter, while the MUI is mainly determined by the high frequency components (>1/NT) of the jitter. For the maximum load, the degradation becomes independent of the spectral contents of the jitter and of the number of carriers. The degradation, which in this case only depends on the jitter variance, is shown in figure 4. The scatter diagrams however, will differ considerably depending of the spectral contents of the jitter. Jitter with mainly low frequency components (<1/NT) gives rise to a random rotation of the useful component and will show an angular displacement in the scatter diagram (figure 5a). Phase jitter with mainly high frequency components (>1/NT) will introduce MUI which causes a circular cloud (figure 5b), as the term of the MUI, which consists of a large number of statistically independent contributions, has uncorrelated real and imaginary parts.

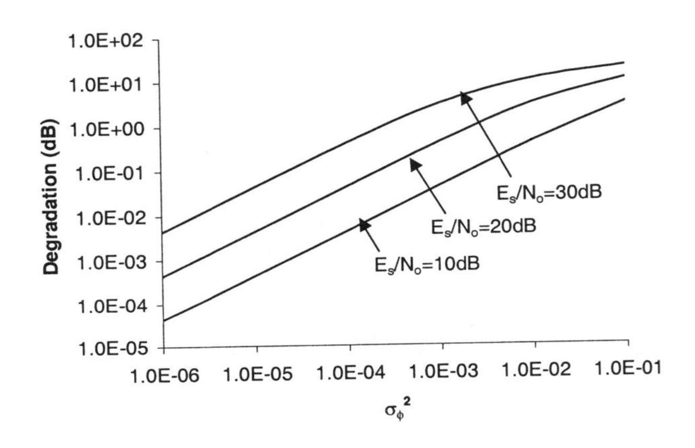

Figure 4. Influence of carrier phase jitter

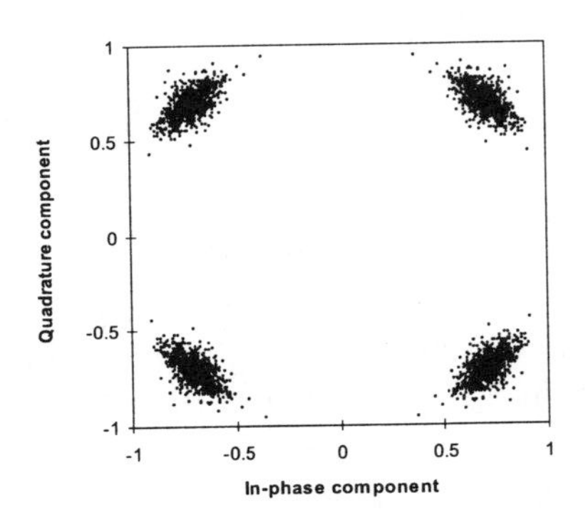

Figure 5a. Scatter diagram for jitter with mainly low frequency components, N=128

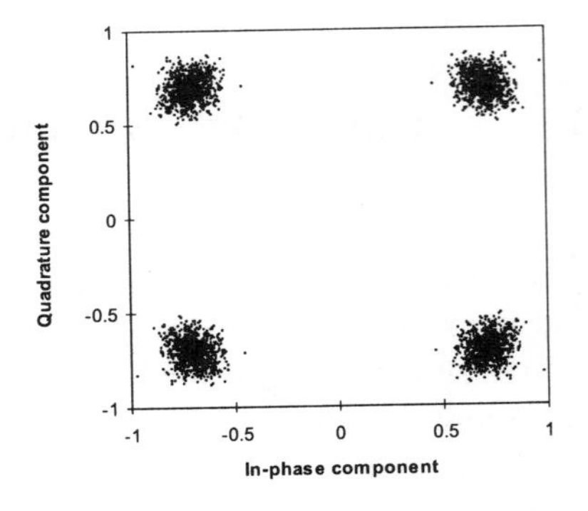

Figure 5b. Scatter diagram for jitter with mainly high frequency components, N=128

4. TIMING ERRORS

4.1 Constant Timing Offset

In the case of a constant timing offset, the coefficients of the spread data at the outputs of the FFT are affected as shown in figure 6. For carriers outside the rolloff area, the constant timing offset has no influence on the amplitude of the coefficient, but only introduces a phase rotation proportional to the carrier index. For carriers inside the rolloff area, the coefficients are rotated over some angle and attenuated as compared to the coefficients of the carriers outside the rolloff area. The equaliser attempts to compensate for the attenuation, caused by the carriers inside the rolloff area and the rotation. However, scaling the FFT outputs affects the noise power level. The MMSE filter therefore makes a compromise between the MUI caused by the carriers inside the rolloff area and the increase of the noise power caused by the scaling. It is clear that the sensitivity of MC-CDMA to the constant timing offset can be eliminated by not using the carriers inside the rolloff area.

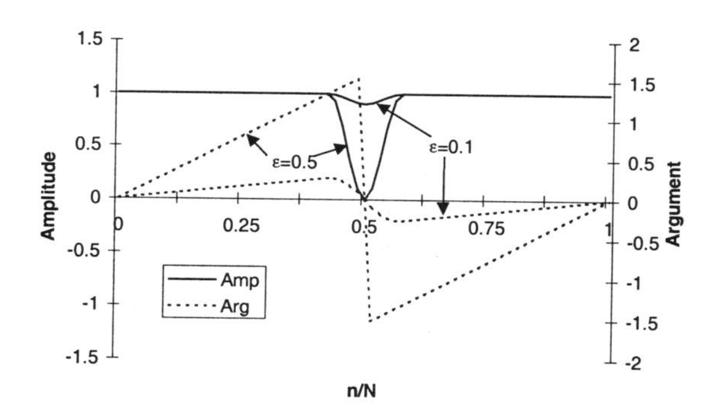

Figure 6. Influence of constant timing error on outputs FFT

4.2 Clock Frequency Offset

When sampling is performed by means of a free-running local oscillator, a clock frequency offset ΔT/T can occur. This clock frequency offset engenders compression (ΔT/T>0) or expansion (ΔT/T<0) of the observed frequencies at the output of the FFT, giving rise to a carrier dependent frequency shift as compared to the transmitted carriers (figure 7). When we focus on the n-th transmitted carrier, we observe an attenuation of the amplitude of the n-th observed carrier, caused by the carrier dependent frequency shift. All other observed carriers are disturbed by non-zero interference caused by the n-th transmitted carrier. Therefore, a clock frequency offset results in an attenuation of the useful component and MUI. It was shown in [13] that a one-tap equaliser is not able to eliminate this MUI, so the clock frequency offset gives rise to performance degradation. This degradation depends on the product of the number of carriers N and the clock frequency offset $\Delta T/T$. From figure 8, where the degradation is shown for the maximum load and $\alpha=0$, we observe that the MC-CDMA system is very sensitive to a clock frequency offset. To obtain small degradations, the clock frequency offset must be limited, i.e. $\Delta T/T \ll 1/N$.

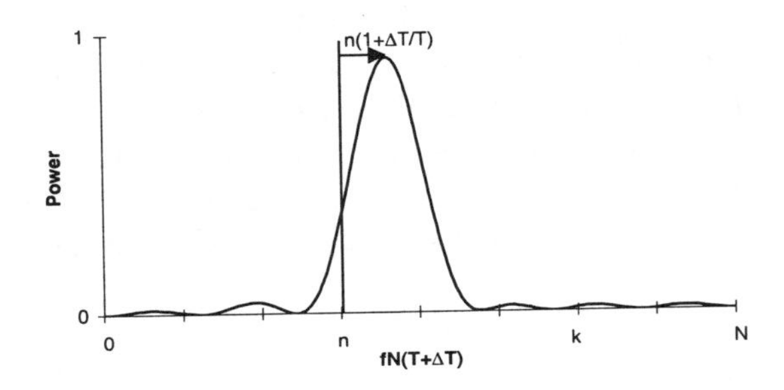

Figure 7. Contribution of the n-th transmitted carrier to the k-th FFT output (N=8, $\Delta T/T=0.2$)

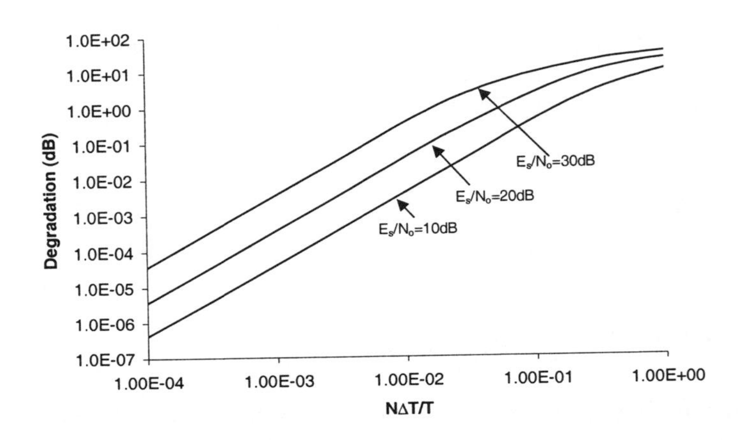

Figure 8. Influence of clock frequency offset, $\alpha=0$

4.3 Timing Jitter

In order to get rid of the constant timing error and the clock frequency offset we can perform synchronised sampling, e.g. by means of a phase-locked sampling clock. The timing error resulting from this PLL can be modelled as a zero-mean stationary process with jitter spectrum $S_\varepsilon(f)$ and jitter variance σ_ε^2. Assuming slowly varying timing errors and small jitter variances, the equaliser coefficients are essentially the same as in the absence of timing jitter. In [13] it is shown that, for the maximum load, $\alpha=0$ and for large N ($N\rightarrow\infty$), the sum of the powers of the fluctuation of the useful component, caused by the random character of the jitter, and the MUI is essentially independent of the number of carriers. Furthermore, this degradation, which is mainly caused by the MUI [13], is independent of the spectral contents of the jitter but only depends on the jitter variance. In figure 9, this degradation is shown as function of the jitter variance.

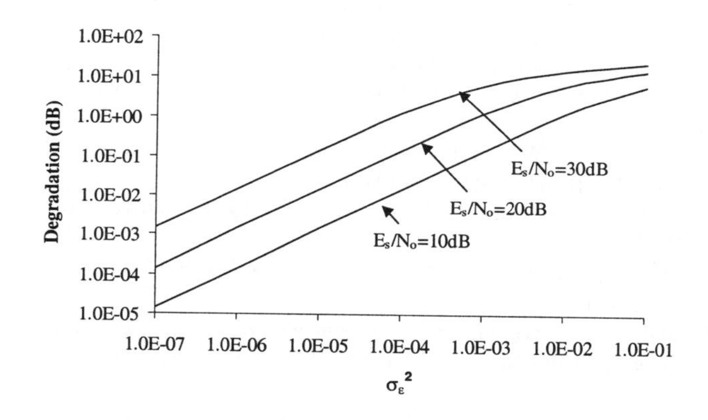

Figure 9. Influence of timing jitter, $\alpha=0$

5. CONCLUSIONS

In this contribution, we have presented an overview of the effect of synchronisation errors on the performance of a MC-CDMA system. A constant phase offset and a constant timing offset can be compensated without loss of performance while the MC-CDMA performance degrades for time-varying timing and carrier phase errors. In the case of a carrier frequency offset or a clock frequency offset, the MC-CDMA performance rapidly degrades and strongly depends on the number of carriers. Stationary carrier phase jitter and timing jitter introduce a degradation that, for the maximum load, is independent of the number of carriers and the spectral contents of the jitter, but only depends on the jitter variance.

REFERENCES

[1] L. Vandendorpe, O. van de Wiel, "Decision Feedback Multi-User Detection for Multitone CDMA Systems", Proc. 7th Thyrrenian Workshop on Digital Communications, Viareggio Italy, Sep 95, pp. 39-52

[2] E.A. Sourour, M. Nakagawa, "Performance of Orthogonal Multicarrier CDMA in a Multipath Fading Channel", IEEE Trans. On Comm., vol. 44, no 3, Mar 96, pp. 356-367

[3] N. Yee, J.P. Linnartz, G. Fettweis, "Multicarrier CDMA in Indoor Wireless Radio Networks", Proc. PIMRC'93, Yokohama, Japan, 1993, pp. 109-113

[4] S. Hara, T.H. Lee, R. Prasad, "BER comparison of DS-CDMA and MC-CDMA for Frequency Selective Fading Channels", Proc. 7th Thyrrenian Workshop on Digital Communications, Viareggio Italy, Sep 95, pp. 3-14

[5] Y. Sanada, M. Nakagawa, "A Multiuser Interference Cancellation Technique Utilising Convolutional Codes and Orthogonal Multicarrier Communications", IEEE J. on Sel. Areas in Comm., vol. 14, no 8, Oct 96, pp. 1500-1509

[6] V.M. Da Silva, E.S. Sousa, "Multicarrier Orthogonal CDMA Signals for Quasi-Synchronous Communication Systems", IEEE J. on Sel. Areas in Comm., vol. 12, no 5, Jun 94, pp. 842-852

[7] N. Yee, J.P. Linnartz, "Wiener Filtering of Multicarrier CDMA in a Rayleigh Fading Channel", Proc. PIMRC'94, 1994, pp. 1344-1347

[8] Multi-Carrier Spread-Spectrum, Eds. K. Fazel and G. P. Fettweis, Kluwer Academic Publishers, 1997

[9] S. Hara, R. Prasad, "Overview of Multicarrier CDMA", IEEE Comm. Mag., no. 12, vol. 35, Dec 97, pp. 126-133

[10] L. Tomba and W.A. Krzymien, "Effect of Carrier Phase Noise and Frequency Offset on the Performance of Multicarrier CDMA Systems", ICC 1996, Dallas TX, Jun 96, Paper S49.5, pp. 1513-1517

[11] H. Steendam, M. Moeneclaey, "The Effect of Carrier Phase Jitter on MC-CDMA Performance", IEEE Trans. on Comm., Vol. 47, No. 2, Feb 99, pp. 195-198

[12] Y. Kim, S. Choi, C. You, D. Hong, "Effect of Carrier Frequency Offset on the Performance of an MC-CDMA System and its Countermeasure Using Pulse Shaping", Proc. ICC'99, Vancouver, Canada, June 6-10, 1999, Paper S5.3, pp. 167-171

[13] H. Steendam, M. Moeneclaey, "The Effect of Synchronisation Errors on MC-CDMA Performance", Proc. ICC'99, Vancouver, Canada, June 6-10, 1999, Paper S38.3, pp. 1510-1514

[14] H. Steendam, M. Moeneclaey, "The Sensitivity of MC-CDMA to Synchronisation Errors", ETT special issue on MC-SS, May-Jun 99, no. 3

[15] S. Nahm, W. Sung, "A Synchronization Scheme for Multi-Carrier CDMA Systems", Proc. ICC'98, Atlanta, GA, June 1998, Paper S37.7, pp. 1330-1334

BLIND FREQUENCY OFFSET/SYMBOL TIMING/SYMBOL PERIOD ESTIMATION AND SUBCARRIER RECOVERY FOR OFDM SIGNALS IN FADING CHANNELS

Shinsuke Hara

Graduate School of Engineering, Osaka University,

2-1, Yamada-Oka, Suita, Osaka 565–0871 Japan

hara@comm.eng.osaka-u.ac.jp

Abstract This paper proposes a fully–blind frequency offset, symbol timing and symbol period estimator for OFDM signals in fading channels. In OFDM system, each symbol is extended with a guard interval, which is the tail of the signal itself, therefore, we can deal with it as *"an unknown pilot signal."* The estimation method is based on the maximum likelihood criterion, making effective use of the unknown pilot signal. This paper also proposes a blind subcarrier recovery method which calculates frequency transfer function essential for subcarrier recovery from the correlation matrix of the unknown pilot signals.

1. INTRODUCTION

In order to maintain orthogonality among subcarriers even in multipath fading environments, each Orthogonal Frequency Division Multiplexing (OFDM) symbol is cyclically extended with the tail of the signal itself, which is called *"guard interval."* The (total) symbol period is written as $T_s = t_s + \Delta$, where t_s is the useful symbol period corresponding to the duration of original OFDM symbol, and Δ is the guard interval (see Fig.1). In other words, OFDM system transmits the same waveform twice in each symbol period, so we can deal with it as *"an unknown pilot signal,"* unlike a normal pilot signal whose waveform we know. Therefore, making effective use of the unknown pilot signal, we can estimate frequency offset, symbol timing and symbol period without any extra pilot signal transmission.

On the other hand, as a suitable coding/detection format, it is common to employ differential encoding/differential detection, because it requires no complicated subcarrier recovery. However, in order to more improve the bit error rate (BER) performance, it is essential to employ coherent detection scheme instead, so it implies the requirement of subcarrier recovery. Here, making effective use of the unknown pilot signal, it is possible to carry out subcarrier recovery, namely, to estimate the amplitude and phase of each subcarrier, without any extra pilot signal transmission as well.

This paper proposes a frequency offset, symbol timing and symbol period estimator[1],[2], and a subcarrier recovery method for OFDM signals, both of which are fully blind in the sense that they do not require any extra pilot signal transmission.

2. FREQUENCY OFFSET/SYMBOL TIMING/SYMBOL PERIOD ESTIMATION

Define $s(t)$,$c(t)$,τ,f_Δ and t_s as the transmitted signal, impulse response of multipath fading channel, propagation delay of the shortest path, frequency offset and symbol period, respectively. Observing the received OFDM signal in the time interval $I_o = [0, MT_s]$ (M is the number of observed symbols), for given $s(t)$,$c(t)$,τ,f_Δ and t_s, the conditional probability density function (pdf) of $r(t)$ can be written as

$$p(r|s, c, \tau, f_\Delta, t_s) = Ae^{-X}, \tag{1}$$

where A is a constant and

$$X = \frac{1}{2N_0} \int_{t \in I_o} \left| r(t) - (c \otimes s)(t - \tau) e^{j(2\pi f_\Delta t + \theta)} \right|^2 dt \tag{2}$$

is the Euclidian distance between a known transmitted signal and the received signal. Here, $\otimes$ denotes the convolution.

Choosing a term which contributes much to the estimation of $s(t)$,$c(t)$, τ,f_Δ and t_s, we can obtain the likelihood function:

$$\Lambda(\tau, f_\Delta, t_s) = \int_{u \in I_o} \int_{v \in I_o} r(u) R_{cs}(u - \tau, v - \tau) r^*(v) e^{j2\pi f_\Delta (u-v)} du dv, \tag{3}$$

where

$$R_{cs}(u, v) = \int_{-\infty}^{\infty} \int_{-\infty}^{\infty} R_c(\xi, \eta) R_s(u - \xi, v - \eta) d\xi d\eta \tag{4}$$

is the autocorrelation function of $(c \otimes s)(t)$. In Eq.(4), $R_c(u, v)$ and $R_s(u, v)$ are the autocorrelation functions of the channel and the transmitted signal, respectively. $R_c(u, v)$ is given by

$$R_c(u, v) = E[c(u)c^*(v)] = g(u)\delta(u - v), \tag{5}$$

where $g(t) = E[c(t)c^*(t)]$ is the delay profile with root mean square (RMS) delay spread T_m. The multipath delay spread means that the impulse response $c(t)$ has non-zero values only for $0 \leq t \leq T_m$. In order to keep the orthogonality between subcarriers, therefore, Δ should be set to be greater than T_m.

On the other hand, since the OFDM signal with guard interval has a cyclostationary property, $R_s(u,v)$ can be written as

$$
\begin{aligned}
R_s(u,v) &= \sum_{m=-\infty}^{\infty} R'_s(u - mT_s, v - mT_s), & (6)\\
R'_s(u,v) &= \delta(u-v) + \delta(u-v+t_s) + \delta(u-v-t_s); & (7)\\
&\quad (-\Delta \leq u \leq t_s \text{ and } -\Delta \leq v \leq t_s).
\end{aligned}
$$

Substituting Eqs.(5), (6) and (7) into Eq.(4), $R_{cs}(u,v)$ becomes

$$
R_{cs}(u,v) = \delta(u-v) + \sum_{m=-\infty}^{\infty} R'_{cs}(u - mT_s, v - mT_S), \quad (8)
$$

where

$$
R'_{cs}(u,v) = \delta(u-v-t_s)g_r(v) + \delta(u-v+t_s)g_r(u), \quad (9)
$$

$$
g_r(t) = \begin{cases} \int_0^{\min(\Delta+t,T_m)} g(\xi)d\xi; & (-\Delta \leq t < 0) \\ \int_t^{\min(\Delta+t,T_m)} g(\xi)d\xi; & (0 \leq t < T_m) \\ 0; & (\text{otherwise}) \end{cases} \quad (10)
$$

Assuming that the propagation delays of the major propagation paths are concentrated on $t = 0$, from Eq.(3), the likelihood function becomes

$$
\begin{aligned}
\lambda(\tau, f_\Delta, t_s) &= \Re\left[e^{j2\pi f_\Delta t_s} \sum_{m=1}^{M} \int_{-\Delta}^{0} r(t+\tau-mT_s)\right.\\
&\quad \left. \times r^*(t+t_s+\tau-mT_s)dt\right]. & (11)
\end{aligned}
$$

The symbol timing τ, the symbol period t_s and the frequency offset f_Δ can be estimated by searching the corresponding values to maximize the likelihood function. Fig.2 (a) shows the block diagram of the estimator. We call the estimator based on Eq.(11) the "optimum estimator."

Furthermore, we can remove the integral operations in Eq.(11). The approximated likelihood function is written as

$$
\lambda'(\tau, f_\Delta, t_s) = \Re\left[e^{j2\pi f_\Delta t_s} \sum_{m=-\infty}^{\infty} r(t+\tau-mT_s)r^*(t+t_s+\tau-mT_s)\right]. \quad (12)
$$

Fig.2 (b) shows the block diagram of the estimator, and we call it the "suboptimum estimator."

3. SUBCARRIER RECOVERY METHOD

Assume that the estimator output is optimally sampled with sampling period t_{smp} after frequency offset compensation. The discrete impulse response of the channel is written as

$$c(t) = \sum_{j=0}^{J-1} q_j \delta(t - jt_{smp}), \tag{13}$$

where J is the number of multipaths in the impulse response. The l–th received signal in the i–th symbol period is written as

$$\begin{aligned} r_l^i &= r^i(t = lt_{smp}) = (c \otimes s)(t = lt_{smp} - \tau) + z(t = lt_{smp}) \\ &= \sum_{j=0}^{J-1} q_j s^i(lt_{smp} - jt_{smp}) + z(lt_{smp}), \end{aligned} \tag{14}$$

where $z(t)$ is the complex-valued white Gaussian noise with zero mean.

Defining L as the number of samples in the guard interval, the received signals in the i–th guard interval $(1 \leq l \leq L)$ and the tail of the i–th useful symbol period $(N + 1 \leq l \leq N + L)$ are written as

$$L \left\{ \begin{array}{l} r_1^i = q_0 s_1^i + q_1 s_{L+N}^{i-1} + \cdots + q_{J-1} s_{L+N-J+2}^{i-1} + z_1, \\ \qquad \vdots \\ r_L^i = q_0 s_L^i + q_1 s_{L-1}^i + \cdots + q_{J-1} s_{L+J+1}^i + z_L, \end{array} \right. \tag{15}$$

$$L \left\{ \begin{array}{l} r_{N+1}^i = q_0 s_{N+1}^i + q_1 s_N^i + \cdots + q_{J-1} s_{N-J+2}^i + z_{N+1}, \\ \qquad \vdots \\ r_{N+L}^i = q_0 s_{N+L}^i + q_1 s_{N+L-1}^i + \cdots + q_{J-1} s_{N+L-J+1}^{i-1} + z_{N+L}, \end{array} \right. \tag{16}$$

where $z_l = z(lt_{smp})$. Define the following vectors:

$$\mathbf{r}_{pre}^i = [r_1^i, r_2^i, \cdots, r_L^i]^T, \tag{17}$$

$$\mathbf{r}_{post}^i = [r_{N+1}^i, r_{N+2}^i, \cdots, r_{N+L}^i]^T, \tag{18}$$

where T denotes the transpose. Furthermore, define the following i–th correlation matrix:

$$\mathbf{R}^i = \mathbf{r}_{pre}^i \cdot \mathbf{r}_{post}^{i\ H}, \tag{19}$$

where H denotes the Hermitian transpose.

Taking into account the fact:

$$s_1^i = s_{N+1}^i, \quad \cdots \quad, s_L^i = s_{N+L}^i, \tag{20}$$

the average correlation matrix can be decomposed as

$$\mathbf{R} =< \mathbf{R}^i >= \frac{1}{I_{av}} \sum_{i=1}^{I_{av}} \mathbf{R}^i = QQ^H \sigma_s^2 + I\sigma_n^2, \tag{21}$$

where σ_s^2 and σ_n^2 are the powers of the OFDM signal and the noise, respectively, I is the $L \times L$ identity matrix, and Q is the lower triangular matrix written as

$$Q = \begin{bmatrix} q_0 & & & & & & \\ q_1 & \ddots & & & & & \\ \vdots & \ddots & \ddots & & 0 & & \\ q_{J-1} & & \ddots & \ddots & & & \\ 0 & \ddots & & \ddots & \ddots & & \\ \vdots & \ddots & \ddots & & \ddots & \ddots & \\ 0 & \cdots & 0 & q_{J-1} & \cdots & q_1 & q_0 \end{bmatrix}. \tag{22}$$

Therefore, with the LU factorization of the correlation matrix of the unknown pilot signals, we can estimate the impulse response. Finally, from the Fourier Transform of the estimated impulse response, we can obtain the frequency transfer function essential for subcarrier recovery, namely, the amplitude and phase of each subcarrier.

4. NUMERICAL RESULTS

Parameters for computer simulation are as follows; The Number of Subcarriers: 32, Modulation: Differentially–Encoded QPSK, Demodulation: Differential or Coherent, Symbol Period: 36 samples, Guard Interval: 4 samples, Useful Symbol Period: 32 samples, Delay Profile: Exponential (3 samples).

Figs.3 (a), (b) and (c) show the estimation performance: the RMS frequency error, the RMS symbol timing error and the RMS period error, respectively. These figures show the excellent of the optimum and suboptimum estimators and the superiority of the optimum estimator over the suboptimum one. The optimum estimator requires only 10–to–20 observation symbols to obtain good estimate for frequency offset, symbol timing and symbol period estimation.

Fig.3 (d) shows the BER performance when DEQPSK/differential detection is employed. This figure clearly shows that 10 observation symbols are sufficient to achieve good BER performance, regardless of values of E_b/N_0.

Fig.3 (e) shows the RMS estimation error of impulse response. Blind estimation of impulse response normally requires longer observation symbols, and in this case, 10^5 observation symbols are required to obtain good estimate.

Finally, Fig.3 (f) shows the BER performance when DEQPSK/coherent detection with differential decoding is employed. For comparison purpose, the BER for DEQPSK/differential detection is also shown in the figure. Here, the BER is evaluated after 20 observation symbols for differential detection whereas 10^5 symbols for coherent detection. The coherent detection is superior to the differential detection, however, the difference is not so large. However, this figure clearly shows that, without any extra known pilot signal transmission, frequency offset, symbol timing and symbol period and furthermore impulse response for subcarrier recovery can be estimated, making effective use of the unknown pilot signal.

5. CONCLUSIONS

This paper has proposed a frequency offset, symbol timing and symbol period estimation method, and a subcarrier recovery method for OFDM signals in multipath fading channels. The methods are fully blind in the sense that they do not require any extra pilot signal transmission.

The optimum frequency offset/symbol timing/symbol period estimator requires only 10–to–20 symbols to achieve good BER performance even in the frequency selective slow Rayleigh fading channel with low E_b/N_0. On the other hand, the subcarrier recovery method requires more than 10^5 symbols for accurate estimation of channel impulse response.

The BER performance of DEQPSK/coherent detection with the proposed subcarrier recovery method is superior to that of DEQPSK/differential detection, however, there can be a lot of rooms to reduce the required number of observation symbols keeping the BER sufficiently low.

References

[1] M. Okada, M. Mouri, S. Hara, S. Komaki and N. Morinaga, "A Maximum Likelihood Symbol Timing, Symbol Period and Frequency Offset Estimator for Orthogonal Multi-Carrier Modulation Signals," *Proc. of IEEE ICT'96*, pp.596–601, April 1996.

[2] M. Okada, S. Hara, S. Komaki and N. Morinaga, "Optimum Synchronization of Orthogonal Multi–Carrier Modulated Signals," *Proc. of IEEE PIMRC'96*, pp.863–867, October 1996.

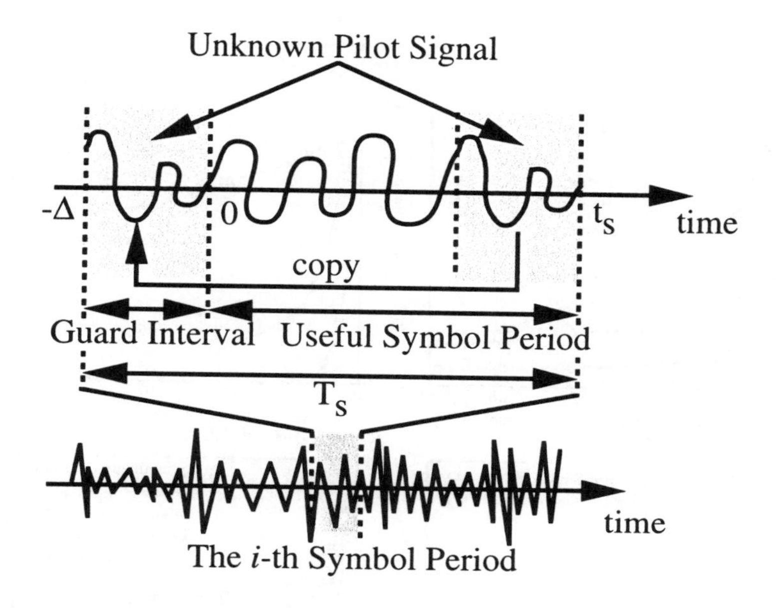

Figure 1 Guard interval insertion

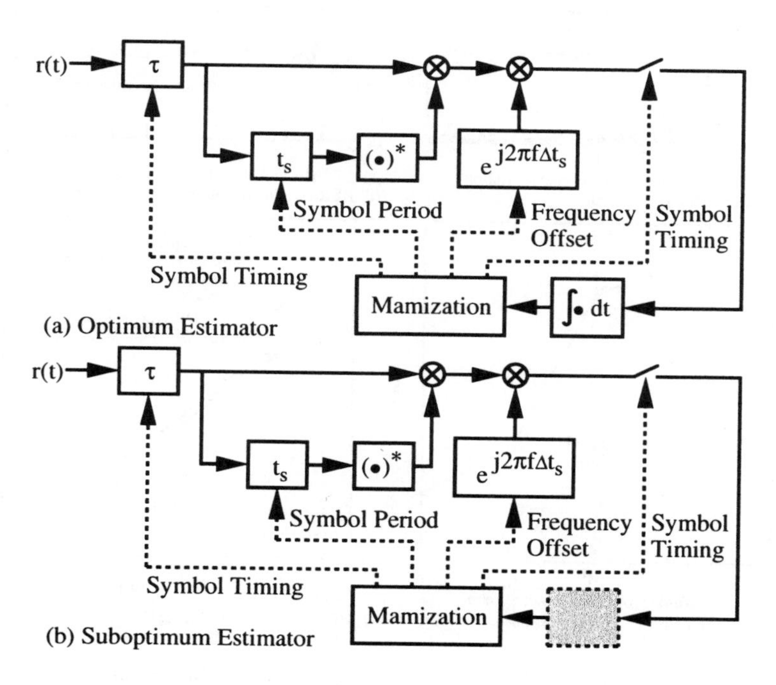

Figure 2 Block diagrams of blind estimators: optimum estimator (a) and suboptimum estimator (b)

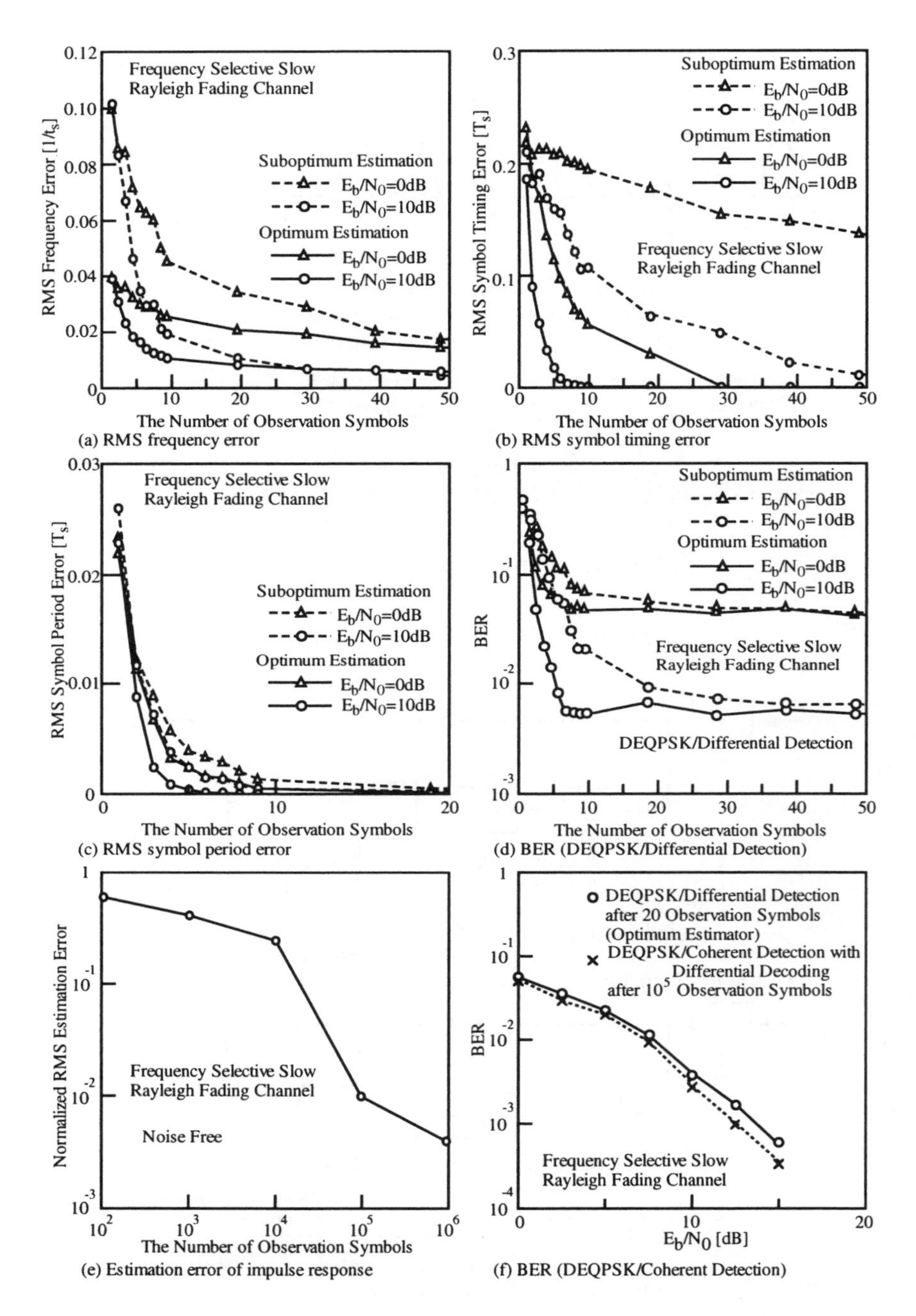

Figure 3 Performance in fading channel: frequency error (a), symbol timing error (b), symbol period error (c), BER (differential detection (d), estimation of impulse response (e), and BER (coherent detection) (f)

DOPPLER SPREAD ANALYSIS AND SIMULATION FOR MULTI-CARRIER MOBILE RADIO AND BROADCAST SYSTEMS

Patrick Robertson and Stefan Kaiser
German Aerospace Center (DLR), Institute for Communications Technology
82234 Oberpfaffenhofen, Germany
Patrick.Robertson@dlr.de; Stefan.Kaiser@dlr.de

Abstract The paper analyses the effects of Doppler spread in time variant channels on the performance of OFDM systems. Several Doppler spectra are analyzed and compared, such as the classical and uniform distributions, as well as the case of two distinct values for the Doppler frequency. We can thus employ the computed variance of the ICI process, together with the channel noise, in system analyses and designs, and avoid lengthy time-domain simulations. We also analyze optimal receiver frequency synchronisation in the case of Doppler spreads, as well as the differences between the up- and downlink in an OFDM(A) mobile radio system.

1. INTRODUCTION

The loss of orthogonality in multi-carrier systems (in particular OFDM systems [1, 2]) due to Doppler spread in the mobile radio channel has significant influence on the system performance. In this paper, we analytically compute the inter carrier interference (ICI) in mobile radio and broadcast systems using multi-carrier modulation. An analytical description of the effect of a classical Doppler spectrum on the C/I ratio for OFDM has been introduced in [3] and employed in methods to combat ICI, [4] [5]. A characteristic of the analysis of [3] is the explicit summation over all sub-carriers in the system. We have used a different approach and have found a closed solution for an infinite number of sub-carriers and classical Doppler distribution, as well as other cases for the Doppler spread, such as the important one of an asymmetrical two path channel. Furthermore, our method allows for the computation of the ICI for individual sub-carriers (such as edge sub-carriers). Finally, we investigate the achievable gain from optimally aligning the local oscillator to the current Doppler spectrum, in order to minimize the ICI level.

2. DERIVATION OF THE ICI LEVEL

In our analysis, we have initially assumed a channel with a known and fixed number of paths; for example a very simple two path model, where the two paths have a different amplitude and Doppler frequency. Fortunately, the delay of each path is irrelevant to our analysis if all echos lie within the guard interval (more correctly referred to as the cyclic extension) employed in OFDM. The individual ICI components which will additively contribute to the total ICI are obtained through varying all sub-carriers and all paths. We can show that we are able to use the central limit theorem for large number of sub-carriers and can model the ICI as zero-mean additive Gaussian noise.

2.1 SUMMATION OF ICI OVER A FINITE NUMBER OF SUB-CARRIERS

Let $L_t(k)$ denote the total ICI energy which results from all $2N$ sub-carriers onto the sub-carrier with index 0, as a function of the Doppler frequency $f_d^{(k)}$, for a single path with index k. To compute $L_t(k)$, we can show [6] that we need to evaluate:

$$\begin{aligned} L_t(k) &= \sum_{i=-N}^{-1} (a^{(k)})^2 \mathrm{sinc}^2 \left(\pi(f_i - f_0 + f_d^{(k)})T\right) \\ &+ \sum_{i=1}^{N} (a^{(k)})^2 \mathrm{sinc}^2 \left(\pi(f_i - f_0 + f_d^{(k)})T\right). \end{aligned} \tag{1}$$

Normalizing the symbol duration T to unity for simplicity, evaluation of these sums yields,

$$\begin{aligned} L_t(k) &= (a^{(k)} \cdot \mathrm{sinc}(\pi f_d^{(k)}))^2 \\ & \left((f_d^{(k)})^2 \cdot \left(\psi'(-f_d^{(k)} - N) - \psi'(1 - f_d^{(k)} + N)\right) - 1\right), \end{aligned} \tag{2}$$

where $\psi'(x)$ is the first derivative of the digamma function [7]. We have used the name "leakage" function for L_t, since it represents the sum of the energy that leaks over from all carriers onto carrier zero, due to a particular channel path k with a particular Doppler frequency $f_d^{(k)}$.

2.2 SUMMATION OF ICI OVER AN INFINITE NUMBER OF SUB-CARRIERS

The easier case is if we let N go to infinity, i.e. using (1) we evaluate $\lim_{N\to\infty} L_t(k)$. The result can be computed straightforwardly to be (again after normalizing the symbol duration T to unity):

$$L_t(k) = (a^{(k)})^2 \cdot \left[1 - \mathrm{sinc}^2(\pi f_d^{(k)})\right]. \tag{3}$$

In practice, we have observed that using the easier case of infinite subcarriers yields virtually the same result as for finite carriers, for N larger than about 60. Using (3) makes the following analysis much easier and often leads to closed-form solutions.

2.3 DOPPLER SPREAD

So far, the "leakage" function $L_t(k)$, expressed in equations (2) and (3), holds for one path with index k, Doppler frequency $f_d^{(k)}$, and amplitude $a^{(k)}$. The total ICI energy experienced by the carrier with index zero is thus just the sum $N_t = \sum_{\forall k} L_t(k)$. Now note that because $L_t(k)$ is proportional to $(a^{(k)})^2$, we can combine all those paths with exactly the same Doppler frequency, and add their energies $(a^{(k)})^2$, to make up a new, hypothetical path of greater energy $(a_{f_d})^2 = a^2 \cdot p(f_d)$, which contributes to $L_t(f_d)$. The new variable a^2 is the sum of the energy of *all* paths of the channel (i.e. summed over all Doppler frequencies). It is formally needed because the integral over $p(f_d)$ has to be unity, but not the integral over $a^2 \cdot p(f_d)$. Therefore, if we are given a statistical distribution for the Doppler frequency, $p(f_d)$, the expression for N_t can be written as:

$$N_t = a^2 \cdot \int_{-\infty}^{\infty} L_t(f_d) \cdot p(f_d) \cdot df_d. \tag{4}$$

The product $a^2 \cdot p(f_d)$ will be denoted in this paper as the weighted PDF; unless otherwise stated it will in fact equal $p(f_d)$.

2.4 USEFUL SIGNAL ENERGY

The useful signal energy $N_u^{(k)}$ which is contributed by one path with Doppler frequency $f_d^{(k)}$ is easily computed by setting $i = 0$, yielding $N_u^{(k)} = (a^{(k)})^2 \text{sinc}^2\left(\pi(f_d^{(k)})T\right)$. Since all paths have independent phases, and hence all useful signal components are uncorrelated, we are able to compute the total useful energy P_u as the sum of $N_u^{(k)}$ over all k.

2.4.1 Definition of Doppler Frequency PDFs.

Discrete. Physically, this distribution of f_d corresponds to a finite number of paths, each with a given Doppler frequency. Note that we do not have to separate physical paths with different delays, but the same Doppler frequency. We denote the number of such distinct paths with P, and use index k' in order to avoid confusion with k, which might be different if some physical paths have the same Doppler frequency.

A special case is a two-path model, which is both simple to compute, and will lend itself to the derivation of the optimal frequency to which the receiver should adjust, in order to minimize the ICI, see Section 3.

Furthermore, it shall serve as an approximation to a classical Doppler spectrum. It is modeled with $a = 1$, and the following PDF for the Doppler frequency:

$$p(f_d) = (1 - p_2)\delta(f_d - f_1) + p_2\delta(f_d - (f_1 + \Delta f)) \tag{5}$$

The variable f_1 denotes the Doppler frequency of one of the paths and p_2 the energy of the other path. The frequency distance between the two paths is given by Δf.

Uniform. It is useful to consider a uniform distribution, with maximal Doppler frequencies $f_{d_{max}}$, as it turns out to be easy to handle analytically, and, moreover, it corresponds to a number of physically interesting cases. Notably, it has been shown in [8], that for scattering environments in three dimensions the PDF of the Doppler frequency follows the uniform, rather than the classical distribution.

Classical. The classical Doppler spectrum results from uniformly distributed angels of arrival at the receiver antenna. The PDF for the classical Doppler case is [9]

$$p(f_d) = \frac{1}{\pi \cdot f_{d_{max}} \cdot \sqrt{1 - \left(\frac{f_d}{f_{d_{max}}}\right)^2}} \; ; \text{ for } |f_d| \leq f_{d_{max}}. \tag{6}$$

The classical Doppler spectrum is a long term average. For considerable lengths of time we will often observe an asymmetrical or biased Doppler spectrum, which we approximate through our discrete two-path model above, with $p_2 \neq 0.5$.

2.5 INTEGRATION FOR FINITE AND INFINITE SUB-CARRIERS

We are now able to proceed with the integration to evaluate (4). The integration is straightforward and leads to closed-form solutions for an infinite number of carriers (albeit resorting to the use of special functions in the solutions).

2.5.1 Discrete. $E^{(k')}$ is the energy of all paths with a distinct Doppler frequency $f_d = f_d^{(k')}$. Of course, equations (2) and (3) are formally computed by simply setting $a^{(k)}$ to unity, and $f_d^{(k)}$ to $f_d^{(k')}$. This yields,

$$N_t = \sum_{k'=0}^{P-1} L_t(f_d^{(k')}) \cdot E^{(k')}. \tag{7}$$

$$P_u = \sum_{k'=0}^{P-1} \text{sinc}^2 \left(\pi(f_d^{(k')})T\right) \cdot E^{(k')}. \tag{8}$$

For the two-path model the useful signal energy is:

$$P_u = (1 - p_2)\text{sinc}^2(f_1\pi) + p_2\text{sinc}^2(\Delta f + f_1\pi). \tag{9}$$

For an infinite number of sub-carriers, the ICI energy is:

$$N_t = (1 - p_2)\left[1 - \text{sinc}^2(f_1\pi)\right] + p_2\left[1 - \text{sinc}^2(\Delta f + f_1\pi)\right]. \tag{10}$$

2.5.2 Uniform. For a uniform distribution, we have found the following solution for an infinite number of carriers,

$$N_t = \frac{\cos(2\pi f_{d_{max}}) + 2f_{d_{max}}\pi\text{Si}(2f_{d_{max}}\pi) - 1 - 2\cdot(\pi\cdot f_{d_{max}})^2}{2\cdot(\pi\cdot f_{d_{max}})^2}, \tag{11}$$

where Si(.) is the sine integral function [7]. Unfortunately, we have had to resort to numerical integration for the finite sub-carrier situation.

The useful energy is always

$$P_u = -\frac{\cos(2\pi f_{d_{max}}) + 2f_{d_{max}}\pi\text{Si}(2f_{d_{max}}\pi) - 1}{2(\pi\cdot f_{d_{max}})^2}. \tag{12}$$

2.5.3 Classical. For infinite sub-carriers, we have used the relation $\sin(x) = \frac{1}{2}(1 - \cos(2x))$, and inserted this into the integral,

$$I_{cd} = \int_{-f_{d_{max}}}^{f_{d_{max}}} \frac{1 - \text{sinc}^2(f_d\pi)}{f_{d_{max}}\cdot\pi\cdot\sqrt{1 - (\frac{f_d}{f_{d_{max}}})^2}}\, df_d. \tag{13}$$

Using [10] to continue, we can express the solution using the generalized hyper-geometric function [7], ${}_p\text{F}_q$, where in our case, $p = 1$, $q = 2$:

$$N_t = 1 - {}_1\text{F}_2\left(\frac{1}{2}; \frac{3}{2}, 2; -(f_{d_{max}}\pi)^2\right). \tag{14}$$

For finite sub-carriers, we again employ numerical integration.

In all cases, the useful energy is

$$P_u = {}_1\text{F}_2\left(\frac{1}{2}; \frac{3}{2}, 2; -(f_{d_{max}}\pi)^2\right). \tag{15}$$

3. OPTIMAL RECEIVER FREQUENCY ADJUSTMENT

3.1 SMALL DOPPLER SPREADS

For small f_d, infinite sub-carriers, and an otherwise arbitrary PDF, the minimal ICI is achieved when the expected value of the Doppler frequency is zero at the receiver, since for small values of x, we can use the approximation $1 - \text{sinc}^2(x) \approx \frac{1}{3} \cdot x^2$ in (3) and in the expression for $N_u^{(k)}$, leading to a quadratic relation between $L_t(k)$ (or $N_u^{(k)}$) and f_d. In other words, a frequency synchronization algorithm which sets the mean Doppler frequency to zero will be close-to-optimal with respect to ICI minimization as long as the Doppler spread is small. For the two path model it is very easy to show that this yields a much easier approximately optimal solution for f_1, namely $f_1 \approx -\Delta f \cdot p_2$.

3.2 LARGE DOPPLER SPREADS

We will restrict ourselves to the two-path model with weighted Doppler PDF. We shall let f_1 be the free variable, and observe the variation of the C/I ratio. When we find the value of f_1 which maximizes this ratio, we have found the best frequency for the receiver's local oscillator, namely the one which results in the value of f_1 for the path in the weighted Doppler PDF with energy $(1 - p_2)$. The values of p_2 and Δf are fixed for one determination of the best f_1.

To do this, we differentiate either the useful signal energy (9) or the ICI energy (10) with respect to f_1: notice that the result is the same except for the sign. Setting this derivative to zero and solving the resulting equation to find f_1 involves either a numerical solution, or a Taylor series expansion followed by an analytical solution of the resulting truncated series, set to zero.

3.2.1 Performance Gain. We shall now compare the gain achieved by using the optimal solution for f_1, compared to the approximation $f_1 \approx -\Delta f \cdot p_2$. We have shown the gain in dB for different values of p_2 and Δf in Fig. 1. As can be seen, there is little to be gained for values of Δf less than 0.5. However, for values greater than 0.5, the gain can be significant even for small values of p_2. This implies that even weak components at high Doppler frequencies (relative to the most powerful path) can have a more serious effect if the sub-optimal frequency adjustment is chosen, and thus has repercussions on the receiver synchronisation design. It is important to remember that the results only apply to the very simple two path model.

3.2.2 Discussion. It appears that a frequency synchronizer which minimizes the ICI directly (for example by observing the C/I of some sub-carriers and maximizing), can provide some gain compared to one which sets the average Doppler frequency to zero. Furthermore, such a synchronizer will per definition find the optimal frequency without having to know the solution for each model for the distribution of the Doppler frequencies.

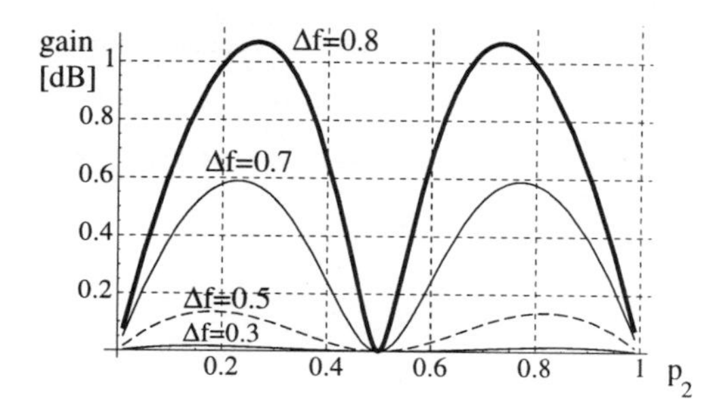

Figure 1 The gain from using the optimal frequency correction for different Δf and p_2 for a two path model.

4. RESULTS

The parameters of the analyzed OFDM system (for a mobile radio scenario) are as follows: The bandwidth is $B = 2$ MHz and the carrier frequency is located at $f_c = 2$ GHz. The number of sub-carries for OFDM is $N = 256$. Thus, the carrier distance is $\Delta f = 7.81$ kHz. We have chosen QPSK for symbol mapping. The maximum number of active users (fully loaded system) is 32, so each user transmits exclusively on its 8 sub-carriers. The 8 sub-carriers of a single user are block interleaved such that they have maximum frequency distance. The data rate per user is 125 kbit/s. We have defined 6 cases with different assumptions about the multi-path propagation in the channel.

The C/I versus the Doppler frequency is shown in Fig. 2. Analytical and simulation results are shown without frequency correction. Additionally, results are presented with optimal frequency correction for those cases where it is applicable (asymmetric PDFs). As is evident from the figure, the analytical results match perfectly with the simulation results.

5. CONCLUSIONS

Summing up, we have analyzed the effects of Doppler spread in a time variant mobile radio channel on the performance of an OFDM system. Analytical results have been derived and verified by simulation results which enable the evaluation of the ICI due to loss of orthogonality. Moreover, an optimal frequency synchronization in the case of Doppler spreads has been analyzed.

References

[1] M. Alard and R. Lassalle, "Principles of modulation and channel coding for digital broadcasting for mobile receivers," *EBU Review*, pp. 47–69, August 1987.

[2] S. Weinstein and P. M. Ebert, "Data transmission by frequency-

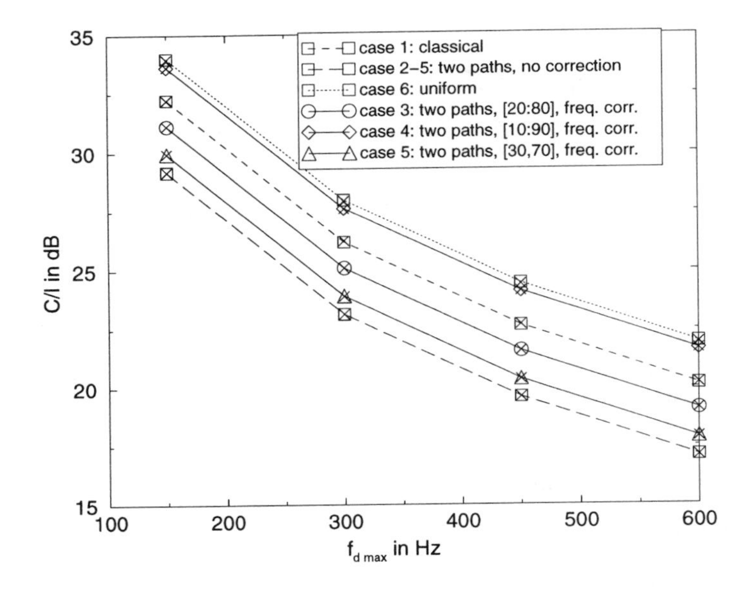

Figure 2 The symbol X shows the analytical results. In parentheses we have shown the path weights in % for the two-path model.

division multiplexing using the discrete fourier transform," *IEEE Trans. Commun.*, vol. 19, pp. 628–634, October 1971.

[3] M. Russel and G. Stüber, "Terrestrial digital video broadcasting for mobile reception suing OFDM," *Wireless Personal Comms, Special Issue Multi-Carrier Comms*, vol. 2, no. 1&2, pp. 45–66, 1995.

[4] K. Matheus and K.-D. Kammeyer, "Optimal design of multicarrier systems with soft impulse shaping including equalization in time or frequency," in *Proc. GLOBECOM '97*, pp. 310–314, 1997.

[5] J. Armstrong, P. Grant, and G. Porey, "Polynomial cancellation of OFDM to reduce intercarrier interference due to doppler spread," in *Proc. GLOBECOM '98*, 1998.

[6] P. Robertson and S. Kaiser, "Analysis of the loss of orthogonality through Doppler spread in OFDM systems," in *Proc. GLOBECOM '99*, December 1999.

[7] I. S. Gradshteyn and I. Ryzhik, *Tables of Integrals, Series and Products.* Academic Press, 1965.

[8] "Final report on RF channel characterization." JTC (AIR) 93.09.23-238R2, 1993.

[9] W. C. Jakes, *Microwave Mobile Communications.* John Wiley & Sons, Inc., 1974.

[10] S. Wolfram, *The Mathematica Book, 3rd ed.* Wolfram Media, Cambridge Univ. Press, 1996.

REDUCED STATE JOINT CHANNEL ESTIMATION AND EQUALIZATION

Isabella De Broeck,
Marco Kullmann
and Uli Sorger
Institute for Network- and Signal Theory
Darmstadt University of Technology,
Merckstraße 25, D–64283 Darmstadt, Germany
bella@nesi.tu-darmstadt.de, marco.kullmann@gmx.de, uli@nesi.tu-darmstadt.de

Abstract In this paper, a joint channel estimation and equalization scheme is proposed and investigated for application to Interleaved Frequency-Division Multiple-Access, a promising spread-spectrum multiple-access schemes for mobile communications. The underlying idea is to perform equalization and channel estimation together in a reduced state diagram, where the channel estimation is adapted continuously. Compared to separately performed channel estimation and equalization, the length of the training sequence for channel estimation as well as the equalizer complexity can be reduced significantly, whereas system performance is improved. Simulation results show a performance gain of 3.3 dB at a bit-error-rate of 10^{-3}. The loss towards equalization with perfect channel estimation is only about 1 dB. The results may be generalized to other systems with interference and a priori unknown channel parameters.

INTRODUCTION

Using spread-spectrum multiple-access (SSMA) schemes in mobile communications a significant performance gain is achieved due to frequency diversity on the one hand. On the other hand, the influence of the channel and, consequently, the intersymbol interferences (ISI) is increased due to the broadband transmission. However, the effects of ISI can be reduced significantly by applying an equalizer at the receiver. Performing an equalization needs knowledge about the channel impulse response at the receiver, so that additionally a channel estimation has to be carried out. Considering channel estimation and equalization separately is not optimal, even if each separate step is optimal. In this paper, a combined scheme is proposed, where channel estimation and equalization are performed together in a reduced state diagram. Compared to separately performed channel estimation and equalization, the additional redundancy for channel estimation as well as the equalizer complexity can be reduced significantly, whereas system performance is improved. In the following, the combined channel estimation and equalization scheme is proposed and investigated exemplarily for application to Interleaved Frequency-Division Multiple-Access (IFDMA), a promising SSMA scheme as described in [SmD99]. The results may be also generalized to other systems with interference and a priori unknown channel parameters.

CHANNEL ESTIMATION

Using an orthogonal SSMA scheme like IFDMA, where the transmission signals of all users are orthogonal to each other, and thus, no multiple-access interference (MAI) is present, the received symbol $y(l)$ in a time discrete model is given by

$$y(l) = \sum_{m=0}^{M} x(l-m) \cdot h(m) + n(l) , \tag{1}$$

where $x(l)$ is the transmitted symbol of any user, $\mathbf{h} = [h(0), h(1) \cdots h(M)]^T$ is the mobile radio channel impulse response vector and $n(l)$ are the additive white Gaussian noise (AWGN) samples with one-sided noise spectral density N_0.

In the case of IFDMA, the symbols $x(l)$ are transmitted block-by-block and the channel impulse response vector $\mathbf{h}$ can be viewed constant over at least one block of IFDMA transmission symbols.

Using the Viterbi algorithm for equalization the channel impulse response vector $\mathbf{h}$ of length $M+1$ determines the number $N_t = 2^M$ of trellis states and, therefore, the complexity of the equalizer. If M is large, the Viterbi algorithm as optimum equalization technique can not be applied due to complexity restrictions, but suboptimum equalization techniques with reduced complexity might be used.

For equalization it is necessary to have an estimate $\hat{\mathbf{h}}$ of the channel impulse response vector $\mathbf{h}$. Normally, a known training sequence is transmitted for channel estimation. In this case, the transmission sequence $x(l)$ as well as the received sequence $y(l)$, $l = 0, \ldots, 2M$ are known at the receiver and, thus, a maximum likelihood (ML) channel estimation can be performed to obtain an estimate $\hat{\mathbf{h}}$ for the channel impulse vector $\mathbf{h}$. This estimate is used for equalization of the subsequently received signals. Compared to perfect equalization, where it is assumed that the exact channel impulse response vector $\mathbf{h}$ is known at the receiver, an equalization with real channel estimation shows a certain performance degradation due to estimation errors.

Channel estimation and equalization can be executed with a relatively low complexity, however, the following disadvantages arise. The training sequence length has to be at least twice as large as the channel impulse response vector $\mathbf{h}$. Therefore, the effective information symbol rate decreases linearly with the length M of the channel impulse response vector $\mathbf{h}$. Moreover, considering channel estimation and equalization separately is not optimal, even if each separate step is optimal. This is due to the fact, that not only the received training sequence but each received symbol $y(l)$ contains information about the channel. Conventional approaches to channel estimation do not take into account this additional information. Therefore, usually long training sequences have to be employed. For example, the "Global System for Mobile Communications" (GSM) uses approximately 20% of the transmitted symbols as training sequences for channel estimation. If the additional information about the channel which is contained in each received signal $y(l)$ is taken into account, an improvement of the system performance and/or a shortening of the training sequence can be realized.

In [Trä98], a new scheme for combined channel estimation and equalization using sequential decoding techniques is described and investigated. The metric adaptation rule derived in [Trä98] is used in the following within a reduced state diagram.

REDUCED STATE EQUALIZER WITH CONTINUOUSLY ADAPTED CHANNEL ESTIMATION

The basic idea of the algorithm for joint equalization and channel estimation is the renewed calculation of the estimated channel impulse response vector $\hat{\mathbf{h}}$ after each received symbol by extending the training sequence with the hypothesis of that symbol. The optimal approach for joint channel estimation and equalization is a tree structure including all possible hypotheses of the symbols at each path. Under the assumption that the hypothesis is correct, for each path the training sequence is extended with the corresponding symbol and a separate channel estimation is calculated, which in turn is used for the metric calculation. At the end a decision is made for the most reliable hypothesis. Due to complexity restrictions this optimal approach is not practical for most applications.

In our approach, a sequential algorithm is applied, where only a maximum number of N_p ($N_p \ll 2^M$) paths is retained at each step after starting the algorithm in an arbitrary initial state in the reduced trellis. The proposed algorithm is initialized with an estimate $\hat{\mathbf{h}}$ of the channel impulse response vector, which is obtained by using a short training sequence. In comparison with the conventional approach with separate channel estimation and equalization the length of this training sequence can be reduced drastically without loss of performance. At each step in the reduced trellis, a separate channel estimation is calculated for each surviving path. This is done by extending the training sequence with the already decided data symbols for that path. Thus, the training sequence virtually grows and leads to a better channel estimation in each further step. The improved estimate $\hat{\mathbf{h}}$ of the channel impulse response vector is used for the metric calculation of the corresponding path. The improved channel estimation as well as the metric calculation can be determined efficiently by using the recursive Kalman algorithm [Trä98, Hay86]. The metric is also used to discriminate the paths of same length at each step in order to retain only the N_p best paths.

Due to the time-dispersive behavior of the channel impulse response large parts of the signal energy of previous symbols are not included in the metric. Thus, for a reliable decision usually many surviving paths have to be considered. To circumvent this problem a new enlarged metric, with an approach comparable to a matched filter, has been developed. The enlarged metric consists of the present general metric and an additional term, which takes into account the missing signal energy by considering for each path the subsequent steps through the trellis. For this the unknown, subsequent symbols are assumed to be equally likely. According to the length of the channel impulse response vector the paths are extended stochastically with the M subsequent symbols. The enlarged metric is calculated from the enlarged path, as described in [Kul99]. Thus, the whole energy of the previous symbols is considered and a more reliable decision is achieved. With that the number of retained paths and, consequently, the complexity can be reduced without affecting the performance.

Using the final, improved channel estimation as initialization for a renewed equalization process the performance can be additionally improved.

In the case, where the channel impulse response changes slowly, i.e. where the Doppler frequency is small, it is possible to take the channel estimation of the previous transmitted block into account. This allows to additionally shorten the training sequence without loosing performance for the equalization process.

For fast changing channels, where the channel impulse response becomes time-variant even during one transmission block, the above described scheme can be used to track the channel impulse response during the transmission block.

SIMULATION RESULTS

For the Monte-Carlo simulations an IFDMA transmission with binary phase shift keying (BPSK) modulation over the GSM test channels TU ("typical urban") and BU ("bad urban") at an approximate transmission bandwidth of $B = 1.25$ MHz is considered. It is assumed that the synchronization at the receiver is perfect. In Fig. 1, the performance

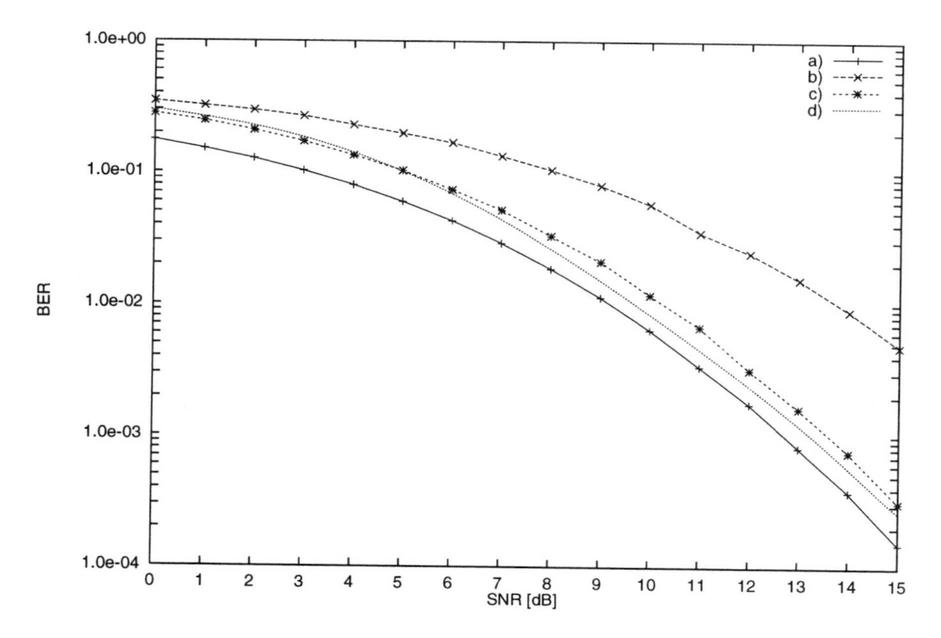

Figure 1 Simulated BER for reduced state equalizer with adaptive channel estimation and training sequence length c) $L_t = 15$, d) $L_t = 4$ in comparison with Viterbi equalizer with a) perfect and b) separate channel estimation, $L_t = 15$.

results for the proposed reduced state equalizer are shown as "Bit Error Rate" (BER) $P_b(\gamma_b)$ versus the "Signal-to-Noise Ratio" (SNR) $\gamma_b = E_b/N_0$, where E_b is the energy per transmitted binary symbol. In this case, the transmission is carried out over the GSM test channels TU with channel impulse response length $M = 7$. For initialization training sequences of length $L_t = 15$ and $L_t = 4$ are used, respectively. In the latter case a Doppler frequency of $f_d = 10$ Hz is assumed and the channel estimation of the previous block is taken into account. For comparison the conventional approach with separate channel estimation and equalization using a training sequence of length $L_t = 15$ as well as equalization with perfect channel estimation are shown. Note, $L_t = 15$ is the minimum length to enable ML channel estimation. In contrast to the optimum equalization with $2^M = 256$ paths in each step the proposed equalizer retains $N_p = 64$ paths, and $N_p = 8$ paths, respectively, by using the enlarged metric. For a BER level of $P_b(\gamma_b) = 10^{-3}$ the gain of the reduced state equalizer in SNR is about 3.3 dB compared to the conventional approach and the loss towards equalization with perfect channel estimation is only about 1 dB. Note, that the additional energy gain due to the shorter training sequence is not considered.

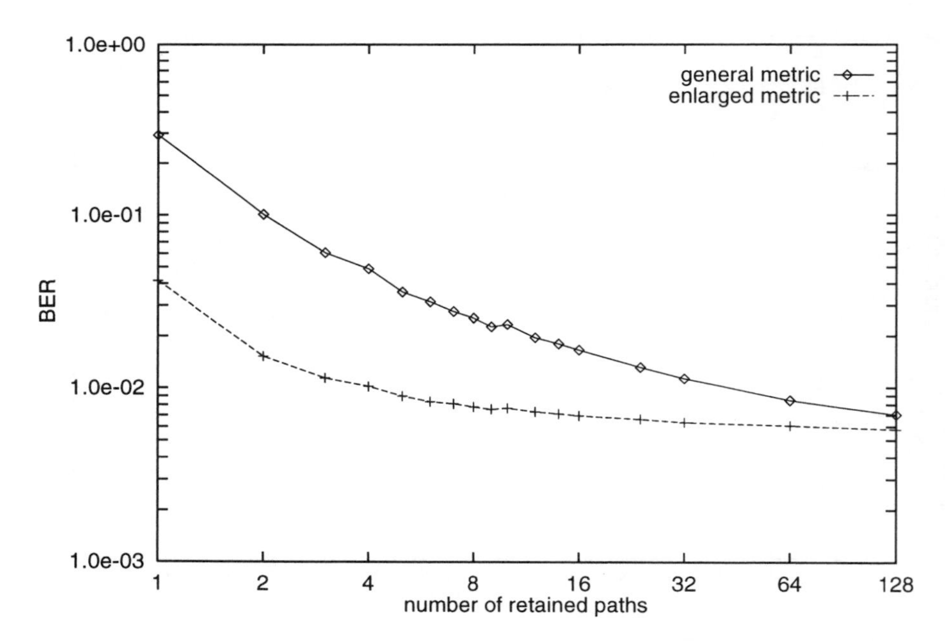

Figure 2 Simulated BER versus the number of retained paths for reduced state algorithm with enlarged metric in comparison with general metric; SNR $\gamma_b = 10$ dB; training sequence length $L_t = 27$

Applying the proposed algorithm with the enlarged metric the same performance is achieved by a significantly smaller number N_p of retained paths. For further clarification of this result, simulations over the GSM test channel BU with channel impulse response length $M = 13$ are carried out. The result of these simulations are shown in Fig. 2, where the BER is depicted versus the number of retained paths at a fixed SNR level of $\gamma_b = 10$ dB. Using the general metric and $N_p = 128$ retained paths a BER of $P_b(\gamma_b) = 7 * 10^{-3}$ is achieved. For achieving the same BER value with the enlarged metric only $N_p = 16$ retained paths have to be considered, which is a significant complexity reduction. Note, for the Viterbi algorithm $N_p = 2^M = 8192$ paths has to be retained.

SUMMARY AND CONCLUSIONS

In this paper, a joint channel estimation and equalization scheme using a reduced state diagram is proposed. Compared to the conventional approach, where channel estimation and equalization are performed separately, the joint approach improves the results of the channel estimation process continuously by taking into account not only the training sequence symbols, but also the information symbols. As a result, the complexity as well as the redundancy can be reduced significantly compared to the optimum equalizer with separate ML channel estimation, whereas system performance is improved. The proposed algorithm is investigated exemplarily for application to IFDMA, a promising SSMA scheme for mobile communication. At a BER of 10^{-3} a performance gain of 3.3 dB is achieved compared to separate channel estimation and equalization, whereas the loss towards equalization with perfect channel estimation is only about 1 dB. The complexity, given by the number N_p of retained path, is reduced significantly from $N_p = 8192$ for the Viterbi algorithm to $N_p = 16$ for the proposed algorithm, if a transmission over the GSM test channel BU is considered.

The proposed scheme may be generalized to other systems with interference and a priori unknown channel parameters.

References

[SmD99] M. Schnell, I. De Broeck, U. Sorger: "A Promising New Wideband Multiple-Access Scheme for Future Mobile Communications Systems", European Trans. on Telecommun. (ETT), to be published

[Trä98] Träger, J.: "Kombinierte Kanalschätzung und Decodierung für Mobilfunkkanäle", Dissertation (Ph.D. Thesis, in German), published in Berichte aus der Kommunikationstechnik, Aachen: Shaker Verlag 1998

[Hay86] Haykin, S.: *Adaptive Filter Theory*, 1st ed. Englewood Cliffs, New Jersey: Prentice Hall, 1986

[Kul99] Kullmann, M.: "Implementierung und Analyse kombinierter Kanalschätzung und Entzerrung für IFDMA", Studienarbeit (in German), Inst. for Network- and Signal Theory, Darmstadt University of Technology, April 1999

CHANNEL ESTIMATION WITH SUPERIMPOSED PILOT SEQUENCE APPLIED TO MULTI-CARRIER SYSTEMS

Peter Hoeher[1] and Fredrik Tufvesson[2]

[1] Information and Coding Theory Lab
University of Kiel, Kaiserstr. 2, D-24143 Kiel, Germany
E-mail: ph@techfak.uni-kiel.de

[2] Department of Applied Electronics
Lund University, P.O. Box 118, SE-221 00 Lund, Sweden
E-mail: Fredrik.Tufvesson@tde.lth.se

ABSTRACT

For the purpose of various synchronization tasks (including carrier phase, time, frequency, and frame synchronization), one may add a known pilot sequence, typically a pseudo-noise sequence, to the unknown data sequence. This approach is known as a spread-spectrum pilot technique or as a superimposed pilot sequence technique.

In previous work, we applied the superimposed pilot sequence technique for the purpose of channel estimation (CE) for single-carrier systems. We proposed and verified a truly coherent receiver based on the Viterbi algorithm, which is optimal in the sense of per-survivor processing. We also suggested a generic low-cost receiver structure based on reduced-state sequence estimation. Here, we extend the receiver structure to cope with multi-carrier systems. Among the distinct advantages compared to conventional pilot-symbol-assisted CE are (i) a lack of bandwidth expansion and (ii) a significantly improved performance in fast fading environments.

1 INTRODUCTION

Consider digital data transmission over time-selective (i.e., non-frequency-selective) fading channels. Assuming coherent demodulation, one of the main problems is carrier synchronization, both in terms of acquisition and tracking, particularly when the channel is fast and when a line-of-sight component is absent [1, 2].

A popular technique to maintain coherent demodulation for a wide class of digital modulation schemes has been proposed by Moher and Lodge [3, 4], and is known as *pilot-symbol-assisted CE*[1]. The main idea of pilot-symbol-assisted CE is to multiplex known pilot symbols (also called training symbols) into an unknown data stream. The receiver firstly obtains tentative channel estimates at the positions of the pilot symbols by means of re-modulation, and then computes final estimates by means of interpolation. Aghamohammadi and Cavers were among the first analyzing and optimizing pilot-symbol-assisted CE given different interpolation filters [5, 6]. Due to the pilot symbols the bandwidth slightly increases.

In this paper, we explore a related technique proposed by Makrakis and Feher [7, 8], originally called spread-spectrum pilot technique. This scheme has also been invented about a dozen years ago, but is less known. The clue is to linearly add a known pilot sequence to the unknown data sequence, see Section 2. Makrakis *et al.* applied the technique to phase synchronization [7, 8]. Later, the technique has been applied for the purposes of frame synchronization [9] and joint time and frequency synchronization of OFDM signals [10], respectively.

Within this paper, we assume a synchronous symbol-by-symbol superposition, where the power of the pilot symbols is typically much less than the power of the data symbols. By construction, there is no increase in bandwidth. Therefore, we use the notion of a *superimposed pilot sequence technique* as suggested in [9]. As opposed to pilot-symbol-assisted CE, no interpolation is necessary. The superimposed pilot sequence technique is therefore more bandwidth *and* power efficient than pilot-symbol-assisted CE, particularly in fast fading conditions.

Our main contribution in [11] was the derivation of a recursive receiver structure for the purpose of CE (phase synchronization). Emphasis was on single-carrier systems. Here, we also treat multi-carrier issues.

2 TRANSMISSION SCHEME

A simplified block diagram of the transmitter, the channel and the receiver is shown in Fig. 1. This picture features only a single carrier of an arbitrary multi-carrier system with orthogonal sub-carriers. In the following, we use the complex baseband notation and assume perfect time, frequency and frame synchronization. In the transmitter a known pilot sequence, typically a pseudo-noise (PN) sequence, is synchronously added to the unknown data sequence.

[1] Channel estimation may be seen as a generalization of carrier synchronization, since estimates of the quadrature components cover the carrier phase as well as reliability information, often called channel state information.

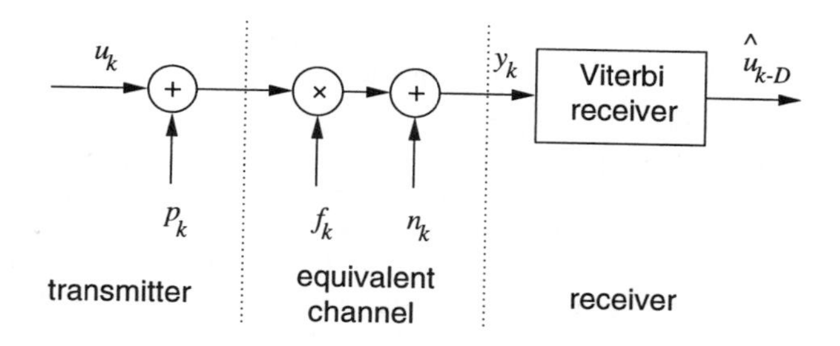

Figure 1: Equivalent block diagram of a single sub-carrier.

The chips of the known pilot sequence and the i.i.d. data symbols are denoted as $p_k \in \mathbb{C}$ and $u_k \in \mathbb{C}$, respectively, where k is the time index. (We assume that the chip duration is equal to the symbol duration.) Our scheme relies on a low correlation between the pilot sequence and the data sequence in order to resolve phase ambiguities. Both sequences are assumed to be M-ary with zero mean: $E[u_k] = E[p_k] = 0$. Let the energy of the pilot symbols be $E[|p_k|^2] = \rho \cdot E_s$ and the energy of the data symbols be $E[|u_k|^2] = (1-\rho) \cdot E_s$, respectively. Hence, the average energy of the composite symbols is E_s, if the sequences are mutually independent: $E[u_k \cdot p_k] = 0$. The average energy per information bit is $E_b = E_s / \log_2 M$.

The normalized power of the pilot symbols, ρ, corresponds to the amount of known information transmitted, and should be optimized. In pilot-symbol-assisted CE systems the related parameter is the spacing of the pilot symbols. In the remaining, ρ is assumed to be constant for all k, since our focus is on tracking. In order to improve the acquisition phase, ρ may be larger for the first symbols within a data frame [10].

Note that in pilot-symbol-assisted CE systems the data and pilot symbols are orthogonal by means of time-division multiplexing, whereas here the cross-correlation between the data and pilot symbols typically is non-zero. However, the superimposed pilot sequence is known and therefore the disturbance can be made small. The negative effect on the bit error rate (BER) is over-compensated by the lack of interpolation, as we will show in Section 4. The transmission system may be interpreted as a two-user system, where one user transmits information unknown to the receiver, and the other user transmits known information. Correspondingly, the transmission technique is related to multi-user systems, and the synchronization/detection problem is related to multi-user detection. Also note the similarities with watermarking schemes.

The matched filter output samples, $y_k \in \mathbb{C}$, can be written as

$$y_k = (u_k + p_k) \cdot f_k + n_k, \tag{1}$$

where $f_k \in \mathbb{C}$ is a multiplicative (time-selective) fading process with $E[|f_k|^2] = 1$, and $n_k \in \mathbb{C}$ are zero-mean white Gaussian noise samples with one-sided power spectral density N_0. Oversampling is not treated within this paper, but could be used to improve the performance in fast fading.

In the transmitter, the major difference between the spread-spectrum pilot technique proposed by Makrakis and Feher and the scheme described above is that in the original proposal the linear addition was done after D/A conversion, whereas we do digital baseband processing. Therefore, we guaranty symbol-synchronous transmission, which is an important property for our novel receiver. As a side effect, the power density spectrum is unaffected by the superimposed pilot sequence when using a PN sequence.

3 RECEIVER STRUCTURE

Instead of using conventional receivers for superimposed pilot sequences [7, 8] or pilot-symbol-assisted CE [3, 4, 5, 6], we will now investigate the recursive receiver based on the Viterbi algorithm proposed in [11], which directly outputs estimates of the data symbols, see Fig. 1. This receiver is primarily designed for the superimposed pilot sequence technique, but is also suitable for the conventional pilot-symbol-assisted CE technique. Different channel estimates are computed for different hypothesis by means of the principles of per-survivor processing [12, 13]. Interference cancellation is done inherently.

The goal is to recursively compute the maximum-likelihood sequence. Since the noise is assumed to be Gaussian, a suitable metric increment is

$$\Lambda_k(\tilde{u}_k; \tilde{u}_{k-1}, \ldots, \tilde{u}_{k-L}) = |y_k - (\tilde{u}_k + p_k) \cdot \tilde{f}_k|^2, \tag{2}$$

where $\tilde{u}_k$ is a hypothesis for the kth data symbol and $\tilde{f}_k$ is a channel estimate, which depends on the hypotheses $\tilde{u}_{k-1}, \ldots, \tilde{u}_{k-L}$. Given these hypotheses, the channel estimate $\tilde{f}_k$ can be computed by linear prediction [12]:

$$\tilde{f}_k = \sum_{l=1}^{L} a_l \cdot y_{k-l} / (\tilde{u}_{k-l} + p_{k-l}), \tag{3}$$

where L is the predictor order. Substituting (3) into (2), we then obtain the desired metric increment. If the channel is wide-sense stationary, the optimal predictor coefficients, a_l, $1 \leq l \leq L$, are the solution of the Wiener-Hopf equations. Instead of prediction, it is also possible to use filtering or smoothing to calculate the channel estimates.

The hypotheses $\tilde{u}_{k-1}, \ldots, \tilde{u}_{k-L}$ belong to the states of a trellis with M^L states, and $\tilde{u}_k$ belongs to the actual branch. The trellis has M branches/state. For large M and L, the complexity can be significantly reduced by applying the principles of reduced-state sequence estimation and set-partitioning [14, 15]. Since the noise is assumed to be white, the metric increments are additive. Therefore, the familiar add-compare-select operation can be applied and the maximum-likelihood path can be found by back-tracing or related operations. A rule of thumb for the decision delay is $D \approx (L+1) \ldots 4(L+1)$. Due to the maximum-likelihood *sequence* estimation, the receiver can cope with a low-power pilot sequence.

Our receiver is suitable to process pilot-symbol-assisted CE too. In contrast to the conventional pilot-symbol-assisted channel estimator, all matched filter output samples, i.e. also the data symbols, are used for CE. The proposed receiver may be extended to accept a priori information and to deliver soft-outputs. These properties are necessary for iterative processing, and they are used for the separable 2-D receiver described in Section 5.

4 PERFORMANCE EVALUATION FOR A SINGLE SUB-CARRIER

For a single sub-carrier, we have verified our receiver given the following set-up: We used 2-PSK modulation ($M = 2$) for the data and pilot symbols (no staggering, no phase offset), i.e. real-valued symbols. For this particular modulation scheme, the performance could be improved by transmitting the pilot sequence as the quadrature component. However, here we want to demonstrate the feasibility of using a common channel instead of two independent channels. The pilot sequence was a long PN sequence known to the receiver, as opposed to a short sequence applied in [8]. Data and pilot symbols were generated by independent pseudo-random generators. We averaged our results over several thousands of possible pilot sequences. The block length was chosen to be 2000 symbols.

The channel model was a flat Rayleigh fading channel per sub-carrier with 2-D isotropic scattering. The performance was studied for a wide range of different fading rates $f_{D_{max}} T_s$, where f_D denotes the Doppler frequency, $f_{D_{max}}$ the maximum Doppler frequency ($-f_{D_{max}} \leq f_D \leq f_{D_{max}}$), and T_s the symbol duration.

In the receiver, the proposed Viterbi receiver with a sufficiently long decision delay was applied. The Doppler spectrum and the signal to noise ratio were assumed to be known in the predictor design. This assumption does not appear to be critical as indicated in related work. The BER was chosen as a performance criteria. Focus was on the tracking phase, i.e. we did not take the first part of the block into account.

In a first set of Monte Carlo simulations we optimized the normalized power, ρ, of the superimposed pilot symbols, see Fig. 2. The signal to noise ratio per information bit was $E_b/N_0 = 15$ dB and the predictor order was chosen to be $L = 6$, which corresponds to $2^L = 64$ states. It appears that the BER is not sensitive with respect to ρ over a wide range. The optimum normalized power is about $\rho = 0.02 \ldots 0.05$ for all Doppler frequencies of interest: The power of pilot symbols should only be about 2 % ... 5 % of the power of the data symbols. In the following, $\rho = 0.05$ is applied. We observed further that a predictor order of about $L = 6$ is sufficient. Given these optimizations, we plotted the BER versus E_b/N_0 for the scheme under investigation, see Fig. 3. As a benchmark, the BER performance of 2-PSK on a flat Rayleigh fading channel given perfect channel estimation is plotted as well. For fading rates up to 10^{-2} the loss is less than 2 dB, and even for fast fading ($f_{D_{max}} T_s = 0.1$) no error floor is visible in the interesting range. This behavior is in contrast to conventional pilot-symbol-assisted CE, where we both have a bandwidth expansion and an error floor in fast fading channels.

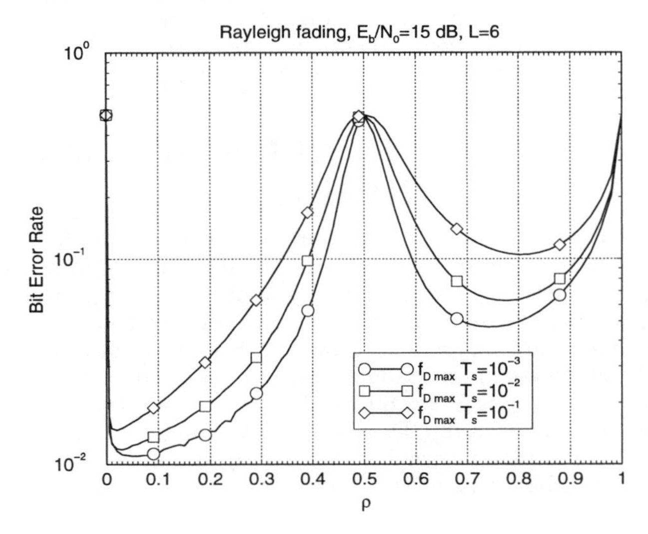

Figure 2: Optimization of the power, ρ, of the superimposed signal.

5 MULTI-CARRIER ISSUES

In multi-carrier systems we can either use single-carrier receivers for each of the sub-carriers, as discussed above, or we can improve the performance by exploiting the correlation between the time and frequency domains. The last aspect could be handled either by using separable receivers or by a truly 2-D receiver, respectively.

In the case of separable receivers, we may first apply a bank of parallel, independent 1-D receivers in one of the directions. These receivers have to be modified to deliver reliability information about the transmitted symbols. In the next stage, we may apply another bank of parallel, independent 1-D receivers in the other direction. These receivers have to be modified to accept a priori information given by the first processing stage. Note the similarities with block interleaving and iterative ("turbo") processing; both iterations are independent.

The truly 2-D receiver can be based on the known 2-D Viterbi algorithm [16], which can be extended to handle per-survivor processing as described above. In this case, one could optionally use 2-D pilot sequences to further improve the performance.

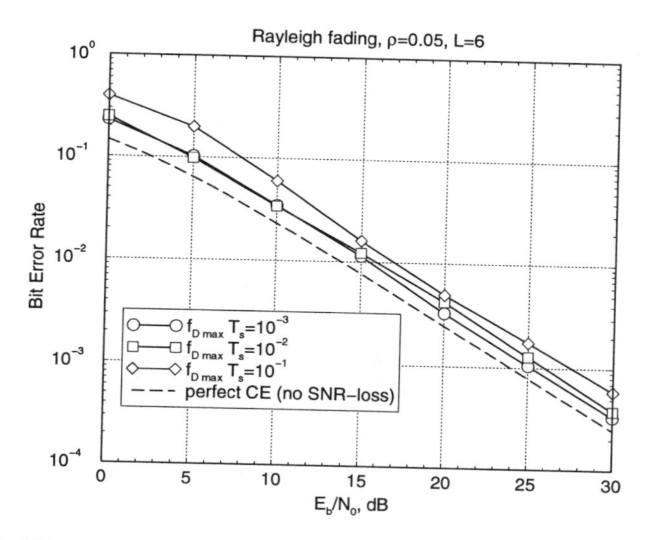

Figure 3: Bit error rate versus signal to noise ratio for channel estimation with superimposed PN pilot sequence and novel receiver.

6 CONCLUSIONS

In this paper, we explored the superimposed pilot sequence technique for the purpose of channel estimation. Due to the redundancy, truly coherent demodulation is achieved. The main contribution was the derivation of a receiver based on the Viterbi algorithm, which is suitable both for single-carrier and multi-carrier systems, and which is optimal in the sense of per-survivor processing.

Compared to pilot-symbol-assisted CE, which is currently state-of-the-art, distinct advantages are as follows:

- No bandwidth expansion
- Better power efficiency in fast fading environments
- The technique is more universally applicable. (The same sequence may be used for time, phase, frequency, and frame synchronization and as a unique word without additional overhead.)
- Data symbols are equally sensitive against transmission errors.

The robustness against fast fading is particularly important for multi-carrier systems, since the normalized Doppler frequency is typically higher due to long symbol durations. Furthermore, the transmission scheme is robust against channel variations, since no worst-case assumptions have to be made. However, our Viterbi receiver is, due to per-survivor processing, much more complex than a single interpolation filter. Therefore, simplified structures based on reduced-state sequence estimation were suggested as well. Future work may be devoted to low-cost receiver structures, optimizations of the pilot sequence, an evaluation of M-ary modulation schemes, receiver structures with oversampling, and a detailed performance analysis including the acquisition behavior.

References

[1] J.G. Proakis, *Digital Communications.* New York: McGraw-Hill, 3rd ed., 1995.

[2] H. Meyr, M. Moneclaey, and S.A. Fechtel, *Digital Communication Receivers.* New York: Wiley, 1998.

[3] M.L. Moher and J.H. Lodge, "A time diversity modulation strategy for the satellite-mobile channel," in *Proc. 13th Biennial Symp. Commun.*, Queen's Univ., Kingston, Canada, June 1986.

[4] M.L. Moher and J.H. Lodge, "TCMP: A modulation and coding strategy for Rician fading channels," *IEEE J. Select. Areas Commun.*, vol. 7, no. 9, pp. 1347-1355, Dec. 1989.

[5] A. Aghamohammadi, H. Meyr, and G. Ascheid, "A new method for phase synchronization and automatic gain control of linearly modulated signals on frequency-flat fading channels," *IEEE Trans. Commun.*, vol. 39, pp. 25-29, Jan. 1991.

[6] J.K. Cavers, "An analysis of pilot symbol assisted modulation for Rayleigh fading channels," in *IEEE Trans. Veh. Techn.*, vol. 40, pp. 686-693, Nov. 1991.

[7] D. Makrakis and K. Feher, "A novel pilot insertion-extraction method based on spread spectrum techniques," presented at *Miami Technicon*, Miami, 1987.

[8] T.P. Holden and K. Feher, "A spread spectrum based system technique for synchronization of digital mobile communication systems," IEEE Trans. Broadcasting, pp. 185-194, Sept. 1990.

[9] A. Steingaß, A.J. Wijngaarden, and W. Teich, "Frame synchronization using superimposed sequences," in *Proc. IEEE ISIT* '97, Ulm, Germany, p. 489, June-July 1997.

[10] F. Tufvesson, M. Faulkner, P. Hoeher, and O. Edfors, "OFDM time an frequency synchronization by spread spectrum pilot technique, in *Proc. Eighth Communication Theory Mini-Conference* in conjunction with *IEEE ICC* '99, Vancouver, Canada, pp. 115-119, June 1999.

[11] P. Hoeher and F. Tufvesson, "Channel estimation with superimposed pilot sequences," to appear in *Proc. Advanced Signal Processing for Communications Symposium* in conjunction with *IEEE GLOBECOM* '99, Rio de Janeiro, Brazil, Dec. 1999.

[12] J.H. Lodge and M.L. Moher, "Maximum likelihood sequence estimation of CPM signals transmitted over Rayleigh flat-fading channels," *IEEE Trans. Commun.*, vol. 38, pp. 787-794, June 1990.

[13] R. Raheli, A. Polydoros, and C.-K. Tzou, "Per-survivor processing: A general approach to MLSE in uncertain environments," *IEEE Trans. Commun.*, vol. 43, no. 2/3/4, pp. 354-364, Feb./Mar./Apr. 1995.

[14] M.V. Eyuboglu and S.U. Qureshi, "Reduced-state sequence estimation with set partitioning and decision feedback," *IEEE Trans. Commun.*, vol. 36, pp. 13-20, Jan. 1988.

[15] A. Duel-Hallen and C. Heegard, "Delayed decision-feedback sequence estimation," *IEEE Trans. Commun.*, vol. 37, pp. 428-436, May 1989.

[16] K. Matheus, *Generalized Coherent Multicarrier Systems for Mobile Communications.* Ph.D. thesis, University of Bremen, Germany, Shaker Verlag, ISBN 3-8265-4382-3, 1998.

A COMPARISON OF CHANNEL ESTIMATION TECHNIQUES FOR OFDM–CDMA AND MC/JD–CDMA SYSTEMS

Bernd Steiner
T-Nova Deutsche Telekom Innovationsgesellschaft mbH
Technologiezentrum
D–64307 Darmstadt
SteinerB@telekom.de

Abstract In the present paper, the problem of joint uplink channel estimation for two special cases of quasi-synchronous MC–CDMA, namely OFDM–CDMA and MC/JD–CDMA, is dealt with. The differences and the common properties of channel estimation for the two concepts are described. In particular, it is shown that the channel estimator structures are similar for both concepts.

1. INTRODUCTION

The first papers investigating the combination of MC techniques with CDMA or SSMA, respectively, were published in 1993, cf. eg. [1]. In the following years, the different issues of MC–CDMA / MC–SSMA including performance aspects, different system concepts and the impact of real–world system impairments were dealt with in a multitude of publications. In this paper, MC–CDMA techniques for mobile radio communications are considered. In order to use the radio spectrum as efficiently as possible, a coherent (or, in particular, joint coherent) demodulation / data detection is mandatory. To facilitate this, however, an accurate channel estimation is necessary. Due to the time variance of the mobile radio channel, the channel estimation can be achieved by training information which is transmitted in addition to the data subcarriers.

In the paper, two MC–CDMA concepts using a slotted quasi–synchronous transmission will be reviewed, namely OFDM–CDMA and MC/JD–CDMA. In an OFDM–CDMA system [2, 3], the number of subcarriers per slot exceeds the total number of subcarriers of a spread data symbol. Similar to a conventional OFDM system, several user data symbols are transmitted in parallel in one OFDM–CDMA symbol. Hence, one slot is represented by a single OFDM symbol. For MC/JD–CDMA [4, 5], the number of subcarriers per slot equals the number of subcarriers per data symbol. Consequently, the different data symbols of a slot are transmitted consecutively rather than in parallel. Therefore, MC/JD–CDMA is closely related to conventional DS–CDMA concepts. A guard interval is favorable in OFDM–CDMA in order to avoid inter carrier interference (ICI) so that the joint data detection is performed in the frequency domain. Such a guard interval to avoid ICI is not feasible for MC/JD–CDMA due to efficiency consideration. Hence, the receiver of an MC/JD–CDMA system has to cope with intersymbol interference as well as multiple access interference. Consequently, a time domain data detection as proposed in [4, 5] is more appropriate in MC/JD–CDMA.

The present paper is organised as follows: In Sect. 2., a system model used to derive the two specific MC–CDMA concepts considered in this paper is described. The following Sect. 3. describes the different channel estimation approaches which can be used for the proposed system concepts. In Sect. 4., it is shown that the simplified channel estimation approach proposed for UTRA–TDD is also applicable for the two concepts considered in this paper. Finally, a conclusion of the paper is given in Sect. 5.

2. SYSTEM MODEL

In this paper, the uplink transmission in slotted, quasi–synchronous MC–CDMA systems similar to UTRA–TDD/TD–CDMA is considered. Here quasi–synchronous means that the delays associated with the reception of bursts originating from different users are small compared to the burst duration T_{bu}. In the following, the number of simultaneously active users is termed K and the number of data symbols per burst is termed N_{d}.

The basic transmitter structure of the MC–CDMA concepts considered in this paper can be described by the block structure shown in Fig. 1. The structure according to Fig. 1 comprises a serial to parallel

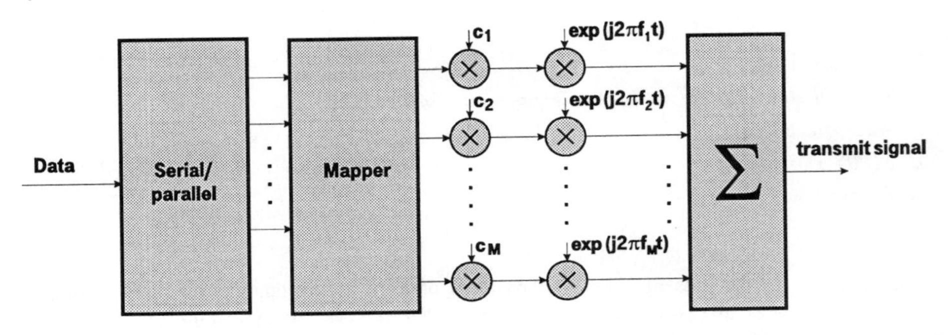

Figure 1 Basic transmitter structure of the considered MC–CDMA concepts

converter, a mapper, a bank of M weighting factor multipliers and a bank of M modulators[1] to form the different subcarriers of the transmit signal. Finally, the different contributions of the subcarriers are summed up.

In the following, the signal synthesis according to Fig. 1 will be described in detail. Here, the equivalent lowpass domain is considered. For convenience, we assume that $B \cdot C = N_{\mathrm{d}}$, with B and C being integer. C is the number of data symbols per output block and B is the number of output blocks of the converter. The N_{d} data symbols $d_n^{(k)}$, $n = 1 \ldots N_{\mathrm{d}}$, of user k are used to form the B data vectors $\boldsymbol{d}^{(k,b)}$, $b = 1 \ldots B$, with $d_i^{(k,b)} = d_{i+(b-1)C}^{(k)}$. The mapping can be described with the user–specific $QC \times C$ matrix $\boldsymbol{H}^{(k)}$. (It is, however, desirable to use the same mapping rule for the K simultaneously active users. In this case, $\boldsymbol{H}^{(k)}$ is the same for all K.) The subsequent scrambling operation can be described by multiplying $\boldsymbol{H}^{(k)}\boldsymbol{d}^{(k,b)}$ by the diagonal matrix $\boldsymbol{\Lambda}^{(k)}$ where $\lambda_i^{(k)} = c_i^{(k)}$ is the ith diagonal element of $\boldsymbol{\Lambda}^{(k)}$. For the weighting factors, $|c_m^{(k)}| = 1$ can be assumed. With the vector

$$\boldsymbol{x}^{(k,b)} = \left(x_1^{(k,b)}, x_2^{(k,b)} \ldots x_M^{(k,b)}\right)^{\mathrm{T}} = \boldsymbol{\Lambda}^{(k)}\boldsymbol{H}^{(k)}\boldsymbol{d}^{(k,b)} \tag{1}$$

the transmit signal is given by

$$s^{(k)}(t) = \sum_{m=1}^{M} \exp(\mathrm{j}2\pi f_m t) \sum_{b=1}^{B} x_m^{(k,b)} \mathrm{rect}\left(\frac{t\,B}{T_{\mathrm{bu}}} + \frac{1}{2} - b\right). \tag{2}$$

[1]The bank of modulators may be implemented as an FFT operation in a real–world system for performance considerations.

As already mentioned, OFDM–CDMA and MC/JD–CDMA are two special cases of MC–CDMA which can be described by the structure shown in Fig. 1. In the following two subsections, both concepts will be described briefly. Note that only the data transmission aspects are dealt with in the present section.

2.1 OFDM–CDMA

In an OFDM–CDMA concept as considered in this paper, all N_d data symbols of a burst are fed into the S/P converter. Therefore, $B = 1$ and $C = N_\mathrm{d}$ holds. With the mapper, the spectral spreading as well as the frequency interleaving are performed. Assuming a spreading gain of Q, the output of the mapper will be $N_\mathrm{d}Q$ values. Consequently, the transmit signal will consist of $M = N_\mathrm{d}Q$ subcarriers. For convenience a mapping example for $Q = 3$ and $N_\mathrm{d} = 4$ is depicted in Fig. 2, see [3].

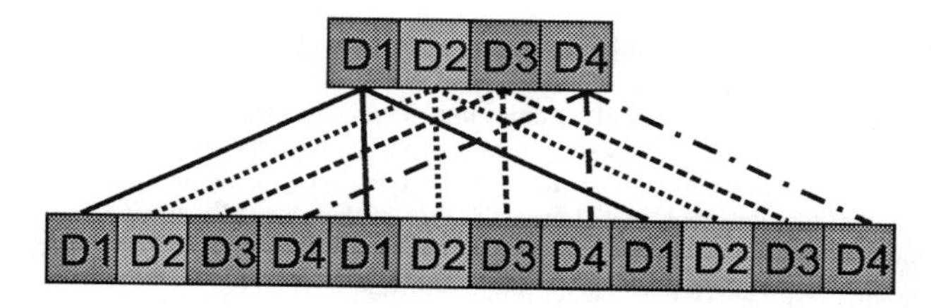

Figure 2 Mapping example for $Q = 3$ and $N_\mathrm{d} = 4$

The transmit signal in an OFDM–CDMA system is therefore given by

$$s^{(k)}(t) = \sum_{m=1}^{N_\mathrm{d}Q} \exp(\mathrm{j}2\pi f_m t) x_m^{(k)} \mathrm{rect}\left(\frac{t}{T_\mathrm{bu}} - \frac{1}{2}\right). \tag{3}$$

The guard interval applied in an OFDM–CDMA system avoids inter carrier interference. With T_g being the duration of the guard interval $\Delta f = f_m - f_{m-1} = \frac{1}{T_\mathrm{bu} - T_\mathrm{g}}$ must hold to avoid inter carrier interference (ICI). In this case only multiple access interference (MAI) occurs in the received signal. Consequently, the joint data detection can be performed in the frequency domain.

2.2 MC/JD–CDMA

In an MC/JD–CDMA concept[2], an S/P conversion is not used, i.e. each segment of the transmit signal only contains contributions from one data symbol [4, 5]. Consequently, the mapper structure is quite simple. Here, the mapper copies the only input value to all Q output branches. With the spreading gain Q being the number of subcarriers, $M = Q$ holds. The transmission in MC/JD–CDMA is similar to the transmission in a conventional TD–CDMA system. In the case of MC/JD–CDMA, the transmit signal is simply given by

$$s^{(k)}(t) = \sum_{m=1}^{Q} c_q^{(k)} \exp(\mathrm{j}2\pi f_m t) \sum_{b=1}^{N_\mathrm{d}} d_b^{(k)} \mathrm{rect}\left(\frac{t\, N_\mathrm{d}}{T_\mathrm{bu}} + \frac{1}{2} - b\right). \tag{4}$$

Due to the limited duration of a data symbol, the introduction of guard intervals to avoid ICI is possible but not reasonable in MC/JD–CDMA. As proposed in [4, 5] it is, however, reasonable to choose the carrier spacing according to $\Delta f = f_m - f_{m-1} = \frac{N_\mathrm{d}}{T_\mathrm{bu}}$. With this assumption,

$$s^{(k)}(t) = \sum_{b=1}^{N_\mathrm{d}} d_b^{(k)} c^{(k)} \left(t + \frac{T_\mathrm{bu}}{N_\mathrm{d}} \left(\frac{1}{2} - b\right)\right) \tag{5}$$

[2]Joint detection (JD) may be applied in all MC–CDMA concepts. In this paper, however, MC/JD–CDMA only refers to the case $B = N_\mathrm{d}$, $C = 1$.

holds for the transmit signal, where $c^{(k)}(t)$ is a user–specific spreading signal. Because of the lack of a guard interval, ICI as well as MAI are present in the received signal in the case of MC/JD–CDMA.

3. CHANNEL ESTIMATION APPROACHES

3.1 PREREQUISITES

At the receiver of an MC–CDMA system, the received signal

$$r(t) = \sum_{k=1}^{K} s^{(k)}(\tau) * h^{(k)}(\tau;t) + n(t) = \sum_{k=1}^{K} r^{(k)}(t) + n(t) \tag{6}$$

contains contributions of all K active users. In eq. (6), $h^{(k)}(\tau;t)$, $k = 1 \ldots K$, are the time–variant impulse responses which characterise the transmission of $s^{(k)}(t)$ to the base station and $n(t)$ is the receiver noise. Without loss of generality, we will assume time–invariant channel impulse responses $h^{(k)}(\tau)$ in what follows. A channel estimation is necessary for the proposed concepts since blind detection techniques are not suited for MC–CDMA because of the time variant mobile radio channel. Depending on the system concept as well as the data detection techniques, the aim of the channel estimation may differ. Typical channel estimation aims are

- a discrete–time estimate of the channel impulse response $h^{(k)}(\tau)$
- a discrete–time estimate of the combined impulse response $b^{(k)}(\tau)$ [3]
- a discrete–frequency estimate of the channel transfer function $H^{(k)}(f)$.

A means to facilitate channel estimation is the transmission of known training information with the data symbols. From a generic point of view the transmit signal of a specific user is given by $s^{(k)}(t) = s_{\mathrm{d}}^{(k)}(t) + s_{\mathrm{t}}^{(k)}(t)$, where $s_{\mathrm{d}}^{(k)}(t)$ is a data signal while $s_{\mathrm{t}}^{(k)}(t)$ is associated with the training signal. In particular, it is desirable to request the signals $s_{\mathrm{d}}^{(k)}(t)$ and $s_{\mathrm{t}}^{(k)}(t)$ to fulfil

$$\left| \int s_{\mathrm{d}}^{(k)}(t) s_{\mathrm{t}}^{(k)}(t) \mathrm{d}t \right|^2 \ll \int |s_{\mathrm{d}}^{(k)}(t)|^2 \mathrm{d}t \int |s_{\mathrm{t}}^{(k)}(t)|^2 \mathrm{d}t, \tag{7}$$

i.e. the training information should not interfere with the data signal. To incorporate training information into the transmit signal, some changes to the block structure, see Fig. 1, will become necessary. Two possible methods to incorporate training information will be considered:

- The usage of *pilot tones*, i.e. within a burst certain subcarriers are transmitted with a known weighting factor for all B output blocks of the S/P converter.
- The usage of *pilot symbols*, i.e. within a burst, a certain number of output blocks of the S/P converter (and, consequently a certain number of input symbols) are known by the receiver.

The first approach is well–suited for OFDM–CDMA since there is only one output block per burst. Therefore, the usage of training symbols would cause the whole OFDM–CDMA symbol to consist of training information. For MC/JD–CDMA, both approaches could be employed. Due to the relatively small number of subcarriers in MC/JD–CDMA and due to the similarity to TD–CDMA, it is more advantageous to use a training symbol based approach for MC/JD–CDMA. The more complex problem of channel estimation for a general MC–CDMA approach, i.e. neither $C = 1$ nor $B = 1$, will not be considered in this paper.

Fig. 3 depicts the two different approaches to incorporate training information into the transmit signal as described above for OFDM–CDMA and MC/JD–CDMA. Note, however, the different orientation of time and frequency axis used for Fig. 3. For OFDM–CDMA it has been assumed that the spacing of the pilot tones — determined by N_{ce} — is the same within the whole OFDM–CDMA symbol. The training signal can therefore be described by

$$s_{\mathrm{t}}^{(k)}(t) = \sum_{m=1}^{M/N_{\mathrm{ce}}} \exp(\mathrm{j}2\pi f_{(N_{\mathrm{ce}}(m-1)+\Delta_{\mathrm{ce}})} t) \left(t_m^{(k)} \right)_{\mathrm{OFDM}} \mathrm{rect}\left(\frac{t}{T_{\mathrm{bu}}} - \frac{1}{2} \right), \tag{8}$$

[3] A combined impulse response is the received signal which occurs in the case that a single user–specific spreading signal $c^{(k)}(t)$ is transmitted.

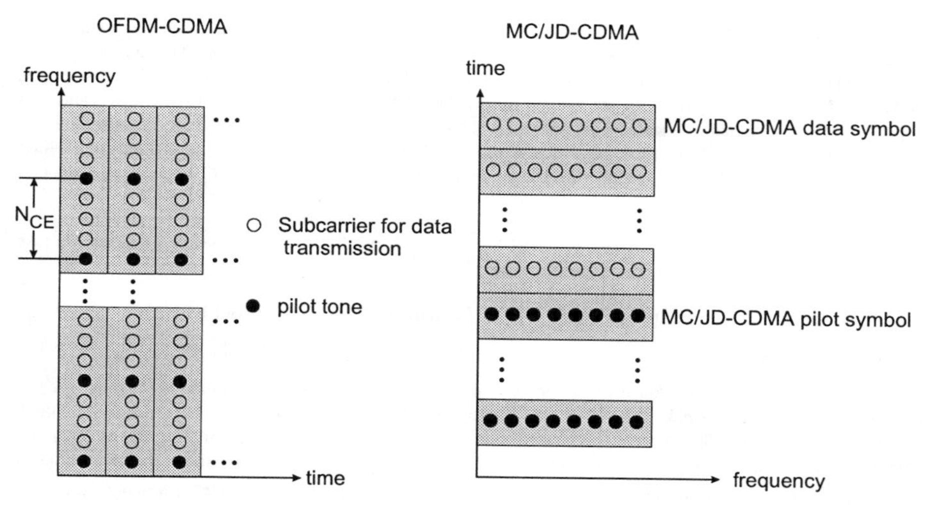

Figure 3 Application of pilot tones for channel estimation in the proposed MC–CDMA concepts

the spacing parameter Δ_{ce} may arbitrarily be chosen between 1 and N_{ce}. For MC/JD–CDMA, the training signal is

$$s_{\text{t}}^{(k)}(t) = \sum_{m=1}^{N_{\text{t}}} \left(t_m^{(k)}\right)_{\text{MC/JD}} c^{(k)}\left(t - \frac{T_{\text{bu}}}{N_{\text{d}}}\left(\frac{1}{2} - m - \Delta_{\text{MC/JD}}\right)\right). \tag{9}$$

In eq. (9) $\left(t_m^{(k)}\right)_{\text{MC/JD}}$ is the so–called training sequence while $\Delta_{\text{MC/JD}}$ is needed to describe the position of the training symbols within a burst.

3.2 ESTIMATOR STRUCTURES

In the following, we assume that a certain section of the received signal $r(t)$ which is solely determined by the training information can be separated. This is the case for OFDM–CDMA when the duration of the guard time is sufficient to avoid ICI. With the vector $\boldsymbol{r} = (r_1, r_2 \ldots r_M)$ containing M properly selected samples (the sampling frequency equals $M\Delta f$) of the received signal, $\boldsymbol{D}$ being the matrix of the discrete Fourier transform, and the diagonal matrix $\boldsymbol{S}$ having only nonzero entries $S_{i,i} = 1$ for those subcarriers which are pilot tones, the vector $\boldsymbol{r}_{\text{t}} = \boldsymbol{D}^{\text{T}}\boldsymbol{R}_{\text{t}} = \boldsymbol{D}^{\text{T}}\boldsymbol{S}\boldsymbol{D}\boldsymbol{r}$ only contains contributions of the pilot tones, furthermore, $\boldsymbol{S}\boldsymbol{D}\boldsymbol{r}$ is the equivalent of $\boldsymbol{r}_{\text{t}}$ in the discrete–frequency domain. In the case of OFDM–CDMA, $\boldsymbol{r}_{\text{t}}$ has the same number of elements as the vector $\boldsymbol{r}$.

In an MC/JD–CDMA system, a certain number of training symbols are transmitted as a midamble or preamble, just as in GSM or TD–CDMA. Therefore the separation of the signal section only containing training information is performed in the time domain, $\boldsymbol{r}_{\text{t}}$ therefore has less elements than $\boldsymbol{r}$. Note, that in the noise–free case, $\boldsymbol{r}_{\text{t}} = \sum \boldsymbol{r}_{\text{t}}^{(k)}$ holds for both OFDM–CDMA and MC/JD–CDMA.

In the following, we will only consider multi–channel estimation. This means that the channel estimator will explicitly take into account that different channels determine $\boldsymbol{r}_{\text{t}}$ and remove cross interference which occurs in the case where a bank of single user channel estimators are applied instead. The structure of the estimator, however, depends on the design of the training information as well as the desired result of the channel estimation.

For OFDM–CDMA, the aim of the channel estimation is obviously a discrete–frequency estimate of the channel transfer function for the subcarriers which are used for data transmission. The number of pilot tones is clearly given by $M/N_{\rm ce}$ while the total number of samples of the desired transfer functions is KM. ($KM(1 - 1/N_{\rm ce})$ of them are needed by the data detector.) Since the number of observations is considerably smaller than the number of unknown values, a reasonable restriction has to be introduced in order to make a channel estimation possible. A reasonable restriction is the assumption that the channel impulse responses $h^{(k)}(\tau)$, associated with the transmission have a finite duration $T_{\rm h}$ and can be modelled by

$$h^{(k)}(\tau) = \sum_{w=1}^{W} h_w^{(k)} \delta\left(\tau - \frac{wM}{\Delta f}\right). \tag{10}$$

Eq. (10) is a reasonable assumption because $T_{\rm h} < T_{\rm g}$ must hold anyway to avoid ICI. With eq. (10), the discrete–time channel impulse response can be modelled by the vector $\boldsymbol{h}^{(k)} = \left(h_1^{(k)}, h_2^{(k)} \dots h_W^{(k)}\right)^{\rm T}$. With the channel impulse response model according to eq. (10), the number of unknown channel parameters has been reduced to KW. $KW < M/N_{\rm ce}$ can be fulfilled by selecting W accordingly. A block structure of an OFDM–CDMA joint channel estimator is shown in Fig. 4. After computing $\boldsymbol{R}_{\rm t}$ as described above, the effect of the pilot tones' weighting factors is removed by multiplying the respective value of $\boldsymbol{R}_{\rm t}$ by $c_m^{(k)*}$. This operation yields K raw estimates of $M/N_{\rm ce}$ samples of the channel transfer functions, which are, however, disturbed by MAI. To remove the MAI, the K raw estimates are transformed into the discrete–time domain. Here, an MAI removal is performed which takes place by performing a matrix vector multiplication. Note, that only a section of W values of the result of the IDFT is needed since (10) is taken into account. After removing the MAI, the discrete–time channel impulse responses are transformed back into the discrete–frequency domain after padding $M - W$ zeros in order to get a vector with M elements.

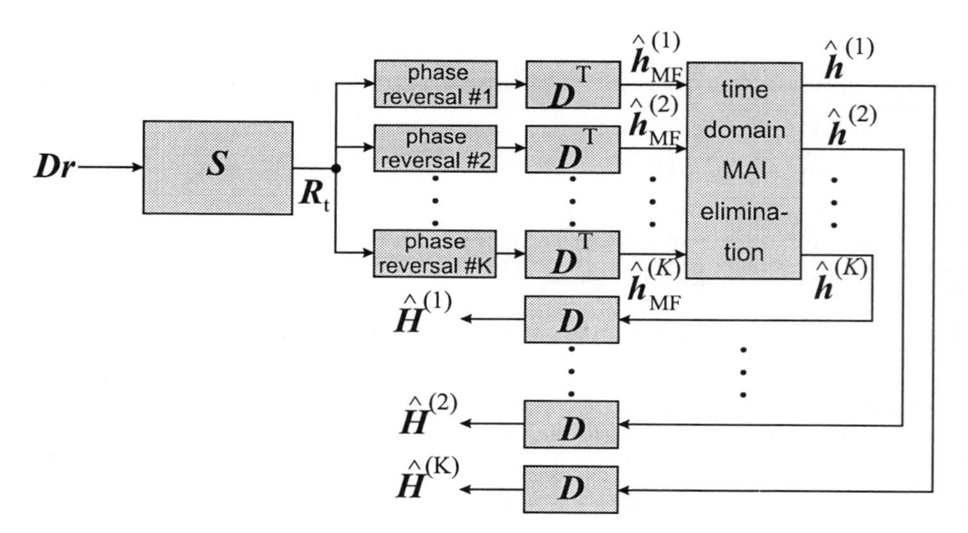

Figure 4 Block structure of an OFDM–CDMA channel estimator

For MC/JD–CDMA, the channel estimation approach is somewhat different. Here, the data detection shall be performed in the time domain rather than in the frequency domain. For MC/JD–CDMA as well as for conventional TD–CDMA, the discrete–time combined impulse responses are an appropriate input for the data detector. As shown in [7], there are two methods to obtain a combined impulse response:

- Direct estimation of a combined impulse response, i.e. the output of the estimator is a discrete–time estimate of the combined impulse response.

- Indirect estimation of the combined impulse response. Here, the result of the estimation is a discrete–time estimate channel impulse response which is convolved with the respective user–specific spreading signal in a consecutive step.

In the case of MC/JD–CDMA where training symbols are used for the channel estimation, the first approach mentioned in the list above is more appropriate. Similar to the issue of channel estimation in OFDM–CDMA, the channel impulse responses are assumed to have a finite duration T_{h}. Due to the finite duration of a user–specific spreading signal $c^{(k)}(t)$ a combined impulse response $b(t)^{(k)}$ also has a finite duration T_{b}. Wit the discrete–time combined impulse responses $\boldsymbol{b}^{(k)} = \left(b_1^{(k)}, b_2^{(k)} \ldots b_{Q+W-1}^{(k)}\right)^{\mathrm{T}}$ is introduced, the aim of the channel estimation is therefore the determination of the vectors $\boldsymbol{b}^{(k)}$, $k = 1 \ldots K$, from $\boldsymbol{r}_{\mathrm{t}}$. To make an unbiased channel estimation possible, the number of elements of $\boldsymbol{r}_{\mathrm{t}}$ must exceed $K(Q+W-1)$. The block structure for an MC/JD–CDMA channel estimator is shown in Fig. 5. Here, the elements of $\boldsymbol{r}_{\mathrm{t}}$ are fed into a bank of filters matched to the individual training symbol patterns, cf. [7]. When estimating the combined impulse responses with a training signal only consisting of training symbols, the knowledge of the user–specific spreading signals is neither needed for the channel estimation nor for the data detection. With respect to the channel estimation, only the training sequences are needed [7]. It can furthermore be shown that the matrix which is needed for the ISI and MAI removal is a sparse matrix. This reduces the computational effort necessary to perform the ISI and MAI removal.

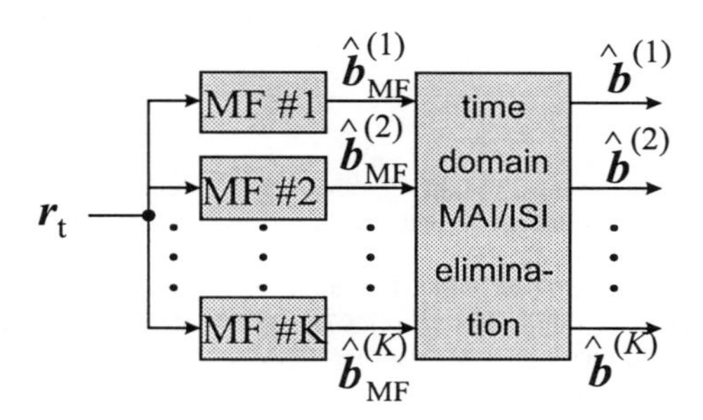

Figure 5 Block structure of an MC/JD–CDMA channel estimator

4. SIMPLIFIED CHANNEL ESTIMATION APPROACH

In the previous section, it has been shown that the channel estimator structures for OFDM–CDMA and MC/JD–CDMA are similar. In particular, both approaches use a MAI or MAI/ISI removal unit which is associated with a matrix vector multiplication and which is typical for crosstalk removal units in MIMO (multiple input / multiple output) systems. This approach has two major drawbacks:

- The SNR (signal to noise ratio) loss associated with the MAI/ISI removal depends on the correlation properties of the training signals and is the price to be paid for the MAI/ISI removal. In order to achieve an acceptable system performance, optimised training signal families have to be determined.
- The estimation process itself necessitates the storage of large matrices and matrix vector multiplications.

To ease the problems which occur in this context, the simplified channel estimation approach proposed by the author for TD–CDMA / UTRA–TDD, [8], may also be used for the considered MC–CDMA concepts. The basic idea described in [8] is to define a "substitute channel" which is the concatenation of the different channel impulse responses or combined impulse responses, respectively. In this case, the problem of multi–channel estimation is reduced to a a single channel estimation. To achieve this complexity reduction, the training signals of the different users are derived as different sections of a single periodic basic sequence.

A prerequisite for this is the quasi–synchronous transmission. When the simplified approach mentioned above is applied to OFDM–CDMA or MC/JD–CDMA, respectively, optimum training signals can be found in both cases. Furthermore, the MAI/ISI removal may be implemented as a correlation which is favorable with respect to implementing the channel estimator in a real–world system.

5. CONCLUSION

In the present paper, the issues of joint uplink channel estimation in quasi–synchronous MC–CDMA mobile radio systems, namely OFDM–CDMA and MC/JD–CDMA, are dealt with. After briefly introducing the two concepts it is shown that a pilot tone based channel estimation approach is appropriate for OFDM–CDMA while a test symbol based one is suited for MC/JD–CDMA. It is furthermore shown that the structure of the channel estimator is similar in OFDM–CDMA and MC/JD–CDMA even though the channel estimation aim is different in both cases. Finally, it is shown that the simplified channel estimation approach which was introduced by the author for UTRA–TDD / TD-CDMA is also applicable for the proposed concepts.

Acknowledgments

The author acknowledges the support of Dr. Dirk von Hugo for carefully proofreading the manuscript.

References

[1] Yee, N.; Linnartz, J.P.; Fettweis, G.: Multi–Carrier CDMA in Indoor Wireless Radio Networks. Proc. International Symposium on Indoor, Wireless and Mobile Radio Communications (PIMRC '93), pp. 109-113, Yokohama, Japan, 1999.

[2] Toskala, A. et al: Cellular OFDM/CDMA Downlink Performance in the Link and System Levels. Proc. Vehicular Technology Conference (VTC '97), pp. 855–859, Phoenix, Arizona, 1997.

[3] Steiner, B.: Uplink Performance of a Multicarrier–CDMA Mobile Radio System Concept. Proc. Vehicular Technology Conference (VTC '97), pp. 1902-1906, Phoenix, Arizona, 1997.

[4] Jung, P.; Berens, F.; Plechinger, J.: Joint Detection for Multicarrier CDMA Mobile Radio Systems — Part I: System Model. Proc. IEEE International Symposium on Spread Spectrum Techniques and Applications (ISSSTA '96), pp. 991-995, Mainz, Germany, 1996.

[5] Jung, P.; Berens, F.; Plechinger, J.: Joint Detection for Multicarrier CDMA Mobile Radio Systems — Part II: Detection Techniques. Proc. IEEE International Symposium on Spread Spectrum Techniques and Applications (ISSSTA '96), pp. 996-1000, Mainz, Germany, 1996.

[6] Steiner, B.: Time Domain Uplink Channel Estimation in Multicarrier–CDMA Mobile Radio System Concepts. Proc. International Symposium on Multi–Carrier Spread–Spectrum, pp. 153–160, Oberpfaffenhofen, Germany, 1997.

[7] Steiner, B.; Valentin, R.: A Comparison of Uplink Channel Estimation Techniques for MC/JD–CDMA Transmission Systems. Proc. IEEE International Symposium on Spread Spectrum Techniques and Applications (ISSSTA 98), pp. 640–646, Sun City, South Africa, 1998.

[8] Steiner, B.; Jung, P.: Optimum and suboptimum channel estimation for the uplink of CDMA mobile radio systems with joint detection. European Transactions on Telecommunications and Related Technologies, vol. 5, pp. 39–50, 1994.

Burst Synchronization for OFDM Systems with Turbo Codes

F. Said, B.Y. Prasetyo, A.H. Aghvami
Centre for Telecommuncations Research, King's College London, Strand, London WC2R 2LS, United Kingdom

Key words: synchronization, OFDM, turbo codes

Abstract: The use of a simplified frame structure has been proposed for fast burst synchronization in Orthogonal Frequency Division Multiplexing (OFDM) systems. A shorter synchronization symbol and a Minimum Mean Square Error (MMSE) algorithm is exploited to estimate the frame starting position. Turbo codes are incorporated in the system to enhance the performance at very low symbol error rates (SER). The performance of the simplified frame is confirmed, by means of simulation, for coherent QPSK systems.

1. INTRODUCTION

OFDM is a parallel data transmission method over a number of parallel sub-channels using frequency division multiplexing [1]. Several OFDM symbol timing estimation techniques have been compared and it is concluded [2] that MMSE yields the most suitable algorithm for burst operations in the multipath channel without the need of a SNR estimate prior to synchronization. The common synchronization methods are based on the inclusion of repetitive synchronization symbols at the beginning of each burst-frame. A shorter synchronization format has been proposed in [3] with higher frequency-offset estimation accuracy albeit with lower detection range. In this paper we use a simplified synchronization format of one symbol only, generated by repeating a selected portion of the first data symbol. Benefits of the proposed structure include significant reduction in frame starting position ambiguity and an estimation accuracy of the introduced carrier frequency offset, equivalent to that in [3]. A further enhancement in terms of lowering SER can be achieved by incorporating channel-coding techniques such as convolutional codes or turbo codes, which have recently gained widespread popularity.

Turbo codes were presented to the coding community in 1993 [4], as a way of dramatically reducing errors in a forward error correction scheme. The novel scheme presented then combined the concepts of parallel concatenation of recursive systematic convolutional codes, non-uniform random interleaving, and iterative soft-in/soft-out decoding. Combining all of these concepts results in an astonishing performance of achieving a bit error rate (BER) of 10^{-5} at an Eb/No of just 0.7 dB when a large interleaver of 65,532 bits is used and a code rate of 1/2. This powerful performance motivated researchers around the world to include turbo codes as an error correcting code in many applications, especially data applications where very low BERs are required and there is no constraint on the delay.

In this paper we investigate one possible technique and some necessary adjustments, in order to incorporate turbo codes into the OFDM system and, by means of simulation, we evaluate the performance enhancement achieved.

2. SYSTEM MODEL

Figure 1 shows a block diagram of the OFDM model considered.

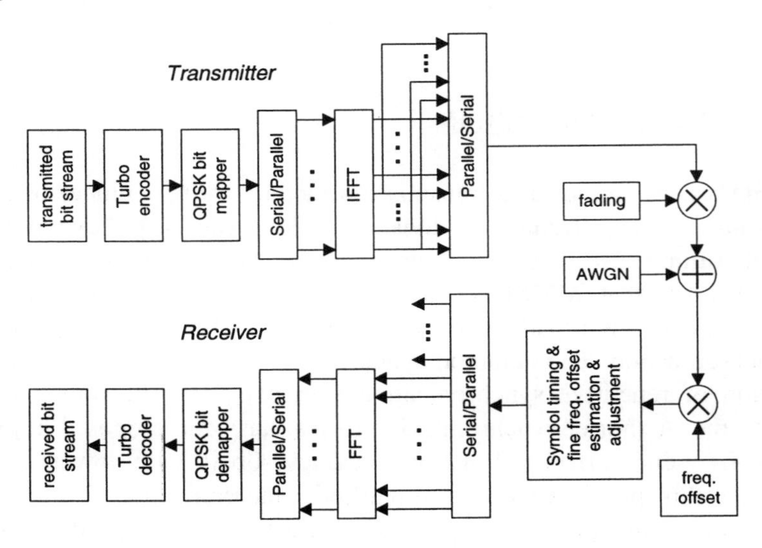

Figure 1. Block diagram of an OFDM system with turbo encoder-decoder

Randomly generated data bits are turbo-coded, yielding a higher number of bits to be transmitted. This is then mapped to obtain a sequence of QPSK symbols. After serial to parallel conversion, an IFFT converts a block of QPSK symbols into its time-domain representation. The transmitted signal is then subject to multipath channel interference, AWGN and carrier frequency offset.

3. TURBO CODING

Turbo codes represent the most important breakthrough in coding since Ungerboeck introduced trellis codes in 1982 [4]. Turbo codes offer near-capacity performance for deep space and satellite channels. A generalised turbo encoder consists of a parallel concatenation of n Recursive Systematic Convolutional Encoders (RSCEs) each separated by a random interleaver. The required rate can be achieved by puncturing the output signals.

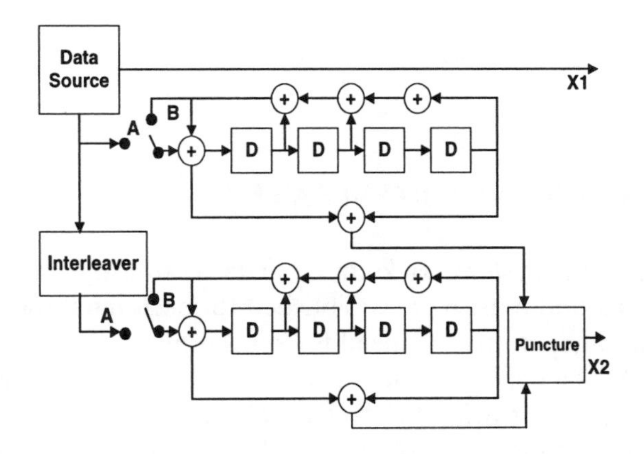

Figure 2. Rate 1/2 $(37,21)_8$ turbo encoder

The turbo encoder used here consists of two RSCE connected to each through a random interleaver. The first RSCE operates directly on the information bit sequence of length *K*. The second RSCE operates on a reordered sequence of information bits, produced by a random interleaver of length *N*. Figure 2 shows an example of the encoder with L=4 memory cells and generator polynomials of octal representation (21,37). Since the component encoders are recursive, it is not sufficient to set the last *L* bits to zero in order to drive the encoder to the all zero-state, i.e. terminate the trellis (turbo codes are block codes and trellis termination is essential at the end of the block). A simple solution to this problem is to move the switch from position A to B at the end of the block, and add a tail of *L* bits to the transmitted signal [5]. In the receiver decoding is performed using a turbo decoder which consists of *n* component decoders one for each encoder. Each component decoder uses the MAP algorithm to produce a soft decision for each bit. After an iteration of the decoding process, every component decoder shares its soft decision output with the other *n-1* component decoders [5]. The block diagram of the turbo decoder whose encoder is given in figure 2 is shown in figure 3.

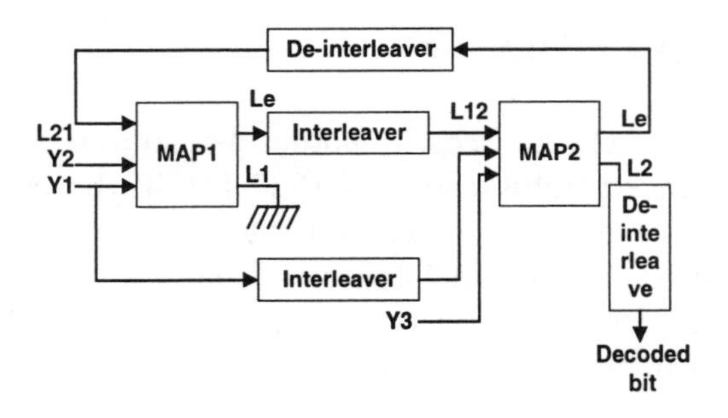

Figure 3. Block diagram of iterative turbo decoder

4. BURST SYNCHRONIZATION

In a wireless LAN system, packetized bursts are transmitted unscheduled, hence synchronization must be re-established for each burst transmitted. To optimize the throughput, the overhead for synchronization purposes must be fully minimised. The conventional and simplified burst formats are given in figure 4 (shaded portions being the cyclic prefix).

As the synchronization symbol, the simplified scheme uses only a portion of the first OFDM data symbol in each frame, excluding the guard interval and the part from which the guard interval is copied. If the lth sample of the kth transmitted OFDM data symbol is denoted as $x_{l,k}$, then the synchronization sample set, s_m, is defined as (1) with N and p indicating the number of sub-carrier and cyclic prefix samples respectively.

$$s_m \underline{\underline{\Delta}} x_{0,m+p}, m \in \{0,1,\ldots,N-2p-1\} \tag{1}$$

Therefore the model uses only one synchronization symbol whose length is the full duration of one OFDM symbol less the period of two guard intervals, i.e. $(N\text{-}p)T_s$ where T_s is the basic sampling period.

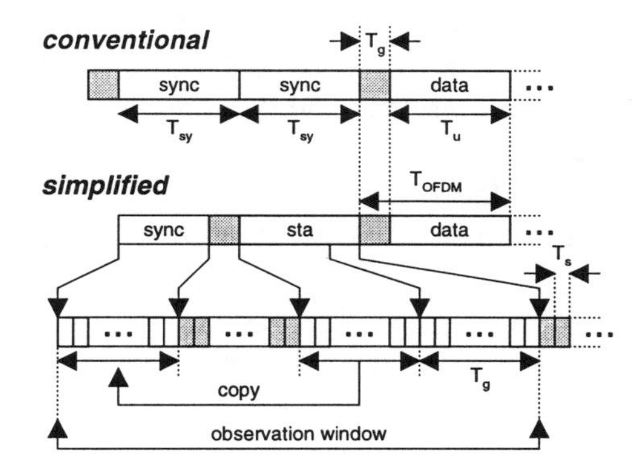

Figure 4. Conventional and simplified OFDM burst frames

We consider an OFDM N-subcarrier baseband transmitted in a dispersive channel with additive white Gaussian noise (AWGN). The sample average power of the transmitted signal above is given by $\sigma_s^2 = \sigma_x^2 = \frac{1}{2}E\{|x_n|^2\}$. Having passed through the channel, the transmitted signal is convolved with the channel impulse response resulting in the signal being received without noise (r_n). This signal then has Gaussian noise added, and hence the time-domain samples of the received OFDM signal sampled at nT_s are $r'_n = r_n + v_n$, where v_n denotes the equivalent lowpass complex white Gaussian noise with zero mean and variance $\sigma_v^2 = \frac{1}{2}E\{|v_n|^2\}$. Any mismatch in the carrier oscillator frequency between transmitter and receiver (f_ε) will cause a frame synchronization offset where the phase of the received sample will be shifted by $\exp\{(j2\pi\delta_f n)/N\}$, where $\delta_f = f_\varepsilon/\Delta f$ and Δf denote the normalised carrier frequency offset and the subcarrier spacing, respectively. These will be modelled in the simulation by modulating r'_n in the baseband with carrier frequency offset Δf.

In the conventional frame structure and using conventional windowing where the averaging filter sweeps the whole synchronization period (T_{sy}), the guard interval placed in front of the burst will cause an ambiguity as to where the actual frame position starts. This is eliminated by the simplified frame structure with appropriate windowing, since there is no guard interval component embedded in the synchronization symbol, and hence will give a sharp peak for maximum position detection.

As a convention let $\hat{n}$ and $r'_{\hat{n}}$, $r'_{\hat{n}+1}$, ... $r'_{\hat{n}+N-1}$ be the estimate of n and the first received N-sample set respectively such that $\hat{n} \in N$ and $r_{\hat{n}+N} = r_{\hat{n}} \exp(j2\pi\delta_f)$. The method of finding the frame starting position is by maximising the similarity probability ($\Pr\{\hat{n} = n\}$) of the sample sequences using the MMSE algorithm. Maximum similarity probability can be achieved when the first N received set correlates with its original copy embedded in the next N-sample set as given by the complex correlation $(S(\hat{n}))$. However this cannot be used as an accurate frame starting point detection method since an OFDM signal does not have a continuous envelope. Instead, by using a similar derivation to that in [2], we arrive at the metric ($M_{\hat{n}}$) to determine the frame starting position based on the MMSE criterion. This gives an estimate of the achieved optimum point using (4).

$$S(\hat{n}) = \sum_{d=0}^{N-1} r'_{d+\hat{n}} r'^{*}_{d+\hat{n}+N} \tag{2}$$

$$M_{\hat{n}} = \sum_{d=0}^{N-1} \left[\left\| r'_{d+\hat{n}} \right\|^2 + \left\| r'_{d+\hat{n}+N} \right\|^2 - 2 \left| r'_{d+\hat{n}} r'^{*}_{d+\hat{n}+N} \right| \right] \tag{3}$$

$$\hat{n} = \arg\{\min(M_{\hat{n}})\} \tag{4}$$

Referring to the simplified frame structure in figure 4, it is obvious that the maximum similarity of the received set is achieved when $n=0$, so that, once the maximum similarity is found, the starting position of useful data is N sample durations towards the time positive direction.

At the maximum value of similarity probability, the carrier frequency offset can be estimated by averaging the phase of the correlated received samples. If the average phase rotation is given by $2\pi\delta_f T_s$ and $|\arg S(\hat{n})|$ can be guaranteed to be less than π, the estimate for the normalised carrier frequency offset is given as,

$$\hat{\delta}_f = \frac{1}{2\pi}\{\arg[S(n)]\} \tag{5}$$

The length of the synchronization symbol determines the accuracy of the estimate since it corresponds to the number of averaging points. Compared with the simplified frame structure in [3], which boasts higher detection accuracy than that in [4], the simplified frame structure in this paper gives exactly the same high accuracy as in [3] since the total number of averaging points is still *N*.

5. SIMULATION RESULTS

Computer simulation for the OFDM system shown in figure 1 has been conducted. The data is error protected using a turbo encoder of constraint length 4 and rate 1/2. The encoder consists of two recursive systematic convolutional encoders of generator polynomials $(37,21)_8$ connected to each other in parallel using a random interleaver with length K=256. The resultant block size is 512 bits, which is modulated using QPSK. The data is transmitted in parallel using OFDM with 64 subcarriers. Each burst is assumed to consist of one synchronization symbol and 4 consecutive data symbols. The number of samples is 64 plus 8 additional time sampling durations ($8T_s$) as the guard interval. In the receiver a turbo decoder which uses the MAP algorithm in an iterative fashion is used with 10 decoding iterations. The dispersive channel consists of a discrete-time 8-tap delay model, which decays exponentially at -3dB/tap. The maximum Doppler spread (f_d) is chosen to be 5 Hz corresponding to a maximum walking speed of 3 km/hr at a carrier frequency of 1.8 GHz.

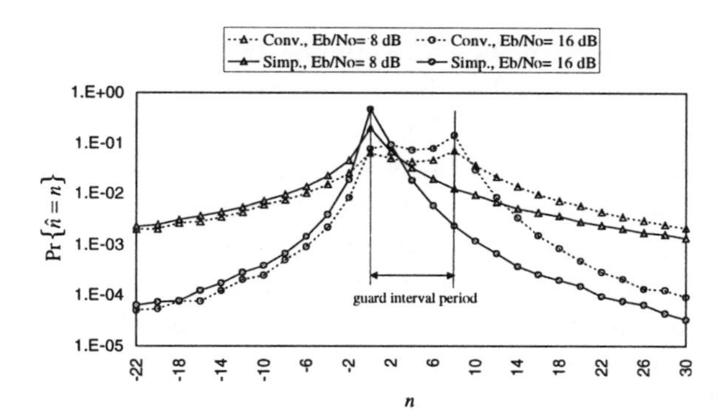

Figure 5. Comparison of simplified (solid) and conventional (dashed) frame structure lock-in probabilities in dispersive channel (fdT=10^{-3}) with $\delta_f = 0.1$ for two Eb/No values (dB)

The lock-in probability graphs in figure 5 show clearly that, due to exclusion of the guard interval in the synchronisation symbol, the simplified frame structure does not have the ambiguous, almost flat region exhibited on the curves for the conventional structure. Figure 6 shows the BER performance of the simplified frame structures owing to the application of turbo codes in several environments with and without frequency offset, for a range of naturally small Eb/No values.

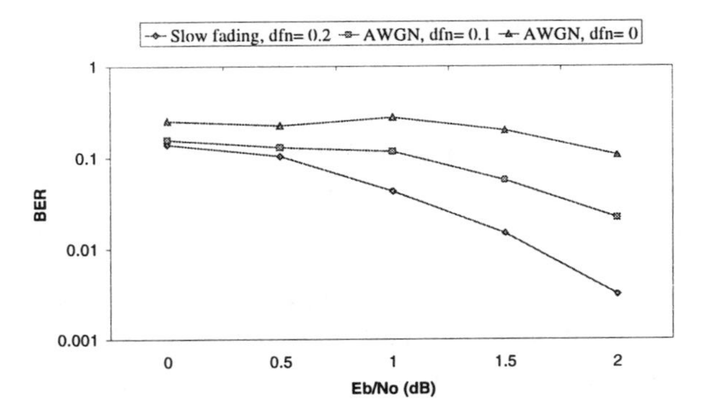

Figure 6. OFDM system performance with turbo codes in AWGN and slow fading environment, with normalized frequency offset (dfn)

6. CONCLUSION

In this work, the performance of a simpler frame format for burst Turbo-coded OFDM transmission has been presented for synchronisation in a multipath slow-fading channel. The use of the simplified frame structure results in a higher lock-in probability with the same carrier-frequency error estimation accuracy as compared to the conventional one. A significant improvement has been achieved by the inclusion of turbo codes in the OFDM system. The turbo encoder employed is a rate 1/2 encoder with RSCE having generator polynomials $(37,21)_8$, and a random interleaver of length K = 256 gave a coding gain of around 4.5 dB at a BER of 10^{-3} compared to the uncoded system. However this gain comes at the expense of some inherent tradeoffs such as the increased delay and complexity.

REFERENCES

[1] R.W. Chang, "Synthesis of Band-limited Orthogonal Signals for Multichannel Data Transmission", Bell Syst. Tech. J., 1966, 45, (12), pp.1775-1796

[2] S.H. Muller-Weinfurtner, "On The Optimality of Metrics for Coarse Frame Synchronization in OFDM: A Comparison", 9th PIMRC Proceedings, 1998, 2, (9), pp.533-537

[3] M. Mizoguchi, T. Onizawa, T. Kumagai, H. Takanashi, M. Morikura, "A Fast Burst Synchronization Scheme for OFDM", ICUPC '98 Proceedings, 1998, 2, (10), pp.125-129

[4] C. Berrou, A. Glavieux, P. Thitimajsshima, "Near Shannon Limit Error-correcting Coding and Decoding: Turbo Codes", IEEE Trans. Commun., Vol.44, No.10, p.1261, Oct.1996

[5] D. Divsalar and F. Pollara " Turbo codes for deep-space communications ", TDA Progress report 42-120.

Acquisition of synchronisation parameters for OFDM using a single training symbol

Dusan Matic*, Nicolas Petrochilos*, Ton A.J.R.M. Coenen*,
Frits Schoute*†, Ramjee Prasad△

* *Telecommunications and Traffic Control System Group*
Delft University of Technology, P.O. Box 5031, 2600 GA Delft, The Netherlands
Email: D.Matic@ITS.TUDelft.NL, Tel: +31-15-278.1782, Fax: +31-15-278-1774
† *Philips Bussines Communications, Hilversum, The Netherlands*
△ *Aalborg University, Aalborg, Denmark*

Abstract This paper presents a technique which acquires the synchronisation parameters for OFDM (Orthogonal Frequency Division Multiplex) and performs channel estimation using a single training symbol. For coherent, microwave, burst-oriented, mobile communication systems, a precise recovery of all synchronisation parameters is required.

The technique proposed uses a new way to estimate the integer part of the carrier frequency offset after the FFT, if the training symbol was beforehand corrected for the fractional part, and can give a good estimate of the symbol timing epoch via multiplication and FFT operation.

OFDM and different kind of synchronisation errors

Let the complex envelope of an OFDM signal after IDFT and up-conversion be written as in the Eq. (1),

$$s(t) = \sum_{k=-\infty}^{+\infty} s_k(t) = \sum_{k=-\infty}^{+\infty} \left[\sum_{i=-N_{FFT}/2}^{N_{FFT}/2-1} x_{i,k} e^{j\left[2\pi(t-kT)\frac{i}{T_{FFT}}+2\pi f_c t+\phi_c\right]} \right] \tag{1}$$

where t is the absolute time, T duration of one OFDM symbol, N_{FFT} is the number of carriers, $x_{i,k}$ the symbol that modulates the i-th carrier of the k-th symbol, T_{FFT} the duration of the useful part (without cyclic prefix) of the OFDM symbol and f_c the carrier frequency at the transmitter. The frequency spacing between sub-carriers is $\Delta f = 1/T_{FFT}$. After down-conversion, assuming correct sampling frequency, proper cyclic prefix removal and that channel does not change during the burst duration, the outputs of the DFT are expressed as in Eq. (2) [1].

$$\widehat{x_{i,k}} = H(i) \frac{\sin(\pi \delta f\, T_{FFT})}{N \sin\left(\frac{\pi \delta f\, T_{FFT}}{N}\right)} x_{i,k}\, e^{j\Psi_{i,k}} + I_{i,k}\ (+noise) \tag{2}$$

where $\delta f = f_c - f_c^{receiver}$ is the frequency offset, $H_{i,k}$ represents the influence of the multipath channel, $I_{i,k}$ describes the intercarrier interference as a function of δf and $\Psi_{i,k}$ the amount of rotation of $x_{i,k}$. $\Psi_{i,k}$ is given in Eq. (3), in which τ is the non-dimensional symbol-timing offset normalised to $\frac{T_{FFT}}{N_{FFT}}$ and ϕ_o is the carrier phase offset.

$$\Psi_{i,k} = 2\pi \left[i \frac{\tau}{N_{FFT}} + k \delta f\, T \right] + \left[2\pi \left(\tau - \frac{1}{2} \right) \frac{\delta f\, T_{FFT}}{N_{FFT}} + \phi_o \right] \tag{3}$$

Introduction

Moose [1] derived the maximum likelihood estimator (MLE) for the carrier frequency offset, which is calculated in the frequency domain, after the FFT. The method is based on the fact that the coressponding samples at the different times differ *only* by the rotating phasor of frequency offset and the influence of noise. Therefore, it was proposed to send two identical symbols and to estimate the frequency offset on the differences of phase rotation between them, since all the sub-carriers undergo same amount of rotation. An MLE function basically averages the offset differences over all carriers. The method needs two symbols and is limited by the acquisition range, which is $\pm\frac{1}{2}$ the sub-carrier spacing Δf, and by the prior knowledge of the symbol timing.

The precise start of a symbol can be determined in combination with the idea proposed by Van de Beek et al [3]. A blind estimation method for tracking time and frequency offsets using the cyclic prefix redundancy was described and an MLE function was derived for AWGN channel. The correlation function is based on the same phenomenon as [1], but is performed in the time-domain. It needs no extra overhead and is very simple for implementation. But, it is not suitable as a help for detection of the start of the burst, since it can be unreliable in case of low SNR or/and multipath, although the variance is small when MLE peaks are averaged over a number of symbols [4].

Schmidl and Cox [2] found an elegant way for both burst start detection and a complete carrier frequency offset estimation. Acquisition is achieved in two separate steps through the use of two OFDM training symbols. The first one is used to indicate the start of the burst and to correct the fractional (between $\pm\Delta f$) carrier frequency offset, while the second provides a way to correct for a carrier frequency offset of an integer number of sub-carrier spacings, by differential coding in respect to the first one. The first symbol is unique, since it consists of two identical halves in time (made by modulating every second sub-carrier) and a cyclic prefix.

The samples that are used for estimation are protected from intersymbol interference (ISI) by the cyclic prefix, which adversly also flattens the top of the correlation function. The width of the plateau is proportional to the length of the cyclic prefix, which leaves ambiguity about the exact timing of the OFDM symbol. The array of samples used for FFT must start within plateau, in order not to have a loss of SNR. The starting point of the array is chosen as a maximum or the middle of this plateau of timing estimation function. The fractional carrier frequency offset is then determined at the same point. The frequency acquisition method is robust and reliable, but needs two symbols and the symbol timing estimation is not accurate.

Proposed algorithm

In the proposed algorithm, the training OFDM symbol has every second carrier modulated with a PN sequence in order to get a symbol that has identical halves in time. These halves are correlated as in [2], in order to detect the beginning of a frame and to acquire fractional carrier frequency offset δf_{frac}, that is between ± 1 subcarrier distance. The performance of this algorithm is the same as the Cramer-Rao bound.

Integer frequency offset estimation It is also possible to estimate the remaining carrier offset δf_{int}, which is an integer number of subcarrier distances, using only the first training symbol.

After the timing point is chosen, within the plateau of cyclic prefix width (or less in case of multipath), and δf_{frac} estimated based on it, the N_{FFT} (number of FFT points) samples of the training symbol are frequency corrected by multiplying the with $\exp(-j2\pi\delta f_{frac}t/T_{FFT})$ and fed into the FFT processor. This results in the PN sequence centered on the sub-carriers, but on the wrong place (shifted by $\delta f_{int} = (2n-1)\Delta f$ from its ideal position) and under influence of multipath channel, timing offset, carrier phase offset, noise and remaining frequency-offset δf_{int}. Let this vector be $\widehat{PN_{\delta fint}}$.

Being in the frequency domain now, the next step is to find n. The method presented by Schmidl [2] is basically summing up the phase differences between subcarriers in two adjacent OFDM training symbols on different positions and finding the maximum. If differential rotations between adjacent sub-symbols are known, as is the case with training symbol, it is possible to do a similar thing within one symbol, using (4). The coefficients of the $\widehat{PN_{\delta fint}}$ (Fourier transform of the samples corrected for the fractional frequency offset) can be expressed as $[\widehat{x_1}, \widehat{x_2}, ..., \widehat{x_{N_{FFT}}}]$, where $\widehat{x}_i$ is given as in the Eq. (2) with k removed for convenience. The original PN sequence is $[P_1, P_2, P_3, ..., P_{N_{FFT}}]$, where $P_{2i} = 0$, and $A = \{-N_{FFT}/2, ..., -2, 0, 2, ..., N_{FFT}/2\}$.The n can be found using the following expression

$$n = \arg\min_{g\in A} f(g) = \arg\min_{g\in A}\left(\sum_{i=1}^{N_{FFT}/2-1} \left| \frac{\widehat{x_{2i-1+g}}}{\widehat{x_{2i+1+g}}} - \frac{P_{2i-1}}{P_{2i+1}} \right| \right) \tag{4}$$

The Eq. (4) can be rewritten as below, in order to avoid the usage of division.

$$n = \arg\min_{g\in A}\left(\sum_{i=1}^{N_{FFT}/2-1} \left| \frac{P_{2i+1}}{P_{2i-1}} \widehat{x_{2i-1+g}} \left(\widehat{x_{2i+1+g}} \right)^* - 1 \right| \right)$$

If δf is substituted with $(2n-1)\Delta f = (2n-1)/T_{FFT}$ in the Eqs. (2) and (3), they simplify significantly. The inter-carrier interference $I_{i,k}$ equals zero, because all sub-carriers have been previously centered. $|P_{2p+1}/P_{2p-1}|$ is always 1. The influence of the carrier frequency and phase offsets, as well as the timing offset get canceled for $g = n$. The remainder is

$$f(n) = \left(\sum_{i=1}^{N/2-1} \left| H_{2i-1+n} H^*_{2i+1+n} \, e^{2j\pi\tau/N_{FFT}} - 1 \right| \right) \tag{5}$$

The crucial assumption is that $H_m \approx H_{m+2}$, what can be done, since adjacent carriers have strongly correlated influence of the channel. Also, since the τ is known to be in the cyclic prefix and T_{FFT} is normalised to the sampling period, then $0 \leq \tau/N_{FFT} \leq T_{cp}/T_{FFT}$, which depends on a system, but it is usually less than 0.25.

Symbol timing estimation - Parallel search technique using the FFT as a noncoherent correlator The correlation of local and received PN sequence, in the time domain, results in the channel impulse response, which peak, in case of microwave communication conditions, indicates the LOS component position. For this purpose, it was chosen to use a maximum length (m) PN sequence. Since their lengths are $2^q - 1$, they have to be extended circularly for one member.

A straightforward calculation requires a large number of operations, which is proportional to N_{FFT}^2. Much processing time can be saved, however, if the Fourier link between convolution and correlation function is employed, by computation of correlation via the frequency-domain [5].

$$R(\bullet) = \sum_{k=0}^{N_{FFT}-1} s(k)p(k+m) = s(k) \star p(-k) = \mathcal{IFFT}\left[x(k)P^*(k)\right] \quad (6)$$

where $s(k)$ is the incoming signal and $p(k) = \mathcal{IFFT}(PN)$ is the training symbol in time, $P^*(\cdot)$ is the complex conjugate of the spectrum of $p(k+m)$, which is the time pair for the function $P(n)$ and $\star$ represents convolution. $p(-k)$ is $p(k)$ with flipped coefficients (as if the direction of axis was reversed). When the PN codes are synchronised, the bin whose range includes the value of the offset will contain the maximum of the correlation function.

The Eq. (6) can be performed on the corrected $\widehat{PN_{\delta fint}}$ and PN. Having the integer n, which describes δf_{int}, the complete carrier frequency offset is known, being $\delta f = \delta f_{int} + \delta f_{frac}$. The position of the PN within $\widehat{PN_{\delta fint}}$ is adjusted as

$$\widehat{PN_\tau} = \left[\widehat{PN_{\delta fint}}(n+1 : N\_FFT) \qquad \widehat{PN_{\delta fint}}(1:n)\right]$$

The Eq. (7) is the (auto)correlation of PN sequences, received and local, but only after correction of the frequency offset. Since multiplying the $\widehat{PN_\tau}$ with conjugated PN sequence has the effect of removing the data from the subcarriers, leaving the channel transfer function, the $TimingCorr$ is also the channel impulse response.

$$TimingCorr = \mathcal{IFFT}(\widehat{PN_\tau}\ (PN)^*) \tag{7}$$

If the start sample for the FFT was somewhere within the cyclic prefix (as estimated in the beginning), it still gives the correct result. The crucial property at this point is that an OFDM symbol, extended with a cyclic prefix, seems as a periodical signal to the FFT ("aliasing").

The *TimingCorr* is an array with one or usually two outstanding members. To determine the precise position of the maximum, which can be between discrete points, we employed a simple, two point inverse interpolator, which calculates the position of the extremum. The largest member index has to be detected and the one next to it in size (on the left or right side). The assumption is that all other members of the array, except these two, are zeros, what can be done in case of microwave channel and m sequences. The used linear inverse interpolator has the following formula

$$d = \frac{\rho_r - \rho_l}{2(\rho_r + \rho_l)} \tag{8}$$

where ρ_l and ρ_r are the indexes of the left and right largest sample, and d is the calculated position of the exact maximum $(-0.5 < d < 0.5)$. The symbol timing offset is estimated as $\widehat{\tau} = \rho_l + 0.5 + d$.

Channel estimation Channel estimation can be done by removing the data and offsets from the carriers. Each member of $(\widehat{PN_\tau}\ (PN)^*)$ has to be corrected with $\exp(-j2\pi i \frac{\widehat{\tau}}{N_{FFT}})$ for the estimated timing offset rotation, where i is the subcarrier index. The remaining errors are assumed as a part of this channel estimation and the following OFDM burst can be coherently demodulated, under assumption that the channel does not change during the burst.

Performance

The simulation results shown exploit an OFDM system with 128 carriers. The overall performance depends very much on the performance of the initial timing detection as defined by Schmidl. If that step is done succesfully, the rest of the system performs well. Otherwise, the estimate was considered as a failure. The Rayleigh channel has an exponential decaying response having 30 reflections. Maximum delay spread was equivalent to $10\tau_{RMS}$.

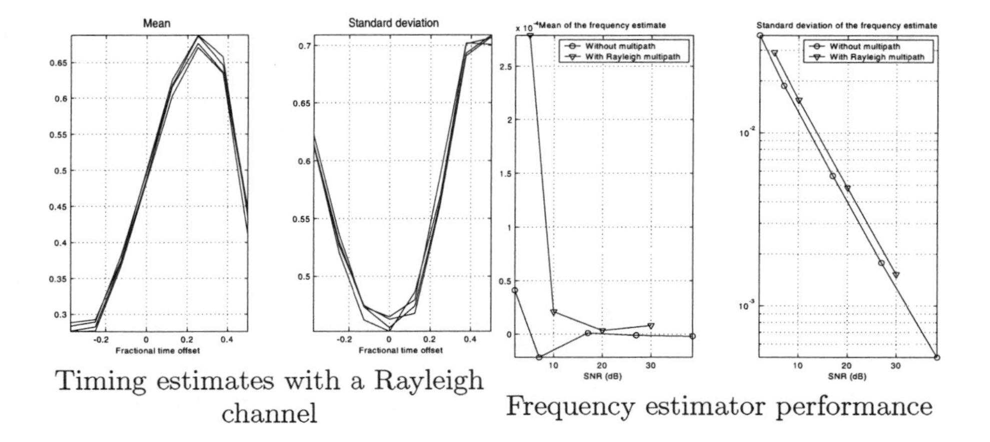

Timing estimates with a Rayleigh channel

Frequency estimator performance

Figure 1. Performance of the timing and frequency estimators

Overall frequency estimator is virtually insensitive to different frequency and timing offsets. In fact, the graphic (1) of the frequency estimation show the performance of the Schmidl frequency estimator, since if the integer frequency estimator performs well, the remaing error is due to the fractional frequency estimator.

The timing estimator performs very good in the AWGN channel, standard deviation is under 1% of sampling period for the SNR of 5dB. The shown performance of the estimator in Rayleigh channel stems from the fact that the interpolator used searches for the maximum, which is moved to the centre of gravity of the channel impulse response. This way of estimation can be accepted in the microwave communication channels, since these are usually Ricean channels and do not have reflections stronger than LOS component. The remaing error in estimation is incorporated and compensated for with a channel transfer function estimate. If a very accurate estimate is needed, other techniques can be used.

Conclusions

A single OFDM training symbol can enable acquisition of synchronisation. A formula was derived for estimating the integer part of the carrier frequncy offset. The overall estimator performance depends on the correct initial timing estimation of the training symbol. This inital estimation itself depends on the number of samples in the correlation.

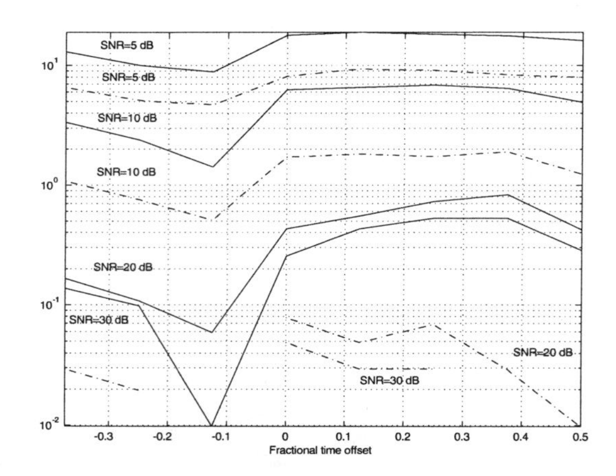

Figure 2 Percentage of overall failure and Schmidl estimator failure

The proposed system can also be used for the channel estimation and refining of the exact postion of the symbol beginning. The technique sustains the combined influence of all offsets, multipath channel and a low SNR.

References

[1] "A technique for orthogonal frequency division multiplexing frequency offset correction", Moose P.H., *IEEE Trans. Comm., vol. 42, no. 10, pp. 2908-2914, Oct 1994.*

[2] "Robust frequency and timing synchronization for OFDM", Schmidl T.M. and Cox C.C, *IEEE Trans. Comm., vol. 45, no. 12, pp. 1613-1621, Dec 1997.*

[3] "ML estimation of Time and Frequency offset in OFDM systems", Van de Beek J-J, Sandell M. and Börjesson P.O., *IEEE Trans.Sig.Proc., vol. 45, no. 7, pp. 1800-1805, July 1997.*

[4] "OFDM timing synchronisation: Possibilities and Limits to the usage of the Cyclic Prefix for Maximum Likelihood Estimation", Matić D., Coenen A.J.R.M. and Prasad R., *Vehicular Technology Conference (VTC '99), September 1999., Amsterdam*

[5] "New fast GPS code-acquisition technique using FFT", Van Nee D.J.R. and Coenen A.J.R.M., *Electronics letters, vol. 27, no. 2, pp. 158-159, Jan 1991.*

Section VII

REALIZATION AND IMPLEMENTATION

Implementation of Multicarrier Systems with Polyphase Filterbanks

Kirsten Matheus
Ericsson Eurolab Deutschland GmbH, Nordostpark 12, 90411 Nürnberg, Germany
kirsten.matheus@eedn.ericsson.se

Karl-Dirk Kammeyer
University of Bremen, FB-1, Department of Telecommunications, P.O. Box 33 04 40, 28334 Bremen, Germany

Ulrich Tuisel
E-Plus Mobilfunk GmbH, P.O. Box 30 03 07, 40403 Düsseldorf, Germany

Abstract This article will emphasize on how to efficiently implement multicarrier (MC-) systems different from and alternative to OFDM. It will show that when using polyphase filterbanks to do so this is a little complex and flexible solution. It will be derived that with polyphase filterbanks not only spacings between adjacent subcarriers in integer multiples of the inverse MC-symbol duration are possible but that also *non-integer* multiples of the inverse MC-symbol duration are easily realizable.

1. INTRODUCTION

With the growing market in mobile communications multicarrier schemes are receiving an increasing amount of attention; mainly because of their capability to handle frequency selective (i.e. mobile radio) channels better than ordinary single carrier systems can.

Figure 1 shows the structure of a MC-system. In the transmitter following a serial to parallel conversion the data symbols are upsampled; generally by the factor $w = N$, with N the number of subcarriers. Afterwards the data is filtered and modulated with the subcarrier's frequency. The values of all subcarriers are then summed up and transmitted as one symbol. The functionality of the receiver is inverse to the one of the transmitter. What distinguishes the different MC-systems is the filtering g_k and h_k.

The multicarrier system of main interest, OFDM, uses rectangular impulses for g_k and h_k. For its implementation the complex structures as shown in Figure 1 can be replaced by an IFFT-unit for the transmitter and an FFT-unit for the receiver. The subcarrier spacing for OFDM is $\Delta F = 1/T_s$ (with ΔF the subcarrier spacing and T_s the MC-symbol duration) unless it is decided to transmit on only e.g. half of the carriers. Nevertheless OFDM always requires that $\Delta F T_s \in \mathbb{N}$. To avoid intersymbol interference an OFDM system generally also comprises the insertion and removal of a guard interval, which increases the OFDM symbol duration to $T_{OFDM} = T_s + T_g$. As no new information is

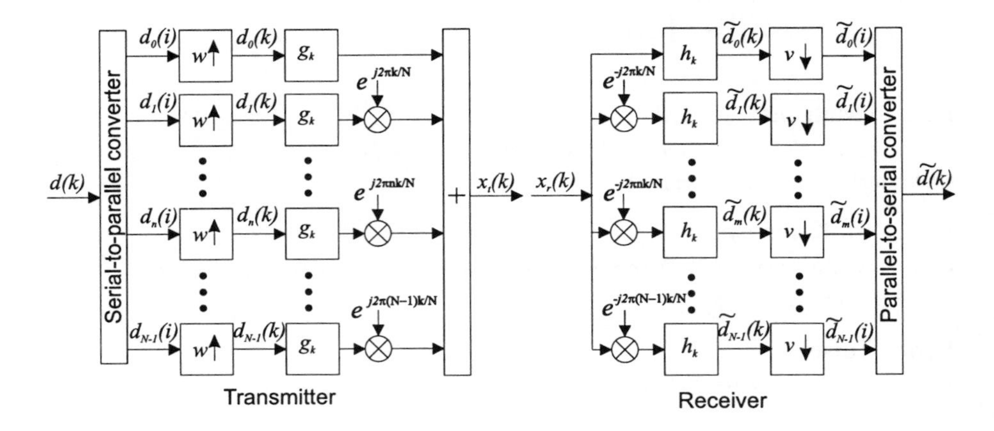

Figure 1 General structure of MC-system

transmitted during the guard interval the bandwidth efficiency is reduced with its insertion.

MC-systems with different impulse shapings for g_k and h_k have been introduced. Some compensate disadvantages of OFDM; e.g.:

- The loss of bandwidth efficiency (e.g. [7, 8]).
- The large out of band radiation (e.g. [14, 6, 2]). Note that with OFDM the symbols are often windowed before transmission to overcome this problem. As will be shown later in Section 2.1, this windowing is very similar to using a different impulse shape in the first place.
- Its sensitivity to frequency shifts (e.g. [5, 3]),
- The non-exploiting of channel diversity when channel coding is not applied (e.g. [10, 11]). Adding a CDMA component to the OFDM system is also a possibility to overcome this disadvantage.
- Its performance loss in strongly time variant environments (e.g. [1, 9]).

2. IMPLEMENTATION OF POLYPHASE FILTERBANKS

2.1 TRANSMITTER

The MC-transmission signal $x_t(k)$ according to Figure 1 is

$$x_t(k) = \underbrace{\sum_{n=0}^{N-1} \sum_{p=0}^{P-1} g_p \, d_n(k-p) \; e^{j2\pi \frac{kn}{N}}}_{x_c(k)} \qquad \text{with } \frac{P}{w} \in \mathbb{N} \tag{1}$$

and P the number of samples of the filter g_k (which can be several times the symbol duration T_s long). The fact that

$$d_n(k) = \begin{cases} d_n(i) & \text{for } k = iw \\ 0 & \text{else} \end{cases} \quad (2)$$

influences $x_c(k)$ (i.e. the convolution with g_k) as only every wth term of $x_c(k)$ differs from 0. To include this in the calculation the index k can be replaced with $k = iw + \lambda$ with $\lambda = 0, \ldots, w-1$. This changes Equation (2) to

$$d_n(k) = d_n(iw + \lambda) = \begin{cases} d_n(i) & \text{for } \lambda = 0 \\ 0 & \text{else} \end{cases} . \quad (3)$$

If p is equally substituted [13, 4] by the polyfilter base $p = p'w + \lambda$ the convolution term $x_c(k)$ becomes

$$x_c(iw + \lambda) = \sum_{p'=0}^{P/w-1} g_{p'w+\lambda} d_n(iw + \lambda - p'w - \lambda) = \sum_{p'=0}^{P/w-1} g_\lambda(p') d_n(i - p'). \quad (4)$$

For the transmission signal of Equation (1) this leads to

$$\begin{aligned} x_t(iw + \lambda) &= \sum_{n=0}^{N-1} \sum_{p'=0}^{P/w-1} g_\lambda(p') d_n(i - p') e^{j2\pi \frac{n}{N}(iw+\lambda)} \\ &= \sum_{p'=0}^{P/w-1} g_\lambda(p') \underbrace{\sum_{n=0}^{N-1} d_n(i - p') e^{j2\pi \frac{n\lambda}{N}} \underbrace{e^{j2\pi \frac{inw}{N}}}_{\text{cyclic shift}}}_{\text{IFFT-base}} \\ &= N \sum_{p'=0}^{P/w-1} g_\lambda(p') \text{IFFT}_N\{d_n(i - p')\}_{(iw+\lambda)\text{mod}N} . \end{aligned} \quad (5)$$

With ideal subcarrierspacing $\Delta F T_s = w/N = 1.0$ the cyclic shift (represented in Equation (5) by the term "$(iw + \lambda)\text{mod}N$") disappears and the implementation of the transmitter looks as shown in Figure 2.

Following the IFFT each N samples of one MC-symbol are multiplied with the respective N samples of the impulse response g_k. Note that when OFDM is implemented with windowing to suppress the large out of band radiation the transmitter structure looks *very* similar to the one shown in Figure 2 when only one branch is considered, i.e. when $P = N$ or resp. $P = N + G$ in the case of OFDM, with G the number of samples in the guard interval.

The possibility to easily implement MC-systems with polyphase filterbanks also when $\Delta F T_s = w/N \notin \mathbb{N}$ (but with $w, N \in \mathbb{N}$) is one of the main advantages of polyphase filterbanks. It means that an additional parameter (the

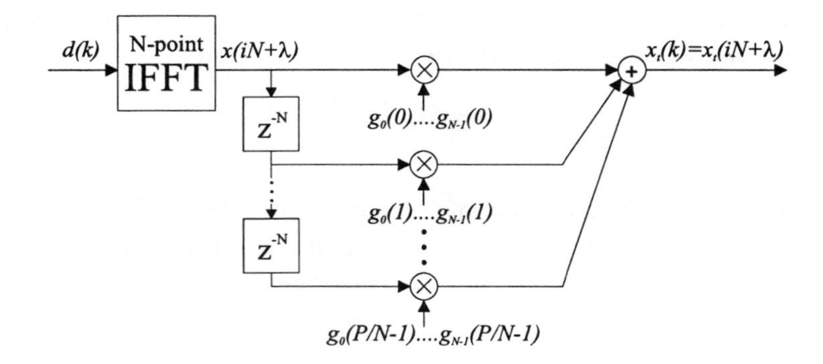

Figure 2 Structur of polyphase filter MC-transmitter for $w = N$

subcarrier spacing) can be exploited to optimally adapt the MC-system to its environment[1]. For OFDM (or any other orthogonal MC-system) $\Delta FT_s \in \mathbb{N}$ has to be maintained and the oversampling w cannot easily be changed.

For the implementation of $\Delta FT_s \notin \mathbb{N}$ the cyclic shift has now to be included in the polyphase filter transmitter as shown in Figure 3. The shift and demultiplexer blocks have the following functionality: Each N input signals $x(iN + \kappa)$ with $\kappa = 0, \ldots, N-1$ of one MC-symbol i are mapped to w output signals $x(iw + \lambda)$ of the same symbol with $\kappa = (iw + \lambda)\text{mod}N$ and "mod" the modulo operation.

An example: The $\lambda = 4^{\text{th}}$ output sample $x(iw + \lambda)$ of the $i = 5^{\text{th}}$ MC-symbol, when $N = 16$ and $w = 17$, is the $\kappa = 9^{\text{th}}$ input signal $x(iN + \kappa)$ also of the $i = 5^{\text{th}}$ MC-symbol.

2.2 RECEIVER

For the receiver the approach is similar to the transmitter one. The recieved signal on subcarrier m before downsampling is

$$\tilde{d}_m(k) = \sum_{q=0}^{Q-1} h_q \, x_r(k-q) \, e^{-j2\pi \frac{(k-q)m}{M}} \qquad \text{with } \frac{Q}{N} \in \mathbb{N} \tag{6}$$

and Q the number of samples of the filter h_k. With the downsampling only every v^{th} sample of the received signal is needed

$$\tilde{d}_m(i) = \tilde{d}_m(k = iv) = \sum_{q=0}^{Q-1} h_q \, x_r(iv - q) \, e^{-j2\pi \frac{(iv-q)m}{N}}. \tag{7}$$

[1]For a given mobile radio environment and symbol rate the main other parameter is the number of subcarriers N [12].

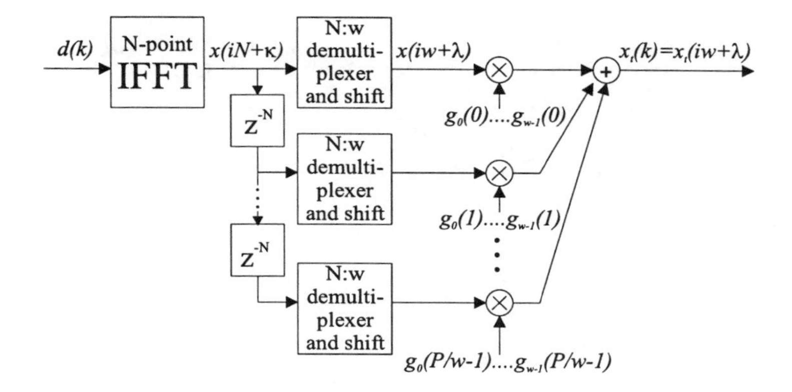

Figure 3 Polyphase filterbank transmitter for multicarrier systems

To enable also a less complex implementation a new polyfilter base [4, 13] is introduced which is now $q = Nq' - \lambda$ with $\lambda = 0, \ldots, N-1$. This leads for the received signal of Equation (7) to

$$
\begin{aligned}
\tilde{d}_m(i) &= \sum_{\lambda=0}^{N-1} \sum_{q'=0}^{Q/N-1} h_{Nq'-\lambda} x_r(iv - Nq' + \lambda) e^{-j\frac{2\pi(iv-Nq'+\lambda)m}{N}} \\
&= \underbrace{e^{-j2\pi i \frac{vm}{N}}}_{\text{rotation}} \underbrace{\sum_{\lambda=0}^{N-1} \sum_{q'=0}^{Q/N-1} h_\lambda(q') x_r(iv - Nq' + \lambda) e^{-j2\pi \frac{\lambda m}{N}}}_{\text{FFT-base}} \\
&= e^{-j2\pi i \frac{mv}{N}} \mathrm{FFT}_N \left\{ \sum_{q'=0}^{Q/N-1} h_\lambda(q') x_r(iv - Nq' + \lambda) \right\}(m). \qquad (8)
\end{aligned}
$$

In case of ideal subcarrier spacing $\Delta F T_s = v/N = 1.0$ the received signal is

$$
\tilde{d}_m(i) = \mathrm{FFT}_N \left\{ \sum_{q'=0}^{Q/N-1} h_\lambda(q') x_r((i - q')N + \lambda) \right\}(m). \qquad (9)
$$

and the implementation is very straight forward as shown in Figure 4. Here the received signal undergoes a similar sort of windowing procedure as it does in the transmitter (see Figure 2) before the FFT is performed.

For $v \neq N$ the receiver structure is shown in Figure 5. In this block diagram the rotation with $e^{-j2\pi i \frac{mv}{N}}$ has been implemented as a cyclic shift before the FFT that merely changes the order of the incoming signals. An incoming signal at the κth position within one MC-symbol $\tilde{x}(iN+\kappa)$ will thus have the position $\lambda = (iv + \kappa) \bmod N$ in the output signal $\tilde{x}(iN + \lambda)$ of the cyclic shift block.

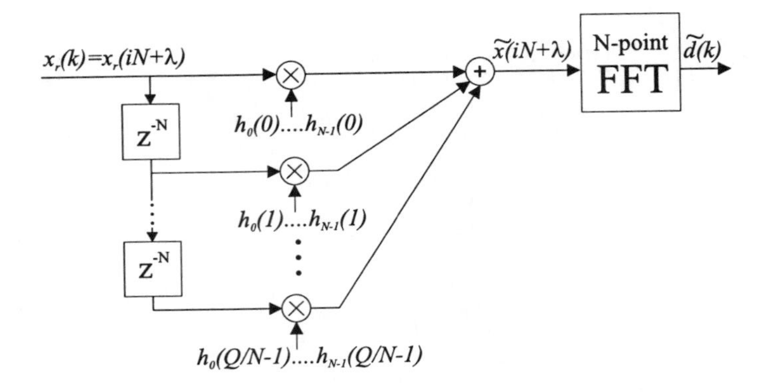

Figure 4 Polyphase filterbank receiver for MC-systems when $\Delta F T_s = 1.0$

Note that while after each N samples in the first branch $v - N$ samples are deleted these values are nonetheless transferred via the memories to the second branch.

An example: In the ith timeslot the sequence that is (samplewise) multiplied with $(h_0(0), \ldots, h_{N-1}(0))$ in the first branch is $(x_r(iv), \ldots, x_r(iv+N-1))$. The sequence that is multiplied with $(h_0(1), \ldots, h_{N-1}(1))$ in the second branch is $(x_r(iv-N), \ldots, x_r(iv-N+N-1) = x_r(iv-1) = x_r((i-1)v+v-1)$. This last sample $x_r(iv-1)$ e.g. has been deleted processed in the first branch (provided $v > N$) in timeslot $i-1$ but is now further processed in the second branch. In the Q/N^{th} branch $(x_r(iv-(Q/N-1)N), \ldots, x_r(iv-(Q/N-1)N+N-1))$ is samplewise multiplied with $(h_0(Q/N-1), \ldots h_{N-1}(Q/N-1))$.

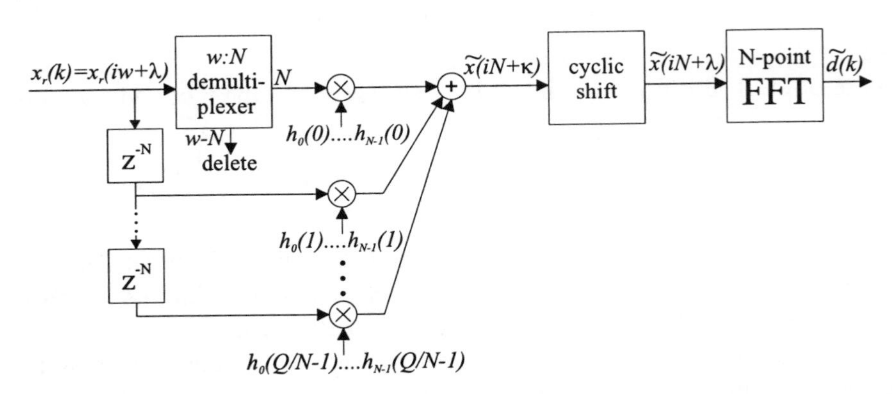

Figure 5 Polyphase filterbank receiver for multicarrier systems

3. AN APPLICATION EXAMPLE

The flexible variation of the subcarrier spacing has been exploited when implementing Gaussian impulse responses for g_k and h_k as done for MCSIS (multicarrier system with soft impulse shaping) [8, 13, 9]. The Gaussian impulses overlap in time and frequency direction (how much depends on the 3dB-Bandwidth parameter). The thus caused intersymbol (ISI) and intercarrier (ICI) interferences influence the performance. In order to be able to apply nevertheless a one-dimensional channel equalization in either time or frequency direction only some residual interference in the respective other direction has to be coped with. The enlargement of the subcarrier spacing can reduce this remaining interference if it is ICI.

How significantly this can improve the performance is shown in Figure 6. Even though the bandwidth efficiency is reduced with the increased subcarrier spacing a performance gain of more than $1dB$ can be achieved[2] at $BER = 10^{-2}$ (compare the solid and the dashed line).

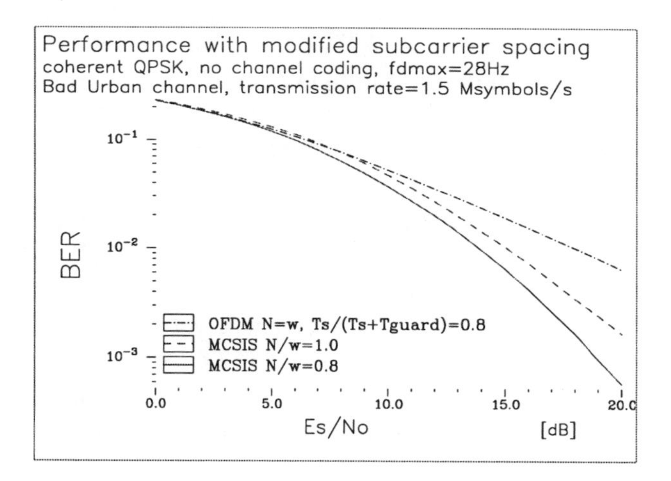

Figure 6 Performance example

4. SUMMARY

It has been shown that also multicarrier systems with non-rectangular impulse shaping can easily implemented with the help of polyphase filterbanks, whose complexity is comparable to the one of OFDM with windowing. Additionally polyphase filterbanks allow the variation of the subcarrier spacing with a fine granularity. Thus polyphase filterbanks provide an additional pa-

[2]The performance gain compared with OFDM is due to diversity effects.

rameter to actively influence the interference situation and the performance of a multicarrier system.

References

[1] J. Armstrong, P. Grant, and G. Povey. Polynomial Cancellation Coding of OFDM to Reduce Intercarrier Interference due to Doppler Spread. In *Proc. of GLOBECOM*, Sydney, 1998.

[2] V. Benedetto, G. D'Aria, and L. Scarabosio. Performance of the COFDM Systems with Waveform Shaping. In *Proc. of ICC*, Montreal, 1997.

[3] D. Dardari. MCM Systems with Waveform Shaping in Multi-User Environments: Effects of Fading, Interference and Timing Errors. In *Proc. of GLOBECOM*, Phoenix, 1997.

[4] N.J. Fliege. *Multirate Digital Signal Processing*. John Wiley and Sons, Chichester, UK, 1994.

[5] M. Gudmundson and P.-O. Anderson. Adjacent Channel Interference in an OFDM System. In *Proc. of VTC*, pages 918–922, Atlanta, 1996.

[6] R. Haas and J.-C. Belfiore. Multiple Carrier Transmission with Time-Frequency Well-Localized Impulses. In *Proc. 2nd IEEE Symposium on Communications and Vehicular Technology in the Benelux*, pages 187–193, Louvain-la-Neuve, Nov. 1994.

[7] B. Hirosaki. An Orthogonally Multiplexed QAM System Using the Discrete Fourier Transform. *IEEE Trans. on Commun.*, 29(7), July 1981.

[8] K.D. Kammeyer, U. Tuisel, H. Schulze, and H. Bochmann. Digital Multicarrier-Transmission of Audio Signals Over Mobile Radio Channel. *European Trans. on Telecommunications (ETT)*, 3(3), May-June 1992.

[9] K. Matheus. *Generalized Coherent Multicarrier Systems for Mobile Communications*. PhD thesis, Dept. of Telecommunications, University of Bremen (FB-1), Shaker Verlag, Aachen, Germany, Jan. 1999.

[10] K. Matheus and K.-D. Kammeyer. Channel Equalization in Multicarrier Systems with Soft Impulse Shaping. In *ITG-Fachb.: Mobile Kommunikation, European Conf. on Personal Mobile Communication*, Sept. 1997.

[11] S. B. Slimane. Performance of OFDM Systems with Time-limited Waveforms over Multipath Radio Channels. In *Proc. of GLOBECOM*, 1998.

[12] H. Steendam and M. Moeneclaey. Optimization of OFDM on Frequency-Selective Time-Selective Fading Channels. In *Proc. of ISSSE*, Sept. 1998.

[13] U. Tuisel. *Multiträgerkonzepte für die digitale, terrestrische Hörfunkübertragung*. PhD thesis, Hamburg University of Technology, May 1993.

[14] Anders Vahlin and Nils Holte. Optimal Finite Duration Pulses for OFDM. In *Proc. of GLOBECOM*, volume 1, pages 258–262, Nov. 1994.

On the improvement of HF Multicarrier modem behavior by using Spread Spectrum techniques to combat fading channels

SANTIAGO ZAZO; FAOUZI BADER; JOSÉ M. PÁEZ-BORRALLO.

ETS Ingenieros de Telecomunicación - Universidad Politécnica de Madrid. Spain.
Phone: 34-91-3367280; Fax: 34-91- 3367350; e-mail: santiago@gaps.ssr.upm.es

Keywords: HF modem, Multicarrier Spreading, Multistage Interference cancellation

Abstract: This paper deals with the application of spread spectrum techniques to Multicarrier HF modems. It is observed that Multicarrier transmission over fading channels with deep nulls show a quite different performance between carriers: those in the vicinity of deep nulls provide a much higher Bit Error Rate (BER) than others far away of the nulls. Our proposal intends to spread different symbols over all the subcarriers simultaneously in order to homogenize performances. This approach, performed by orthogonal codes, shows an evident link with MC-CDMA techniques. In addition, joint detection strategies as Interference Cancellation (IC) with coherent detection are also applied in order to improve its performance. Several computer simulations support the feasibility of our proposal.

1. INTRODUCTION

HF transmission through ionospheric refraction is generally accepted as a very difficult task due to the time variant high dispersive channel characteristic. The ionosphere is a dispersive medium that spreads the signal in both time and frequency. In addition, received radio signals have usually been reflected from more than one ionosphere layer, a phenomenon known as multipath propagation. On the other hand, the ionization characteristic of this atmosphere layer also depends on solar conditions, season, latitude and local weather and time are in fact which the responsible of the time variant characteristic of this medium [1].

This behavior of the physical medium constrains the maximum feasible bandwidth to 3 KHz. Transmission rate over 1.8-2.4 Kbps. requires enhanced techniques with also complexity increase.

Most of the current modems are based on single carrier modulation including very powerful codes and long interleavers to compensate long burst errors [1]. Probably, the common basis of these modems is the Military Standard MIL-STD-188-110A [2] which specifies a continual transmission system to the maximum of 2400 bps.

One of the main civil applications of these modems is the HFDL system (High Frequency Data Link) developed by ICAO (International Civil Aviation Organization). HFDL is a TDMA system dealing with data transmission between aircraft and ground stations providing a data link by HF propagation meanwhile the airplane is far away the coast line [3]. Although this is a burst transmission instead of the continuous stream, the physical layer is almost identical to the single carrier modem of the MIL-STD-188-110A. In the current standard [3], long delays is not a critical point because it is devoted only for data transmission; even, if one slot is lost due to the improper behavior of the equalizer+interleaver+decoder, a request strategy provide the retransmission in probably better conditions. By our own, we are working in this application, but with a much wider purpose: we are focused on the capability of the system to carry out data and digital voice signals in an interactive operation mode. This approach requires a minimum delay also with data transmission rate over 2400 bps. These features do not apply for single carrier techniques but can be obtained by multicarrier modulation (OFDM) due to its more robust behavior against frequency selective fading channels [4, 5,6].

Let us describe briefly our proposal for the next generation of HFDL system in the next section. After that, we will focus on a particular improvement related with the use of multicarrier spreading; in fact, our approach can be formulated as a synchronized MC-CDMA where all the 'users' must be demodulated.

2. THE MC-HFDL SYSTEM

This section is devoted to the presentation of the block diagram of the HFDL system. We will like to point out that the multicarrier version MC-HFDL is not a current standard but it has been presented at the ICAO panel on April 1999 at Montreal (Canada) as a potential alternative for the next definition of the standard [5]. Let us also remark the most critical constraints in our design:

- Our application is concerned with interactive communications where longer delays of 50-80 msec. are not acceptable.
- We will incorporate a digital voice coder: of course, the maximum data rate achievable will provide the better voice quality. We expect to guarantee data rate up to 3600 bps even in bad channel conditions.

Figure 1 shows the block diagram of the OFDM data modem where all the main devices are presented both in transmission and reception.

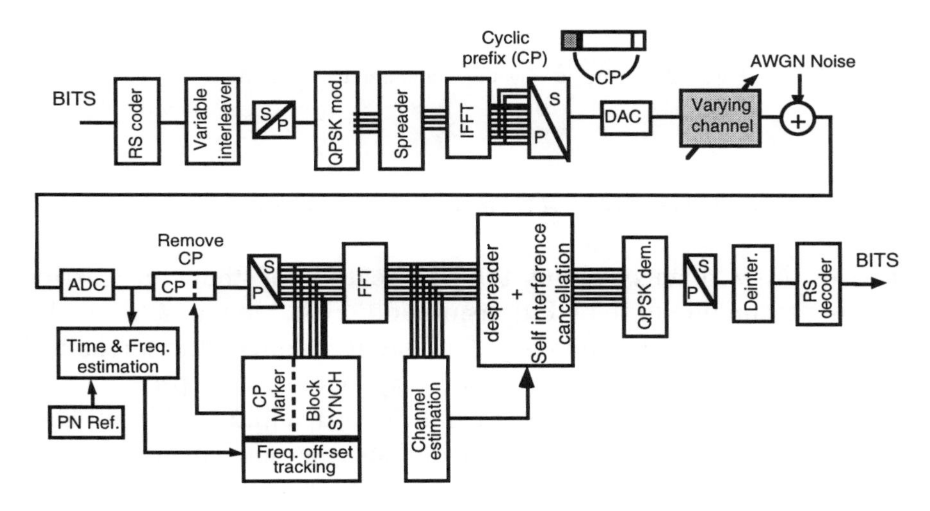

Figure 1. Block diagram of the proposed HF SS-OFDM modem

We have labeled it a HF Spread-Spectrum OFDM modem. Due to space reasons, details about channel coding (Reed-Solomon (45, 63)), interleaving (frequency domain), time and frequency estimation schemes, pilot insertion, channel estimation…must be referred to previous works [5, 6, 7]. On the other hand, spreader/despreader and interference cancellation is the main purpose of the present paper and will be discussed in detail in the next section.

Figure 2 shows an schematic view of the data structure to clarify the previous explanations. In this case, a pilot 16-OFDM symbol is inserted between couples of data 16-OFDM symbols just as an example. Data symbols of 32-OFDM and 64-OFDM are also under analysis. Also, the period insertion of the pilot symbol must be also determined; our experience shows that the channel variability requires channel estimation each 250-300 msec.

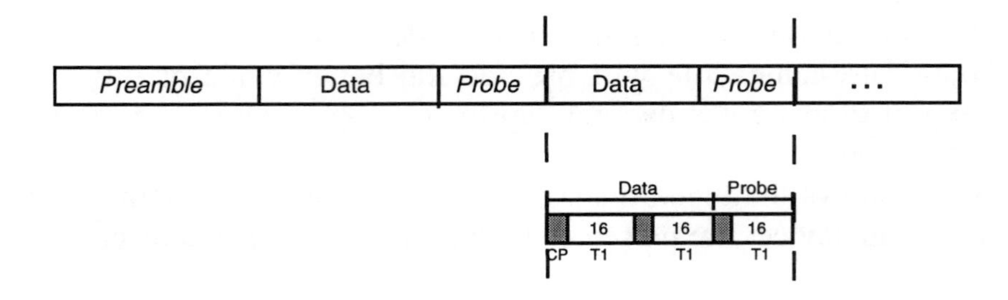

Figure 2. Schematic view of preamble and pilot insertion

3. THE MULTICARRIER SPREADING

Let us now discuss about the application of the multicarrier spreading technique to the HF transmission. First, we will address the motivation and intuitive approach that may recommend the spreading. After that, a formulation of our approach as a problem of synchronized multiple access will link both disciplines: HF-OFDM and MC-CDMA. Finally, some computer simulations will confirm the significant improvement of the spreading technique.

Channel estimation can be easily accomplished by known pilots insertion. Several pilot patterns can be proposed: rectangular, hexagonal... which show a satisfactory performance in terms of its density and the final estimation error. In our application, delay is a critical point that unable pilot patterns employing several OFDM symbols. By our experience, the insertion of a pilot patter only in the frequency dimension allows a delay-free estimation with a computational very low cost by using the FFT algorithm. In fact, pilot symbol may use shorter time duration using alternate carriers and the final estimation for all the carriers is performed by interpolation.
However, in spite of the flat fading characteristic, a severe performance degradation can be observed at the subcarriers at the vicinity of deep nulls. Figure 3 shows a schematic view of the estimation procedure remarking the fact that at high SNR subcarrier, proper channel estimation is obtained and therefore an accurate demodulation is provided. On the other hand, at deep nulls, the local

SNR decreases and coarse estimation may decrease the system performance dramatically. This is a typical observation in HF multicarrier modems: several carriers provide a low BER meanwhile other ones present a much higher BER which definitely decreases the expected performance.

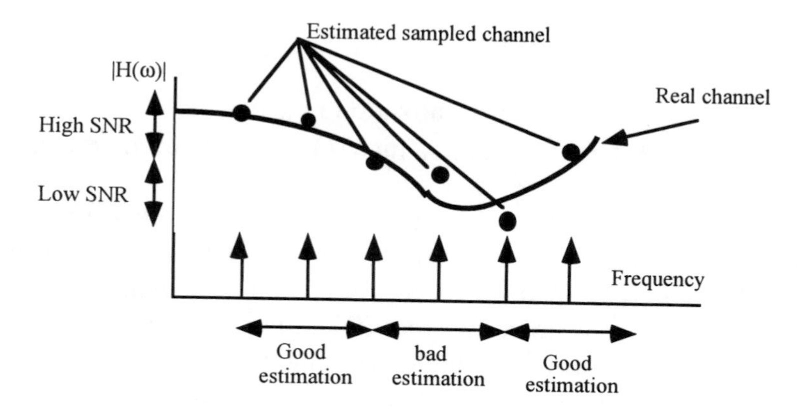

Figure 3. Behavior of the channel estimation technique

Our proposal intends to improve the robustness of the OFDM system by using spread spectrum techniques: the main idea that arises of figure 2 is that subcarriers located in the vicinity of the spectral null will provide a worse estimation accuracy meanwhile the remainder will provide a very accurate measure (their local SNR is upper the mean in band SNR). Therefore, we propose to spread all the symbols by all the subcarriers (through an orthogonal code) in order to uniform the channel distortion. By this approach all the subcarriers will behave almost identically because the real effect of the channel is now devoted by the mean SNR instead of local SNR (the latter depends of the particular null locations).

The idea stated in the last paragraph can be analyzed as a problem of multiple access in a synchronized MC-CDMA system but with a different motivation. In our application each symbol performs as a different 'user' which must be demodulated. The concept of Multiple Access Interference (MAI) also appears in terms of Intercarrier Interference (ICI). Assuming perfect time and frequency synchronization and negligible Intersymbol Interference (ISI) due to the CP insertion, the received signal **r** at one OFDM symbol can be expressed as follows:

$$\mathbf{r} = \mathbf{HCb} + \mathbf{n} \tag{1}$$

matrix **C** is the Walsh-Hadamard coding matrix, **b** is the transmitted data vector and **H** is a diagonal matrix that represents the samples of the channel frequency response at each subcarrier frequency; **n** is the additive gaussian noise. At the receiver, it is obtained:

$$\mathbf{y} = \mathbf{C}^H\mathbf{Gr} = \mathbf{C}^H\mathbf{GHCb} + \mathbf{C}^H\mathbf{Gn} \tag{2}$$

matrix **G** now represents the equalization matrix.

If $\mathbf{C}^H\mathbf{GHC} \neq \mathbf{I}$(**I** is the identity matrix) ICI appears increasing the expected BER.

In addition, the performance of direct despreading can be improved significantly by using Multiuser Detection Techniques (MUD). In fact eq.(2) just performs a matched filter which should be optimum in Gaussian noise, perfect ZF equalization and orthogonal codes (it would perform as a single user channel). By this motivation, we propose to include a simple and effective MUD technique: a multistage Interference Cancellation [8].

The implementation of this technique in terms of the spreader / despreader and multistage IC is quite simple because the number of subcarriers in this application is usually in the range 16-64.

4. COMPUTER SIMULATIONS

The objective of this section is to evaluate the potential capability of the spreading and interference cancellation techniques to the application of HF transmission. We are not concerned in this paper with the performance of the MC-HFDL because many parameters and techniques must analyzed and contrasted. Also, the evaluation and final performance is out of the scope of this paper.

Therefore, we have simulated a uncoded and no interleaving transmission through an static channel; this scenario is composed by and initial pilot symbol and a continuos data stream. The choice of an static channel is due to the desire of analyze the performance in a perfect known situation. The channel impulse response is:

$$h[n] = \delta[n] + 0.8\delta[n-1] \tag{3}$$

Impulse response in eq. (3) shows a two paths model with one very deep null (let us observe the vicinity of the zero of the transfer function to the unit circle). Of course, a line of further research will include simulations in a more realistic HF channel.

We have implemented a phase equalizer as a tradeoff between performance and complexity. However, MMSE technique should be investigated with expected improved performance. Our first approach used ZF criterium but it had to be neglected due to the dramatic noise enhancement in the vicinity of the spectral null.

We have tested this simple scenario for several number of carriers 16, 32 and 64, and we found that the observed behavior is almost independent of the number of carriers. In a first approach we expected that increasing the number of carriers will increase also the spreading gain also improving the performance. The reason of this unexpected behavior is that when increasing the number of subcarriers you are also increasing the number of subcarriers seriously affected by the deep null conditions. This situation is in fact compensated with the increased spreading gain providing similar results.

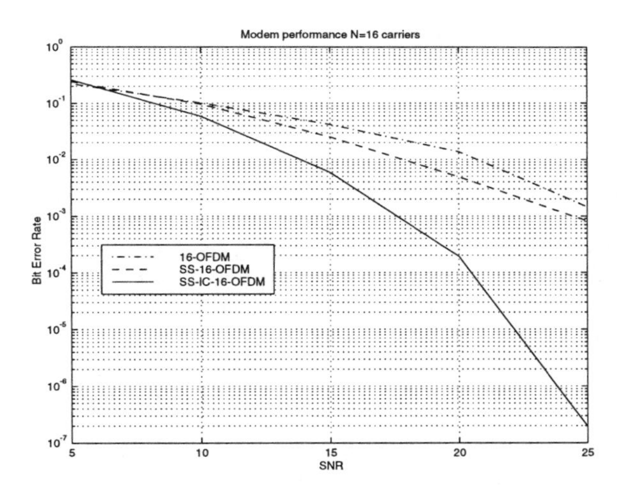

Figure 4. BER for several modems. Dashdot line a 16-OFDM, dashed line is a SS-16-OFDM and solid line is a SS-IC-16-OFDM

Figures 4, and 5 show the comparative BER for different number of subcarriers. In each graphic three curves are presented: solid line represents the SS technique with multistage (3 stages) IC; dashed line represents the SS technique as a bank of matched filters (the FFT is in fact a bank of matched filters) and the dashdot line represents the behavior of conventional OFDM modem. A significant improvement is observed which points out the convenience of a more detailed study with a complete set of simulations in more realistic scenarios.

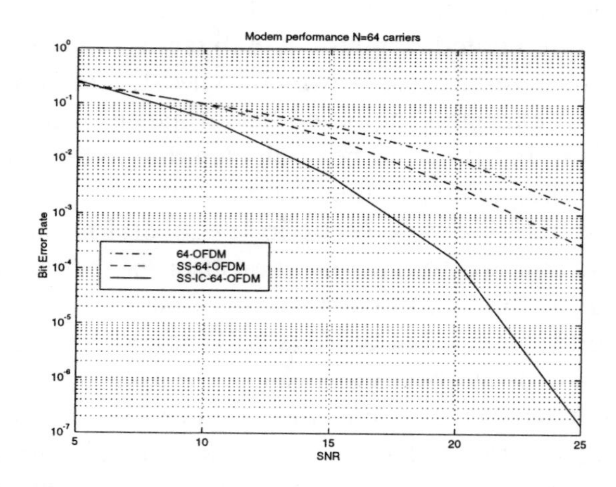

Figure 5. BER for several modems. Dashdot line is a 64-OFDM, dashed line is a SS-64-OFDM and solid line is a SS-IC-64-OFDM

References

[1]: Johnson, E.E., et al. *Advanced High Frequency Radio Communications*. Artech House 1997.

[2]: Department of the Army, Information Systems Engineering Command, *MIL-STD-188-110A: Interoperability and Performance Standards for Data Modems*. Philadelphia, PA. Naval Publications and Forms Center, Attn. NPODS, September 1991.

[3]: SARPs for HFDL. Validated by AMCP-Memo/8, September 1998, Montreal (Canada).

[5]: Pennington J. Techniques for medium speed data transmission over HF channels. *IEE Proceedings Vol.136, Part I, No. 1*, February 1989, pp. 11-19.

[5]: Zazo, S. *Comparison between single carrier and multicarrier techniques for the HFDL system*. Aeronautical Mobile Communications Pannel (AMCP), Montreal, (Canada), March 1999..

[6]: Zazo, S., Páez-Borrallo, J.M., Fernández-Getino García, M.J. High Frequency Data Link (HFDL) for Civil Aviation: a comparison between single and multitone voiceband modems. *Proceedings IEEE Vehicular Technology Conference VTC'99*, Houston (Texas USA) 1999.

[7]: Fernández-Getino García, M.J., Páez-Borrallo, J.M., Zazo, S. "Novel pilot patterns for channel estimation in OFDM mobile systems over frequency selective fading channels". *Proceedings IEEE International Symposium on Personal Indoor, Mobile and Radio Communications, PIMRC'99*, Osaka, Japan 1999.

[8]: Kaiser, S. Multicarrier CDMA mobile radio systems – Analysis and Optimization of detection, decoding and channel estimation. Ph. D. Thesis 1998.

AN IMPROVED MMSE BASED BLOCK DECISION FEEDBACK EQUALIZER

Andreas Bury, Jürgen Lindner
Department of Information Technology
University of Ulm, Germany
bury@it.e-technik.uni-ulm.de

Abstract We present a generalization and an extension of the block decision feedback equalizer described by Reinhardt in [1]. The original equalizer was derived for a class of transmission which is described by an orthogonal spreading over a group of non-interfering fading channels. We generalize this concept for block transmissions with arbitrary interference; the generalized description also includes reception with multiple antennas using joint diversity combining and equalization. In extension, we introduce soft feedback for all individual symbols, in contrast to the single soft value used in [1], offering an improved performance. We describe the proposed equalization method, provide examples for application, and illustrate the performance by simulation results.

1. INTRODUCTION

In many practical cases it is convenient to transmit data in blocks which can be processed independently. Examples for block transmission methods are MC-CDM(A) on the basis of OFDM [3], Interleaved FDMA as described in [2], or serial transmission with termination to assure that there is no interference between blocks [4]. If there is interference within a block, the optimum detector performs a highly complex search of the most probable combination of transmit symbols. A less complex method is to employ equalization together with single symbol detection. Linear equalization allows simple implementation.

The proposed decision feedback equalizer offers an improvement over linear equalization – however, at the cost of some complexity increase.

Subsequently we use bold print for random variables, single underscores for vectors and double underscores for matrices.

2. TRANSMISSION MODEL

We consider a transmission where blocks of symbols are transmitted independently without interference between them. However, within a block there is arbitrary interference among individual symbols. Figure 1 shows a model for transmission of a single block.

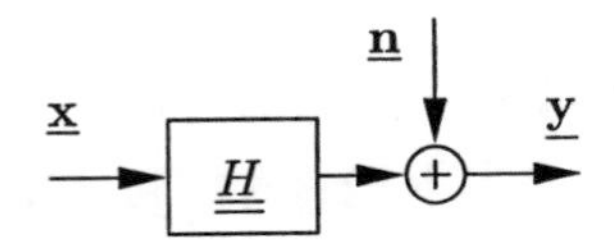

Figure 1 Model of block transmission

A length-N column vector of transmit symbols, $\underline{\mathbf{x}}$, is chosen randomly from an alphabet of M different values, $A_{\mathbf{x}} = \{\chi_1, \chi_2, \cdots, \chi_M\}$. The vector is transmitted over a channel which is described by the arbitrary $K \times N$ matrix $\underline{\underline{H}}$. This matrix reflects how different received symbols interfere with each other; it is a conjunction of all transforms that occur within the chain from transmitter to receiver. To obtain sufficient statistics for estimation of the transmitted symbols, the rank of $\underline{\underline{H}}$ must be N, and therefore $K \geq N$. A noise vector, $\underline{\mathbf{n}}$, is added and a length-K vector, $\underline{\mathbf{y}} = \underline{\underline{H}} \cdot \underline{\mathbf{x}} + \underline{\mathbf{n}}$, is received. The noise vector follows a zero-mean multivariate Gaussian distribution which is uniquely determined by its covariance matrix $\underline{\underline{\Phi}}_{\mathbf{nn}}$.

3. EQUALIZER STRUCTURE

Figure 2 depicts the structure of the proposed equalizer. There are multiple stages of filtering and decision. The first stage is a linear equalizer, $\underline{\underline{A}}^{(0)}$; all other stages combine the linearly filtered received signal (through $\underline{\underline{A}}^{(r)}$) with a linearly filtered version of the decided symbols (through $\underline{\underline{B}}^{(r)}$) after the previous filtering stage. Since decided symbols are fed back for another decision, the filters $\underline{\underline{B}}^{(r)}$ are called feedback filters. The decision DEC is performed on individual symbols according to the maximum a-posteriori (MAP) criterion which, for the case of equal a-priori probabilities of all transmit symbols, simplifies to the maximum likelihood (ML) rule; ML decision is implemented by selecting the element in $A_{\mathbf{x}}$ whose squared Euclidean distance to $\tilde{\mathbf{x}}^{(r)}$ is minimum.

In general, filters of different stages may have different coefficients. For the proposed equalizer, all filter matrices except for stage 0 depend on the values of the received sample vector $\underline{y}$. Furthermore, special types of channel matrices $\underline{\underline{H}}$ may lead to simple filter matrices.

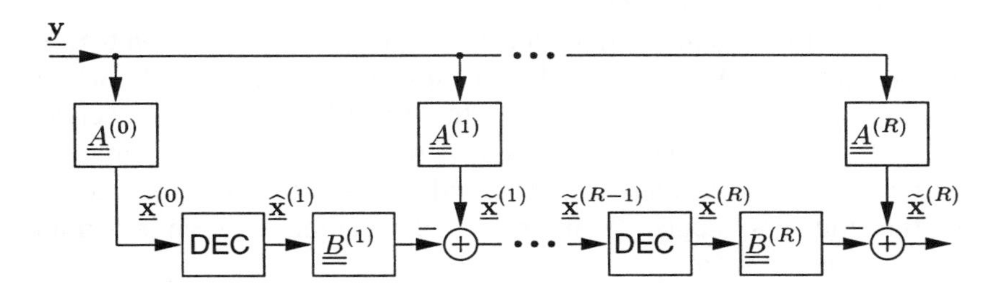

Figure 2 Model of equalizer

4. MMSE BASED FILTER DERIVATION

The decision performed by DEC is reliable if an estimated symbol $\tilde{\mathbf{x}}^{(r)}$ is "as close as possible" to the corresponding transmitted symbol $\mathbf{x}$. Therefore, a heuristic method for derivation of the filters $\underline{\underline{A}}^{(r)}$ and $\underline{\underline{B}}^{(r)}$ is to apply the MMSE criterion

$$J^{(r)} = E\left\{\left(\underline{\tilde{\mathbf{x}}}^{(r)} - \underline{\mathbf{x}}\right)^H \cdot \left(\underline{\tilde{\mathbf{x}}}^{(r)} - \underline{\mathbf{x}}\right)\right\} \longrightarrow \min_{\underline{\underline{A}}^{(r)}_{\text{opt}}, \underline{\underline{B}}^{(r)}_{\text{opt}}} .$$

When assuming zero-mean transmit symbols, the MSE for element $\mathbf{x}_i$ of the transmitted vector $\underline{\mathbf{x}}$ is (the stage index r is omitted here for brevity)

$$\begin{aligned} J_i = {} & (\underline{a}_i\underline{\underline{H}} - \underline{d}_i)\underline{\underline{\Phi}}_{\mathbf{xx}}(\underline{a}_i\underline{\underline{H}} - \underline{d}_i)^H - \underline{b}_i\underline{\underline{\Phi}}_{\mathbf{x}\hat{\mathbf{x}}}{}^H(\underline{a}_i\underline{\underline{H}} - \underline{d}_i)^H + \underline{a}_i\underline{\underline{\Phi}}_{\mathbf{xn}}{}^H(\underline{a}_i\underline{\underline{H}} - \underline{d}_i)^H \\ & -(\underline{a}_i\underline{\underline{H}} - \underline{d}_i)\underline{\underline{\Phi}}_{\mathbf{x}\hat{\mathbf{x}}}\underline{b}_i{}^H \quad + \underline{b}_i\underline{\underline{\Phi}}_{\hat{\mathbf{x}}\hat{\mathbf{x}}}\underline{b}_i{}^H \quad - \underline{a}_i\underline{\underline{\Phi}}_{\mathbf{n}\hat{\mathbf{x}}}\underline{b}_i{}^H \\ & +(\underline{a}_i\underline{\underline{H}} - \underline{d}_i)\underline{\underline{\Phi}}_{\mathbf{xn}}\underline{a}_i{}^H \quad - \underline{b}_i\underline{\underline{\Phi}}_{\mathbf{n}\hat{\mathbf{x}}}{}^H\underline{a}_i{}^H \quad + \underline{a}_i\underline{\underline{\Phi}}_{\mathbf{nn}}\underline{a}_i{}^H \end{aligned} \tag{1}$$

In this equation, $\underline{\underline{\Phi}}_{\mathbf{xx}}$, $\underline{\underline{\Phi}}^{(r)}_{\hat{\mathbf{x}}\hat{\mathbf{x}}}$, $\underline{\underline{\Phi}}_{\mathbf{nn}}$, $\underline{\underline{\Phi}}^{(r)}_{\mathbf{x}\hat{\mathbf{x}}}$, $\underline{\underline{\Phi}}_{\mathbf{xn}}$, and $\underline{\underline{\Phi}}^{(r)}_{\mathbf{n}\hat{\mathbf{x}}}$ are the respective auto covariance and cross covariance matrices of transmitted symbols, decided symbols, and noise; a vector $\underline{d}_i$ is the i-th row of an identity matrix $\underline{\underline{I}}$; $\underline{a}_i$ and $\underline{b}_i$ are the i-th rows in the respective matrices $\underline{\underline{A}}^{(r)}$ and $\underline{\underline{B}}^{(r)}$.

For the feedback filter matrices $\underline{\underline{B}}^{(r)}$, we restrict the diagonal elements to be zero, in order to avoid feeding decided symbols through to their next decision. Furthermore we assume that noise is statistically independent from transmitted symbols, that transmitted symbols within a block are statistically independent, that decided symbols are statistically independent, and that there is no cross correlation between decided symbols and noise:

$$\underline{\underline{\Phi}}_{\mathbf{xn}} = \underline{\underline{0}}, \quad \underline{\underline{\Phi}}_{\mathbf{xx}} = \sigma_{\mathbf{x}}^2 \cdot \underline{\underline{I}}, \quad \underline{\underline{\Phi}}_{\hat{\mathbf{x}}\hat{\mathbf{x}}}^{(r)} = \sigma_{\mathbf{x}}^2 \cdot \underline{\underline{I}}, \quad \underline{\underline{\Phi}}_{\mathbf{n}\hat{\mathbf{x}}}^{(r)} = \underline{\underline{0}}\,.$$

Here $\sigma_{\mathbf{x}}^2$ is the variance of the transmit symbols.

Depending on the variance of noise and interference, the correlation between transmitted symbols and decided symbols varies; for low variance of noise and interference, there is a low bit error probability, and a high correlation between transmitted and decided symbols. We assume that decided symbols are correlated only with the corresponding transmitted symbols and uncorrelated with all other transmitted symbols. Then the cross correlation between transmitted and decided symbols writes

$$\underline{\underline{\Phi}}_{\mathbf{x}\hat{\mathbf{x}}}^{(r)} = E\left\{\underline{\mathbf{x}}^H \underline{\hat{\mathbf{x}}}^{(r)}\right\} = \sigma_{\mathbf{x}}^2 \cdot \underline{\underline{D}}_{\varrho}^{(r)}\,, \tag{2}$$

with $\underline{\underline{D}}_{\varrho}^{(r)}$ being a diagonal matrix with elements ϱ_i on the main diagonal; these values have influence on the equalizer performance. As will be explained subsequently, they are estimated according to the channel matrix, and the received vector $\underline{\mathbf{y}}$.

With these assumptions, we find for the optimum filter coefficients

$$\underline{\underline{A}}_{\text{opt}}^{(0)} = \underline{\underline{H}}^H \cdot \left(\underline{\underline{H}} \cdot \underline{\underline{H}}^H + \frac{1}{\sigma_{\mathbf{x}}^2} \cdot \underline{\underline{\Phi}}_{\mathbf{nn}}\right)^{-1}, \tag{3}$$

and for all other stages $r > 0$

$$\underline{\underline{A}}_{\text{opt}}^{(r)} = \underline{\underline{D}}_{p}^{(r)} \cdot \underline{\underline{H}}^H \cdot \underline{\underline{Q}}^{(r)}\,, \tag{4}$$

$$\underline{\underline{B}}_{\text{opt}}^{(r)} = \underline{\underline{A}}_{\text{opt}}^{(r)} \cdot \underline{\underline{H}} \cdot \underline{\underline{D}}_{\varrho}^{(r)} - \text{diag}\left(\underline{\underline{A}}_{\text{opt}}^{(r)} \cdot \underline{\underline{H}} \cdot \underline{\underline{D}}_{\varrho}^{(r)}\right). \tag{5}$$

with

$$\underline{\underline{Q}}^{(r)} = \left(\underline{\underline{H}}\left(\underline{\underline{I}} - \underline{\underline{D}}_{|\varrho|^2}^{(r)}\right)\underline{\underline{H}}^H + \frac{\underline{\underline{\Phi}}_{\mathbf{nn}}}{\sigma_{\mathbf{x}}^2}\right)^{-1}, p_i^{(r)} = \frac{1}{1 + \underline{h}_i{}^H \underline{\underline{Q}}^{(r)} \underline{h}_i \left|\varrho_i^{(r)}\right|^2}\,.$$

Here, $\underline{h}_i$ is the i-th column of the channel matrix $\underline{\underline{H}}$. The function diag$(\cdot)$ extracts the main diagonal of a matrix and sets all other elements to zero.

This solution is applicable for arbitrary block transmissions.

5. ADJUSTMENT OF SOFT FEEDBACK

The values $\varrho_i^{(r)}$ are required for calculation of equalizer filters for the higher stages $r > 0$; compare equations (2), (4), and (5). We use the model depicted in Figure 3 to find expectations for these values.

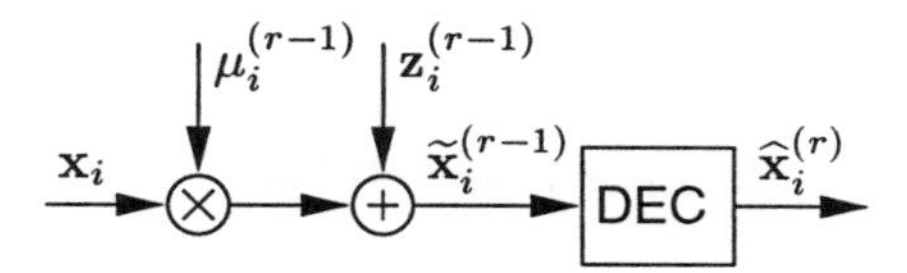

Figure 3 Transmission chain for single-symbol detection

This model illustrates the chain from a transmitted symbol to the corresponding decided symbol after one of the equalizer's decision devices. The transmitted symbol, $\mathbf{x}_i$, is multiplied by a factor $\mu_i^{(r-1)}$; a random variable $\mathbf{z}_i^{(r-1)}$ is added to obtain the estimated symbol $\tilde{\mathbf{x}}_i^{(r-1)}$, and the decision device delivers the decided symbol $\hat{\mathbf{x}}_i^{(r)}$. The random variable $\mathbf{z}_i^{(r-1)}$ contains additive noise from the channel as well as the random interference from other symbols; they are modeled as a single source of zero-mean Gaussian noise. We now consider two approaches to find expectations for the corresponding value $\varrho_i^{(r)}$; for simplicity of notation, we leave out the symbol index i and the equalizer stage index r:

Global expectation

The cross-correlation of a transmitted and the corresponding decided symbol is the expectation

$$E\left\{\mathbf{x}^*\hat{\mathbf{x}}\right\} = \sum_{l=1}^{M}\left(\text{prob}\left\{\mathbf{x} = \chi_l\right\}\chi_l{}^* \sum_{k=1}^{M}\text{prob}\left\{\hat{\mathbf{x}} = \chi_k|\mathbf{x} = \chi_l\right\}\chi_k\right). \quad (6)$$

In this equation, $\text{prob}\left\{\mathbf{x} = \chi_l\right\}$ is the a-priori probability for occurrence of symbol χ_l, and $\text{prob}\left\{\hat{\mathbf{x}} = \chi_k|\mathbf{x} = \chi_l\right\}$ is the conditional probability for decision in favor of χ_k when given that χ_l has been transmitted. These probabilities are calculated as area integrals of the probability density $f_{\tilde{\mathbf{x}}|\mathbf{x}=\chi_l}(\tilde{x})$ over the decision region for $\hat{\mathbf{x}} = \chi_k$; therefore they depend on the transmit symbol alphabet, the multiplicative factor μ, the noise variance $\sigma_{\mathbf{z}}^2$, and the decision rule. We then calculate $\varrho = E\left\{\mathbf{x}^*\hat{\mathbf{x}}\right\}/\sigma_{\mathbf{x}}^2$. In general, a two-dimensional lookup table may be used to find $\varrho = \varrho(\sigma_{\mathbf{z}}^2, \mu)$. For all PSK alphabets the decision only depends on the angle of an estimated symbol; therefore the lookup is only one-dimensional, depending on the ratio $\sigma_{\mathbf{z}}^2/\mu^2$.

Conditional expectation

For an estimated sample value $\tilde{x}$ we can compute a-posteriori probabilities $\text{prob}\left\{\mathbf{x} = \chi_l|\tilde{x}\right\}$. When decision $\hat{x}$ is in favour of the symbol with maximum a-posteriori probability, χ_l, we estimate the symbol error probability for that particular symbol:

$$\widetilde{P}_S(\tilde{x}) = 1 - \text{prob}\{\hat{\mathbf{x}} = \chi_l|\tilde{x}\} \; . \tag{7}$$

To obtain a conditional expectation $E\{\mathbf{x}^*\hat{\mathbf{x}}|\tilde{x}\}$, we think of an equivalent transmission over an AWGN channel, which exhibits the same *average* symbol error probability as the *estimated* symbol error probability $\widetilde{P}_S(\tilde{x})$ when deciding for $\hat{x}$. In this equivalent transmission, when the noise variance is adjusted such that $P_S = \widetilde{P}_S(\tilde{x})$, we may calculate the corresponding expectation $E\{\mathbf{x}^*\hat{\mathbf{x}}|P_S\}$, according to equation (6). Therefore, a simple one-dimensional lookup table is used to find $\varrho = \varrho(\widetilde{P}_S)$.

For the case of 2-PSK, $A_\mathbf{x} = \{+1, -1\}$, we may – instead of using a lookup table – simplify to the direct computation $\varrho = 2 \cdot \text{prob}\{\hat{\mathbf{x}} = \chi_l|\tilde{x}\} - 1$.

6. SPREAD TRANSMISSION

We now consider the special case of transmission over a set of interference-free fading channels, where orthogonal symbol spreading is used to achieve a diversity gain. Figure 4 shows a model for this type of transmission.

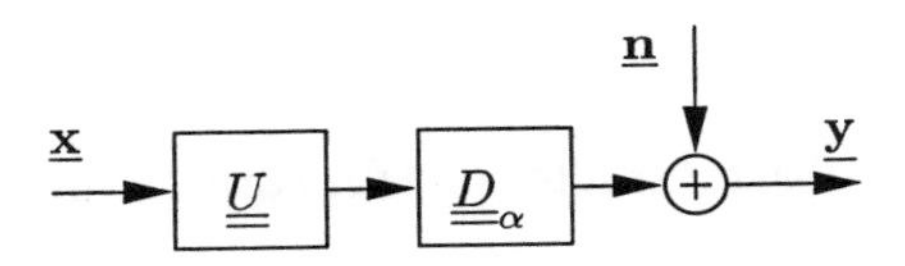

Figure 4 Transmission over non-interfering channels with orthogonal spreading

In this figure, $\underline{\underline{U}}$ is an orthogonal matrix for spreading, $\underline{\underline{D}}_\alpha$ is a diagonal matrix with the channel amplitudes on the main diagonal and zeros elsewhere, and we assume white noise with $\underline{\underline{\Phi}}_{\mathbf{nn}} = \sigma_\mathbf{n}^2 \cdot \underline{\underline{I}}$. The orthogonal spreading matrix has the property $\underline{\underline{U}}^H = \underline{\underline{U}}^{-1} = \underline{\underline{U}}$. For the cases of a Hadamard transform or a Fourier transform, multiplication with $\underline{\underline{U}}$ may be implemented as a fast algorithm (Fast Fourier Transform, FFT; Fast Hadamard Transform, FHT). By inserting $\underline{\underline{H}} = \underline{\underline{D}}_\alpha \cdot \underline{\underline{U}}$ into equations (3), (4) and (5), we find

$$\underline{\underline{A}}^{(0)}_{\text{opt}} = \underline{\underline{U}}^H \cdot \underline{\underline{D}}_{\tilde{\alpha}} \quad \text{with} \quad \tilde{\alpha}_i = \frac{{\alpha_i}^*}{|\alpha_i|^2 + \frac{\sigma_\mathbf{n}^2}{\sigma_\mathbf{x}^2}} \tag{8}$$

$$\underline{\underline{A}}^{(r)}_{\text{opt}} = \underline{\underline{D}}^{(r)}_p \cdot \underline{\underline{U}}^H \cdot \underline{\underline{D}}_\alpha{}^H \cdot \underline{\underline{Q}}^{(r)} \, , \tag{9}$$

$$\underline{\underline{B}}^{(r)}_{\text{opt}} = \underline{\underline{A}}^{(r)}_{\text{opt}} \cdot \underline{\underline{D}}_\alpha \cdot \underline{\underline{U}} \cdot \underline{\underline{D}}^{(r)}_\varrho - \text{diag}\left(\underline{\underline{A}}^{(r)}_{\text{opt}} \cdot \underline{\underline{D}}_\alpha \cdot \underline{\underline{U}} \cdot \underline{\underline{D}}^{(r)}_\varrho\right) \, . \tag{10}$$

$$\underline{\underline{Q}}^{(r)} = \left(\underline{\underline{D}}_\alpha \cdot \underline{\underline{U}} \cdot \left(\underline{\underline{I}} - \underline{\underline{D}}^{(r)}_{|\varrho|^2} \right) \cdot \underline{\underline{U}}^H \cdot \underline{\underline{D}}_\alpha{}^H + \frac{\sigma_\mathbf{n}^2}{\sigma_\mathbf{x}^2} \cdot \underline{\underline{I}} \right)^{-1},$$

$$\frac{1}{p_i^{(r)}} = 1 + \underline{u}_i{}^H \cdot \underline{\underline{D}}_\alpha{}^H \cdot \underline{\underline{Q}}^{(r)} \cdot \underline{\underline{D}}_\alpha \cdot \underline{u}_i \cdot \left| \varrho_i^{(r)} \right|^2.$$

For the special case where for arbitrary i and fixed r the feedback values $\varrho_i^{(r)}$ are all identical, computation of the filter coefficients can be further simplified, such that no matrix inversion is required; this has been derived in [1].

This model of spread transmission is applicable to, e. g., the following transmission schemes:

MC-SSMA (Multi-Carrier Spread Spectrum Multiple Access) as described in [3] uses a Hadamard transform to spread a vector of transmit symbols over a set of OFDM subchannels; each OFDM subchannel may be considered as a single Rayleigh fading channel. OFDM-CDMA is another name for essentially the same method.

IFDMA (Interleaved Frequency Division Multiple Access) as introduced in [2] can be described as an OFDM transmission where Fourier spreading is used to transmit over sets of equally spaced OFDM subchannels.

Serial block transmission over a multipath channel, using cyclic extension of transmitted blocks and corresponding extraction at the receiver, may be considered as an OFDM transmission with Fourier spreading over all subchannels. The equalizer can be implemented in the frequency domain by assuming a Fourier matrix for spreading. A linear equalizer for frequency domain equalization (FDE) has been presented in [4].

7. SIMULATION RESULTS

Figure 5 shows simulation results for 2PSK transmission over 8 independent Rayleigh fading channels using Hadamard spreading over all channels. The following equalizers are compared: ML is a maximum likelihood equalizer employing a brute force search, COND is the proposed equalizer with conditional feedback adjustment (equation (7)), GLOB is the proposed equalizer with global feedback (equation (6)), ISBDFE is the equalizer described in [1], LIN is the linear MMSE equalizer, and INV is the linear inversion (zero forcing) equalizer. For iterative equalizers, the number of iterations is provided. Simulations have shown that the described rules for adjustment of soft feedback $\varrho_i^{(r)}$ can be slightly improved since the assumptions made for equalizer derivation are not perfectly true; a more pessimistic adjustment of the coefficients leads to better results: for COND we assumed a higher noise variance by the factor of $\sqrt{2}$, and for GLOB we adjusted according to the rule $\varrho = 1/\sqrt{1 + 2 \cdot \sigma_{\mathbf{z}}^2/(\mu^2 \cdot \sigma_{\mathbf{x}}^2)}$ which is based on assuming a Rayleigh channel instead of AWGN. For the GLOB equalizer all feedback coefficients of one stage are identical and it does not require any matrix inversion.

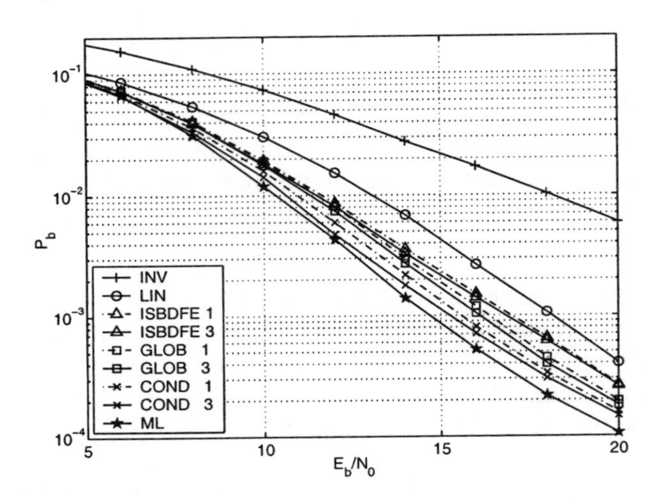

Figure 5 Simulation results: 2PSK, Hadamard spreading over 8 Rayleigh fading channels

From the results we see that conditional feedback adjustment (COND) delivers better performance than global adjustment (GLOB). For global feedback the proposed adjustment rule delivers better results than adjusting according to the SNR of a received block (ISBDFE). Multiple decision feedback stages give a slight improvement.

References

[1] M. Reinhardt: *Kombinierte vektorielle Entzerrungs- und Decodierverfahren*, Dissertation, Ulm, 1997

[2] U. Sorger, I. De Broeck, M. Schnell: *Interleaved FDMA – A New Spread-Spectrum Multiple-Access Scheme*. Proc. MC-SS '97, K. Fazel and G. P. Fettweis (eds.), Kluwer, Dordrecht 1997

[3] K. Fazel, S. Kaiser, M. Schnell: *A Flexible and High Performance Cellular Mobile Communications System Based on Orthogonal Multi-Carrier SSMA*. Wireless Personal Communications, Vol. 2, Nos. 1&2, 1995, pp. 121–144

[4] A. Czwylwik: *Comparison between adaptive OFDM and single carrier modulation with frequency domain equalization*. Proc. IEEE Veh. Tech. Conf., VTC '97, Phoenix, pp. 865–869

MC-CDMA vs. DS-CDMA IN THE PRESENCE OF NONLINEAR DISTORTIONS

K. Fazel* and S. Kaiser**

** Digital Microwave Systems, Bosch Telecom GmbH, D-71522 Backnang, Germany*

*** Institute for Communications Technology, German Aerospace Center (DLR), D-82234, Wessling, Germany*

Abstract In this paper we analyze the influence of the non-linearity effects due to high power amplifiers in MC-CDMA scheme. The results are compared to those obtained with the classical DS-CDMA scheme. It is shown that the total degradation for the up-link of a cellular system is about 2 dB more for the MC-CDMA system compared to the DS-CDMA system. In the down-link the MC-CDMA system outperforms the DS-CDMA system, where, with predistortion, the HPA non-linearities may have the same influence on the performance of the up-link of both systems, i.e. MC-CDMA and DS-CDMA systems.

1. INTRODUCTION

Broadband transmission schemes are usually proposed in cellular mobile radio communications that allow to exploit the multi-path diversity. An interesting approach is based on Direct-Sequence Code-Division-Multiple-Access (DS-CDMA) system [1]. On other hand, the Multi-Carrier Code-Division-Multiple-Access (MC-CDMA) system offers many additional advantages compared to the conventional DS-CDMA scheme [2]. However, it is known that multi-carrier systems using OFDM are more sensitive to the High-Power-Amplifier (HPA) non-linearities than single carrier systems. This is due to the complex Gaussian distribution of the OFDM signal, where, its amplitude is Rayleigh distributed, which results in severe clipping-effects.

The aim of this paper is to analyze the influence of the effects of the non-linearity due to the travelling waves tube (TWTA) and the solid state power amplifiers (SSPA) in MC-CDMA for a cellular mobile communications system. The system performance is compared with the classical DS-CDMA scheme. In addition, the performance of an analytical predistortion technique is evaluated.

2. TRANSMISSION MODEL

The transmission scheme under study, based on MC-CDMA is illustrated in Figure 1. The transmitter consists of a spreader, interleaver, OFDM modulator and High-Power-Amplifiers (HPA).

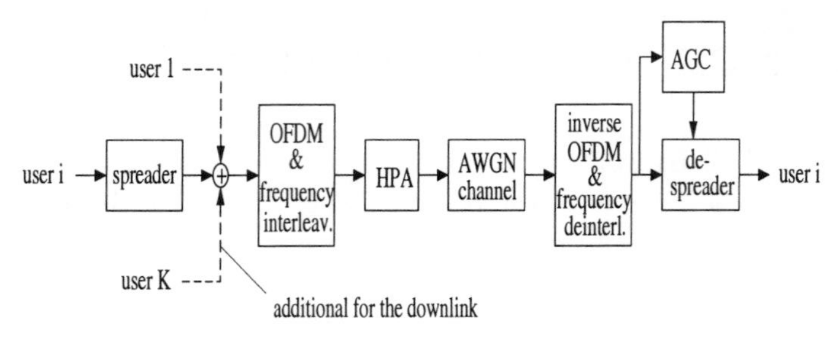

Figure 1 The transmission scheme based on MC-CDMA

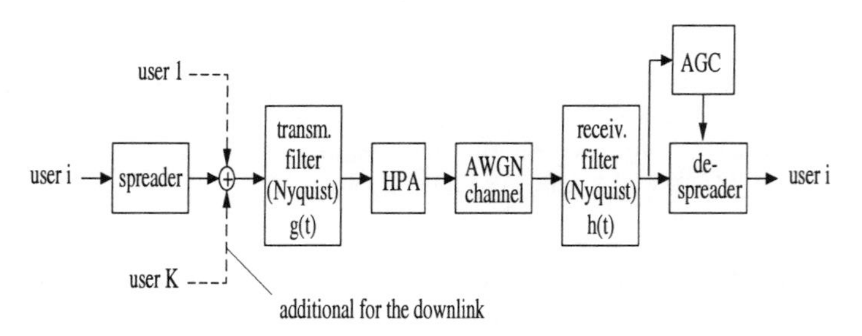

Figure 2 The transmission scheme based on DS-CDMA

In order to have a good reference for system comparisons a conventional DS-CDMA is considered (see Figure 2). Its transmitter consists of a spreader, transmit filter and the HPA. For both systems, a conventional correlative detection is used. As disturbance, we consider only the effects of HPA in the presence of an AWGN. In order to compensate partially the effects of the HPA non-linearities, an automatic gain control (AGC) with phase compensation is used in both systems.

Both schemes use BPSK modulation. The processing gain for MC-CDMA is up to $G_{MC} = 64$, and for DS-CDMA $G_{DS} = 63$. The user bit rate $R_b = 1/T_b$ is chosen to be $R_b = 16.0$ kbit/s for MC-CDMA and $R_b = 15.6$ kbit/s for DS-CDMA. By devoting about 10-15% of the whole bandwidth for channel sounding and synchronization and by considering the loss due to the DS-CDMA filtering and the MC-CDMA guard-time, the resulting transmission bandwidth B is equal to 1.25 MHz.

3. EFFECTS OF NON-LINEAR DISTORTIONS

Before analyzing the non-linearties effects, we look briefly at different HPA models and introduce some definitions.

3.1 HIGH-POWER-AMPLIFIER (HPA) MODELS

The non-linear HPA can be modeled as a memoryless device [3]. Let $x(t) = r(t)\, e^{-j\phi(t)}$ be the HPA complex input signal with amplitude $r(t)$ and phase $\phi(t)$. The corresponding output signal can be written as:

$$y(t) = R(t)\, e^{-j\Phi(t)}, \tag{1}$$

where, $R(t) = f(r(t))$ describes the AM/AM conversion. The AM/PM distortion $\Phi(t) = g(r(t))$ produces an additional phase modulation [3, 4].

Travelling Wave Tube Amplifier (TWTA): For these type of amplifiers the above functions are [3]:

$$R_n(t) = \frac{2\, r_n(t)}{1 + r_n^2(t)}, \quad \text{and} \quad \Phi(t) = \phi(t) + \frac{\pi}{3}\, \frac{r_n^2(t)}{1 + r_n^2(t)}, \tag{2}$$

where, in the above expressions, the input and the output amplitudes are normalized by the saturation amplitude A_{sat}. This kind of amplifier has the most critical characteristics.

Solid State Power Amplifier (SSPA): For these amplifiers the above functions are given with some specific parameters as follows:

$$R_n(t) = \frac{r_n(t)}{(1 + (r_n(t))^{10})^{1/10}}, \qquad \text{and} \qquad \Phi(t) = \phi(t). \tag{3}$$

Input- and Output Back-offs: The non-linear distortions of HPA depend on the input (IBO) and the output back-offs (OBO): $IBO = \frac{P_{in}}{P_{sat}}$, and $OBO = \frac{P_{sat}}{P_{out}}$, where, $P_{sat} = A_{sat}^2$ represents the saturation power, $P_{in} = E(|x(t)|^2)$ the mean power of the input signal $x(t)$ and $P_{out} = E(|y(t)|^2)$ is the mean power of the transmitted signal $y(t)$.

3.2 MC-CDMA

The binary information bits b_i, $b_i \in \{-1, 1\}$ with duration T_b, of each user $i = 1, \ldots, K$ are spread by the corresponding spreading code $c_i(t)$. The Walsh-Hadamard codes of length L are used which result in zero cross-correlation. For the up-link case, after spreading, the signal is mapped to N_c sub-carriers of the OFDM [5]. However, for the down-link, the spread data of all active users are added synchroneousley and mapped to the N_c sub-carriers.

Assuming a perfect frequency interleaving, we suppose that the input signal of OFDM is statistically independent. By using a high number of carriers, the OFDM signal $x(t)$ (being a complex signal) can be approximated by a complex Gaussian distribution with zero mean and variance σ^2. Hence, the amplitude of the OFDM signal is Rayleigh distributed and the phase is uniform distributed. Figure 3 shows the amplitude distribution of a MC-CDMA (derived by simulations) signal for the up-link case for a chip energy $E_c = -8$ dB, $L = 64$, $N_c = 512$. It has to be noticed that the probability is presented in logarithmic scaling. The MC-CDMA signal is Rayleigh distributed. We will see later, that compared to the up-link of a DS-CDMA system, the MC-CDMA signal has much higher peak amplitudes. This high signal amplitude leads to a higher degradation for small output back-off values and results in severe clipping effect.

In the same Figure, the signal amplitude distribution (derived by simulations) for the down-link case with $L = 64$, $N_c = 512$ and $K = 32$ (resp. $K = 64$) active users is presented. This signal is also Rayleigh distributed with average power $\sigma^2 = K \times E_c = 7$ dB (resp. 10 dB). Comparing this distribution to that of the down-link of a DS-CDMA system (see later), one can notice that it has a lower peak amplitude. Therefore, it will be more resistant than the DS-CDMA signal.

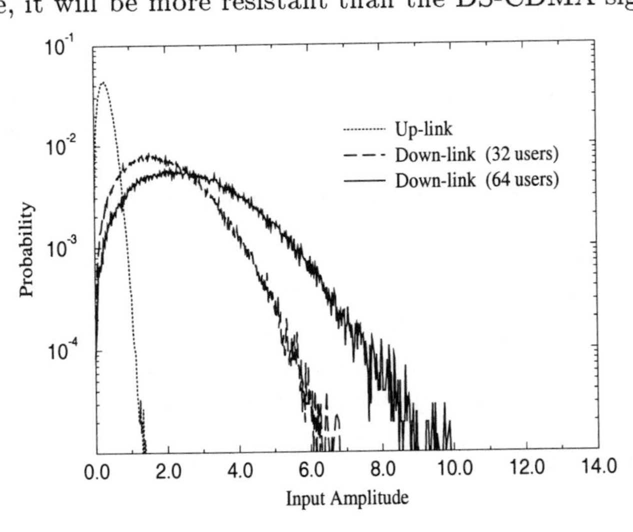

Figure 3 The amplitude distribution of a MC-CDMA signal before HPA

Hence, for low OBO values, the signal $y(t)$ at the output of the amplifier will be highly disturbed, where after the inverse OFDM it leads to a non-linear channel with memory characterized by wrapped output chips containing clusters, which results in inter-chip-interference. The received bit of user i after de-correlation is given by (see [6]):

$$\hat{b}_i = \text{sign}(\sum_{l=1}^{L} \hat{S}(t)\, c_i(l)\, P_{T_c}(t - lT_c)) = \text{sign}(b_i + I_i(OBO) + n_i) \qquad (4)$$

where, $I_i(OBO)$ is the interference term depending on the OBO, and n_i represents the additive white Gaussian noise samples.

3.3 DS-CDMA

After spreading by preferentially phased Gold codes $c_i(t)$ the overall transmitted signal $x(t)$ of all K added synchroneous users signals (down-link) $S(t) = \sum_{i=1}^{K} x_i(t)$ results in $x(t) = S(t) * g(t)$ where, $g(t)$ represents the transmit filter impulse response and " $*$ " the convolution operation. For $K = 1$, the above expression will be also valid for the up-link scenario.

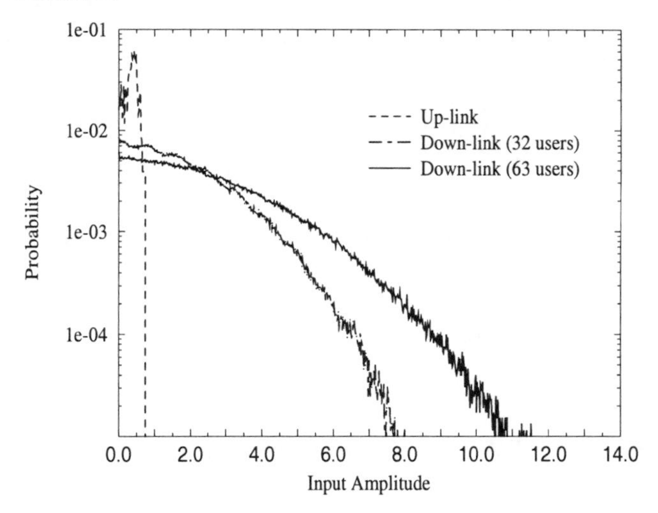

Figure 4 The amplitude distribution of a DS-CDMA signal before HPA

The real signal $x(t)$ will be submitted to the HPA. For low OBO values, the signal $y(t)$ at the output of the amplifier will be highly disturbed, where after the receive filter $h(t)$ it leads to a non-linear channel with memory characterized by a wrapped output chips containing clusters, which results in inter-chip-interference.

Figure 4 shows the amplitude distribution (derived by simulations) of the filtered signal before the HPA for the up-link with $L = 63$ and $E_c = -8$ dB. It has to be noticed that the probability is presented in logarithmic scaling. One can notice that this signal has lower peak amplitudes than the MC-CDMA signal.

However, for high number of users, the amplitude distribution of the transmitted signal at the base station can be approximated by a Gaussian distribution with zero mean and variance σ^2. In the same Figure the amplitude distribution (derived by simulations) for the down-link case for $K = 32$ (resp. $K = 63$) active users with $L = 63$ and $\sigma^2 = 7$ dB (resp. $\sigma^2 = 10$ dB) are presented. One can notice that for high number of active users, it results in a high peak-amplitude values that will cause high degradation in the transmitted signal for low OBO values.

4. PREDISTORTION TECHNIQUES

Several methods of data predistortion techniques for non-linearity compensation have been introduced for single carrier systems [4]. Here, we will consider a simple method based on the analytical inversion of the HPA characteristics.

Let $z(t) = |z(t)|e^{-j\psi(t)}$ be the signal that has to be amplified and $y(t) = R(t)e^{-j\Phi(t)}$ the amplified signal. Since the HPA is a non-linear device, the resulting output will be distorted. To limit this distortion, a device with output signal $x(t) = r(t)e^{-j\phi(t)}$ can be inserted in baseband before the HPA in order that the HPA output $y(t)$ is as close as possible to the original signal $z(t)$. Hence the predistortion function will be chosen such that the global function between $z(t)$ and $y(t)$ will be equivalent to an idealized amplifier:

$$y(t) = \begin{cases} z(t), & \text{if} \quad |z(t)| < A_{sat} \\ A_{sat} \frac{z(t)}{|z(t)|}, & \text{if} \quad |z(t)| \geq A_{sat} \end{cases} \tag{5}$$

4.1 TWTA

The inversion of the TWTA leads to the following equations:

$$r_n(t) = \begin{cases} \frac{1}{z_n(t)}[1 - \sqrt{1 - z_n^2(t)}], & \text{if } |z_n(t)| < 1 \\ \frac{z(t)}{|z(t)|}, & \text{if } |z_n(t)| \geq 1 \end{cases} \tag{6}$$

and

$$\phi(t) = \begin{cases} \psi(t) - \frac{\frac{\pi}{3} z_n^2(t)}{1 + z_n^2(t)}, & \text{if} \quad |z_n(t)| < 1 \\ \psi(t) - \frac{\pi}{6}, & \text{if} \quad |z_n(t)| \geq 1 \end{cases} \tag{7}$$

4.2 SSPA

In the same way the inversion of the SSPA equations leads to:

$$r_n(t) = \frac{z_n(t)}{(1 - (z_n(t))^{10})^{1/10}}, \quad \text{and} \quad \phi(t) = \psi(t). \tag{8}$$

5. SYSTEM PERFORMANCE

For performance evaluation, the total degradation T_D for a given bit-error-rate (BER) is considered as the main criterium:

$$T_D = SNR - SNR_G + OBO, \tag{9}$$

where, $SNR = \frac{E(|y(t)|^2)}{N_o}$ (N_o is the noise spectral density) is the signal to noise ratio in the presence of non-linear distortions, the SNR_G is the signal to noise ratio in the case of linear channel. One can notice that as the OBO decreases, the HPA will be more efficient. But on the other hand the non-linear distortion effects will be increased, and therefore higher SNR is needed to compensate this effect compared to a linear channel. For high OBO values, the HPA will work in its linear zone and there will be no distortions. However, the loss of the HPA efficiency through the high OBO value is taken into account in the T_D expression. Since the T_D changes from a decreasing to an increasing function, it is reasonable to expect an optimal value for OBO which minimizes the T_D. Here, in order to be independent from the channel coding, we consider an uncdoed BER of the order of $BER = 10^{-2}$ for our analysis.

We use a direct simulation approach based on signal oversampling as it is proposed in [4]. Here, an oversampling factor of 8 is considered.

For MC-CDMA system $N_c = 512 \times 8$ points FFT with useful carrier of $N_u = 512$ is used. The sub-carrier distance is 2.441 kHz with $T_s = 409.6\mu$s. The achievable net chip rate is 1.0236M-chip/s. The guard interval is $\Delta = 16\mu$s. In order to reduce the correlation of adjacent sub-carriers a pseudo random frequency interleaver is applied.

For DS-CDMA system both transmit and receive filters are squared raised cosine filter with roll-off $\alpha = 0.15$. The achievable net chip rate is 0.9828 M-chip/s.

In Table 1 the minimum total degradation $T_{D_{min}}$ at $BER = 10^{-2}$ for both systems are summarized. The lower degradation of the MC-CDMA system for the down-link (DL) is due especially to its lower peak signal amplitude compared to the DS-CDMA signal and to the presence of frequency interleaving. In the case of TWT amplifiers, the gain provided by predistortion for MC-CDMA (MC-CDMA+Pred.) is more than 1.5 dB for up- (UL) and down-link (DL).

6. CONCLUSIONS

In this article we analyzed the performance of a MC-CDMA scheme for a cellular mobile communications system by taking into account the effects of the non-linearity due to the travelling waves tube amplifier (TWTA) and solid state power amplifiers (SSPA). The results are com-

Parameters	Minimal total degradation T_{min}					
	DL, K=32		DL, K=64		UL	
	SSP	TWT	SSP	TWT	SSP	TWT
MC-CDMA	2.6	4	3.2	4.7	1.25	3
DS-CDMA	5	5.5	5.3	5.8	≤ 0.9	≤ 1.1
MC-CDMA+Pred.	-	2.5	-	3.1	-	≤ 1.2

Table 1 $T_{D_{min}}$ for $BER = 10^{-2}$

pared to those obtained with the classical DS-CDMA scheme. In addition, the performance of an analytical predistortion technique is evaluated. It is shown that with no predistortion technique the total degradation for the up-link is about 2 dB more for the MC-CDMA system compared to the DS-CDMA system. However, for the down-link case (assuming a full system load) the MC-CDMA system outperforms about 1.5 dB the DS-CDMA system. This is due to the lower peak amplitude of the MC-CDMA signal which is Rayleigh distributed and to the presence of a frequency interleaving which offers a higher diversity for the MC-CDMA system. The results with analytical predistortion showed that for the up-link of a MC-CDMA one can achieve about 1.8 dB gain with respect of a non-predistorted scheme for the TWTA. Similar gains for the down-link case are also obtained. Hence, with predistortion the HPA non-linearities may have the same influence on the performance of the up-link of both systemes, i.e the MC-CDMA and DS-CDMA systems.

References

[1] G. L. Turin, "Introduction to spread-spectrum antimultipath techniques and their application to urban digital radio," *Proceedings of the IEEE*, vol. 68, pp. 328–353, March 1980.

[2] K. Fazel and L. Papke, "On the performance of convolutionally-coded CDMA/OFDM for mobile communications system," in *IEEE, Proc. Conf. PIRMC'93*, September 1993.

[3] A. M. Saleh, "Frequency-independent and frequency-dependent non-linear models of TWTA," *IEEE Trans. Commun.*, vol. 29, pp. 1715–20, Nov. 1981.

[4] G. Karam, *Analysis and compensation of non-linear distortions in digital microwaves systems.* PhD thesis, ENST-Paris, 1989.

[5] S. Kaiser and K. Fazel, "A flexible spread-spectrum multi-carrier multiple-access system for mulimedia applications," in *IEEE, Proc. Conf. PIRMC'97*, Sept. 1997.

[6] K. Fazel and S. Kaiser, "Analysis of Non-Linear Distortions on MC-CDMA," in *IEEE, ICC'98 Proc.*, May 1998.